Normal Partitions and Hierarchical Fillings of N-Dimensional Spaces

Gennadiy Vladimirovich Zhizhin
Russian Academy of Natural Sciences, Russia

A volume in the Advances in Chemical and Materials Engineering (ACME) Book Series

IGI Global
PUBLISHER of TIMELY KNOWLEDGE

Published in the United States of America by
IGI Global
Engineering Science Reference (an imprint of IGI Global)
701 E. Chocolate Avenue
Hershey PA, USA 17033
Tel: 717-533-8845
Fax: 717-533-8661
E-mail: cust@igi-global.com
Web site: http://www.igi-global.com

Library of Congress Cataloging-in-Publication Data

Names: Zhizhin, G. V. (Gennadiĭ Vladimirovich), author.
Title: Normal partitions and hierarchical fillings of n-dimensional spaces / by Gennadiy Vladimirovich Zhizhin.
Description: Hershey, PA : Engineering Science Reference, [2021] | Includes bibliographical references and index. | Summary: "This book pays considerable attention to biological problems, including a mathematical model of plant populations based on Mendel's experiments, complex problems of nucleic acid interactions, problems of the genetic code, and the formation of quaternary structures of living matter"-- Provided by publisher.
Identifiers: LCCN 2020028648 (print) | LCCN 2020028649 (ebook) | ISBN 9781799867685 (hardcover) | ISBN 9781799867692 (paperback) | ISBN 9781799867708 (ebook)
Subjects: LCSH: Plant populations--Mathematical models. | Mendel's law. | Genetic code.
Classification: LCC QK910 .Z45 2021 (print) | LCC QK910 (ebook) | DDC 581.7/88--dc23
LC record available at https://lccn.loc.gov/2020028648
LC ebook record available at https://lccn.loc.gov/2020028649

This book is published in the IGI Global book series Advances in Chemical and Materials Engineering (ACME) (ISSN: 2327-5448; eISSN: 2327-5456)

British Cataloguing in Publication Data
A Cataloguing in Publication record for this book is available from the British Library.

Advances in Chemical and Materials Engineering (ACME) Book Series

J. Paulo Davim
University of Aveiro, Portugal

ISSN:2327-5448
EISSN:2327-5456

Mission

The cross disciplinary approach of chemical and materials engineering is rapidly growing as it applies to the study of educational, scientific and industrial research activities by solving complex chemical problems using computational techniques and statistical methods.

The **Advances in Chemical and Materials Engineering (ACME) Book Series** provides research on the recent advances throughout computational and statistical methods of analysis and modeling. This series brings together collaboration between chemists, engineers, statisticians, and computer scientists and offers a wealth of knowledge and useful tools to academics, practitioners, and professionals through high quality publications.

Coverage

- Computational methods
- Fatigue and Creep
- Wear of Materials
- Ductility and Crack-Resistance
- Sustainable Materials
- Heat Treatments
- Nanomaterials
- Biomaterials
- Thermo-Chemical Treatments
- Chemical Engineering

IGI Global is currently accepting manuscripts for publication within this series. To submit a proposal for a volume in this series, please contact our Acquisition Editors at acquisitions@igi-global.com or visit: https://www.igi-global.com/publish/.

The Advances in Chemical and Materials Engineering (ACME) Book Series (ISSN 2327-5448) is published by IGI Global, 701 E. Chocolate Avenue, Hershey, PA 17033-1240, USA, www.igi-global.com. This series is composed of titles available for purchase individually; each title is edited to be contextually exclusive from any other title within the series. For pricing and ordering information please visit https://www.igi-global.com/book-series/advances-chemical-materials-engineering/73687. Postmaster: Send all address changes to above address. ©

Titles in this Series

For a list of additional titles in this series, please visit: www.igi-global.com/book-series

Technology Development for Adsorptive Heat Energy Converters Emerging Research and Opportnities
Kostyantyn M. Sukhyy (Ukrainian State University of Chemical Engineering, Ukraine) Elena A. Belyanovskaya (Ukrainian State University of Chemical Engineering, Ukraine) and Mikhailo P. Sukhyy (Ukrainian State University of Chemical Engineering, Ukraine)
Engineering Science Reference • © 2021 • 305pp • H/C (ISBN: 9781799844327) • US $195.00

Nanotechnologies and Clusters in the Spaces of Higher Dimension Emerging Research and Opportunities
Gennadiy Vladimirovich Zhizhin (Independent Researcher, Russia)
Engineering Science Reference • © 2021 • 286pp • H/C (ISBN: 9781799837848) • US $195.00

Advanced Surface Coating Techniques for Modern Industrial Applications
Supriyo Roy (Haldia Institute of Technology, India) and Goutam Kumar Bose (Haldia Institute of Technology, India)
Engineering Science Reference • © 2021 • 342pp • H/C (ISBN: 9781799848707) • US $215.00

Handbook of Research on Advancements in Supercritical Fluids Applications for Sustainable Energy Systems
Lin Chen (Institute of Engineering Thermophysics, Chinese Academy of Sciences, China & University of Chinese Academy of Sciences, China)
Engineering Science Reference • © 2021 • 821pp • H/C (ISBN: 9781799857969) • US $415.00

Inspiration and Design for Bio-Inspired Surfaces in Tribology Emerging Research and Opportunities
Hisham Abdel-Aal (Drexel University, USA)
Engineering Science Reference • © 2021 • 367pp • H/C (ISBN: 9781799816478) • US $195.00

Advanced Applications of 2D Nanostructures
Subhash Singh (National Institute of Technology Jamshedpur, India) and Kartikey Verma (Indian Institute of Technology, Kanpur, India)
Engineering Science Reference • © 2020 • 300pp • H/C (ISBN: 9781799829911) • US $225.00

Innovative Applications of Layered Materials and Structures
Sarmila Sahoo (Heritage Institute of Technology, Kolkata, India)
Engineering Science Reference • © 2020 • 300pp • H/C (ISBN: 9781799835875) • US $225.00

Using Lasers as Safe Alternatives for Adhesive Bonding Emerging Research and Opportunities
Barbara Ewa Ciecińska (Rzeszow University of Technology, Poland)
Engineering Science Reference • © 2020 • 248pp • H/C (ISBN: 9781799846345) • US $175.00

IGI Global
PUBLISHER of TIMELY KNOWLEDGE

701 East Chocolate Avenue, Hershey, PA 17033, USA
Tel: 717-533-8845 x100 • Fax: 717-533-8661
E-Mail: cust@igi-global.com • www.igi-global.com

Table of Contents

Preface

In 1885, the book of Efgraf Fedorov "Began the doctrine of figures" (Fedorov, 1885) went out of print. It classifies two - and three - dimensional figures that can fill the Euclidean plane and three - dimensional space adjacent to each other along whole edges and, accordingly, flat faces without gaps. These figures were called parallelogons and, correspondingly, parallelohedrons, due to the fact that their sides are pairwise parallel. It gives the concept of stereohedrons as equal figures that make up a parallelohedron together. In 1889 and in 1891 the following works of Fedorov, "Symmetry of finite figures" and "Symmetry of regular systems of figures" (Fedorov, 1889, 1891), went out of print. In these works, Fedorov deduced 230 spatial groups, i.e. 230 unique geometric laws of arrangement of elementary particles of crystalline structures in three -dimensional Euclidean space and 17 two - dimensional groups of arrangement of elementary particles in the Euclidean plane. In 1891, Schönflis's book "Crystal Systems and the Structure of Crystals" was also published (Schönflis, 1891). Her author many times has quoted Fedorov, pointing to his primacy in a number of questions on theoretical crystallography. Fedorov's extensive correspondence with Schönflis (Shafranovsky, 1963) has been preserved, from which the leading role of Fedorov in the derivation of 230 spatial symmetry groups follows. The studies were reflected in a report by David Hilbert (Hilbert, 1901) at the International Congress of Mathematicians in Paris, in which a number of mathematical problems were formulated that determined the development of mathematics for the next century. One of these problems (the eighteenth) raised the question of constructing spaces from congruent polyhedrons. The report expressed the desire to investigate the question of constructing spaces of higher dimension from congruent figures (Aleksandrov, 1969). The answer to this request was a series of researchers' work in subsequent years. However, many of them tried to consider spaces of higher dimension and figures of higher dimension by mechanically extending to a space of higher dimension ideas about three - dimensional Euclidean space without taking into account the essential properties and features of spaces of higher dimension (Minkowski, 1911; Delaunay, 1929, 1937, 1961; Delaunay, & Sandakova,1961; Voronoi, 1952; Alexandrov, 1934, 1954; Venkov, 1954, 1959). Although back in 1880, the work of the American mathematician Stringham appeared on figures in n - dimensional space (Stringham, 1880). This work was apparently unknown to these researchers, as well as the work of Stott from Amsterdam (Stott, 1900, 1910). Much later, the works of Coxeter and Grünbaum, widely known at present, on figures of higher dimensions appeared (Coxeter, 1963; Grünbaum, 1967). From the analysis of geometric figures of higher dimensions, it would be necessary to begin the solution of the question of constructing spaces of higher dimensions with the help of these figures. That is exactly what the ingenious crystallographer Fedorov did when he studied the question of constructing three - dimensional space with the help of three - dimensional figures. However, this did not happen. As a result, mechanically extending Fedorov's teaching on parallelohedra in three - dimensional spaces to the four - dimensional space, Delaunay in 1929 systematized, as he believed, four - dimensional parallelohedrons (Delaunay, 1929). Moreover, no evidence was given for any of these 51 parallelohedrons that they have dimension 4. Such proof may be the proof of the fulfillment of the Euler - Poincaré equation for a given figure with dimension equal to 4 (Poincaré, 1895). The principle that Delaunay used in the construction, as he believed, four - dimensional parallelohedrons consisted in the construction of three - dimensional parallelohedrons from three - dimensional figures. Now we can make sure that for the figures constructed by Delaunay, by counting the number of vertices, edges, planar faces and three - dimensional faces included in any of the figures, the Euler - Poincaré equation is not fulfilled for them. This means that none of the 51 parallelohedrons are convex figures at all.

Also, Delaunay's teaching on higher - dimensional stereohedrons (Delaunay, 1961), which, according to Fedorov's ideas, should be part of four - dimensional parallelohedrons (Zhizhin, 2019 a), remains incorrect. After the publication of the Delaunay theory of stereohedrons in 1961, not a single concrete stereohedron of higher dimension was given, nor was there a concrete example of constructing spaces of higher dimension using these stereohedrons.

In the study of the structure of substances in recent decades, phenomena were discovered that were unknown at the time of David Hilbert. These include spontaneous zooming (scaling processes), for example, during second - order phase transitions (Landau, 1937; Kadanoff, 1966; Wilson, 1971); discovery of crystals with the absence of translational symmetry in three -dimensional space (Shechtman, et al., 1984); detection of the fractal nature of matter (Mandelbrott, 1982); hierarchical filling of space with polytopes of higher dimension (Zhizhin, 2012); the highest dimension of most molecules of chemical compounds (Zhizhin, 2018).

This forces us to expand the formulation of the question of constructing n - dimensional spaces, posed by David Hilbert in 1900, and to abandon the methods of considering the construction of spaces by geometric figures that do not take into account the accumulated discoveries in the physics of the structure of substances. Here we should take into account the new paradigm of the discrete world built in the latest works of Zhizhin (Zhizhin, 2018, 2019 a, b).

This monograph aims to consider the construction of spaces of various dimensions from 2 to any finite dimension n, taking into account the indicated conditions, including zooming in on shapes, properties of geometric figures of higher dimensions, which have no analogue in three -dimensional space.

In the monograph we consider the conditions of conditions of existence of polytopes of higher dimension, clusters of chemical compounds as polytopes of the highest dimension, higher dimensions in the theory of heredity, the geometric structure of the product of polytopes, the products of polytopes on clusters and molecules, parallelohedrons and stereohedrons of Delaunay, parallelohedrons of higher dimension and partition of n - dimensional spaces, hierarchical filling of n - dimensional spaces, joint normal partitions and hierarchical fillings of n - dimensional spaces. All these problems are extremely important for modern nanotechnology, which plays a significant role in the processes of obtaining new substances and their practical use in various fields of human activity (Zhizhin, 2020).

The monograph pays considerable attention to biological problems. These include a mathematical model of plant populations based on Mendel's experiments, the most complex problems of nucleic acid interactions, problems of the genetic code, the formation of quaternary structures of living matter, and others.

REFERENCES

Alexandrov, A. D. (1934). The derivation of four - dimensional abnormal parallelohedrons. *Proceedings of the U.S.S.R. Academy of Sciences. Atmospheric and Oceanic Physics*, *6*, 803–817.

Alexandrov, A. D. (1954). On filling space with polyhedrons. *Bulletin of the Leningrad University. Mathematics Series.*, *1*, 33–43.

Alexandrov, P. S. (Ed.). (1969). *Hilbert's problems*. Science.

Coxeter, Y. S. M. (1963). Regular Polytopes. Academic Press.

Delaunay, B.N. (1929). Sur la partition ré gulière de l`espace A4 dimensions. *Bulletin de L`Académie des Sciences de L`Urss.*, *1*, 79 - 110, 147 - 164.

Delaunay, B. N. (1937). Geometry of positive quadratic forms. *Progresses of Mathematical Sciences*, *3*, 16–62.

Delaunay, B. N. (1961). Proof of the main theorem of the theory of stereohedrons. *Reports. Academy of Sciences of the USSR*, *138*(6), 1270–1272.

Delaunay, B. N., & Sandakova, N. N. (1961). Theory of stereohedrons. *Proceeding Mathematical In. name V.A. Steklov of Academy Sciences of the USSR*, *64*, 28–51.

Fedorov, E.S. (1885). Beginning of the teachings on the figures. *Notes of the St. Petersburg Mineralogical Society. 2nd series*, *21*, 1 - 277.

Fedorov, E.S. (1889). Symmetry of finite figures. *Notes of the St. Petersburg Mineralogical Society. 2nd series, 25*, 1 - 52.

Fedorov, E.S. (1891). Symmetry of the correct systems of figures. *Notes of the St. Petersburg Mineralogical Society. 2nd series, 28*, 1 - 146.

Grünboum, B. (1967). *Convex Polytopes*. Springer.

Hilbert, D. (1901). Gesammelte Abhandlangen. *Archiv f. Mathematics and Physics*, *3*(1), 44–63, 213–237.

Kadanoff, L. P. (1966). Scaling Laws For Ising Models Near . *Phsics.*, *2*, 263–272.

Landau, L. V. (1937). Theory of phase transitions 1. *Journal of Experimental and Theoretical Physics*, *7*, 19–38.

Mandelbrott, B. B. (1982). *The Fractal Geometry of Nature*. Freeman.

Minkowski, H. (1911). Diskontinuitätshereich für arithemetische Äquivalenz. *Gesammelte Abhandlangen*, *II*, 70–72.

Poincaré A. (1895). Analysis situs. *J. de é Ecole Polytechnique, 1,* 1 – 121.

Schoenflies, A. (1891). *Krystallsysteme und Krystallstruktur*.

Shafranovsky, I. I. (1963). *Efgraf Stepanovich Fedorov*. Publishing House of the USSR Academy of Sciences.

Shechtman, D., Blech, I., Gratias, D., & Cahn, J. (1984). Metallic phase with longerange orientational order and no translational symmetry. *Physical Review Letters*, *53*(20), 1951–1953.

Stott A. B. (1900). On certain series of sections of the regular four - dimensional hyper - solids. *Verhandelingen der Koninklijke Akademie van Wetenschappen te Amsterdam*, 3 - 7.

Stott A. B. (1910). Geometrical deduction of semiregular from regular polytopes and space fillings. *Verhandelingen der Koninklijke Akademie van Wetenschappen te Amsterdam,* 1 – 11.

Stringham, W. I. (1880). Regular figures in *n* - dimensional space. *American Journal of Mathematics*, *3*, 1–14.

Venkov, B. A. (1954). On a class of Euclidean polyhedrons. *Bulletin of the Leningrad University. Mathematics.*, *2*, 11–31.

Venkov, B.A. (1959). On the design of parallelohedra. *Mathematical Collection, 49*(2), 207 -224.

Voronoi, G. F. (1952). *Studies on primitive parallelohedra. In Collected Works, 2*. Publishing House of the Academy of Sciences of the Ukrainian SSR.

Wilson, R. G. (1971). Renormalization group and critical phenomena 1. Renormalization group and the Kadanoff scaling picture. *Physics Rev. B.*, *4*, 3174–3183. doi:10.1103/PhysRevB.4.3174

Zhizhin, G. V. (2012). Hierarchical filling of spaces with polytopes. In *Science and human progress. 7th St. Petersburg meeting of Nobel Prize winners*. St. Petersburg: Publishing House of the Polytechnic University.

Zhizhin, G. V. (2018). *Chemical Compound Structures and the higher Dimension of Molecules: Emerging research and Opportunities*. IGI Global. doi:10.4018/978-1-5225-4108-0

Zhizhin, G. V. (2019a). *The Geometry of Higher - Dimensional polytopes*. IGI Global. doi:10.4018/978-1-5225-6968-8

Zhizhin, G. V. (2019b). *Attractors and Higher Dimensions in Population and Molecular Biology: Emerging Research and Opportunities*. IGI Global. doi:10.4018/978-1-5225-9651-6

Zhizhin, G. V. (2020). *Nanotechnologies and Clusters in the Spaces of Higher Dimension: Emerging Research and Opportunities*. IGI Global.

Chapter 1
On the Conditions for the Existence of Higher–Dimensional Polytopes

ABSTRACT

For irregular n-simplex, n-cross-polytope, n-cube (n-prismahedron) analytic expressions are obtained for calculating the number of faces of different dimensions included in these polytopes. It is shown that the expressions obtained for each of these types of polytopes lead to the Euler-Poincaré equation, regardless of its general topological conclusion. The fulfillment of the Euler-Poincaré equation is the main condition for the existence of polytopes of higher dimension. It is proved that from the obtained analytical expressions for the numbers of faces of different dimensions, the necessary condition for the existence of polytopes follows, which determines the incidence coefficients of low-dimensional faces with respect to high-dimensional faces. It was found that a cross-polytope of any dimension is the result of the rotation of a simplex around the helicoid axis.

INTRODUCTION

As early as 1880, the American mathematician Stringham published a paper on the correct figures in n - dimensional space (Stringham, 1880). Moreover, a multidimensional figure is considered correct if all the polyhedral angles in the vicinity of all vertices of the figure are correct and compatible with the motion. Polyhedral angles are considered correct if all the edges emanating from the vertex have the same length and the angles between the edges are equal. In this work, for a special type of multidimensional figures, a generalized Euler equation was obtained that relates the numbers of elements of different dimensions in this figure into one equation. For the first time, images of regular four - dimensional figures are shown in projection onto a two - dimensional plane. The conditions for the existence of regular n - dimensional figures are formulated. Figures 1, 2, 3 respectively show the projections of four - dimensional regular figures of three types in the Stringham image, which exist for any finite n: 4 - simplex, 4 - cube, 4 – cross - polytope.

DOI: 10.4018/978-1-7998-6768-5.ch001

Figure 1. Simplex

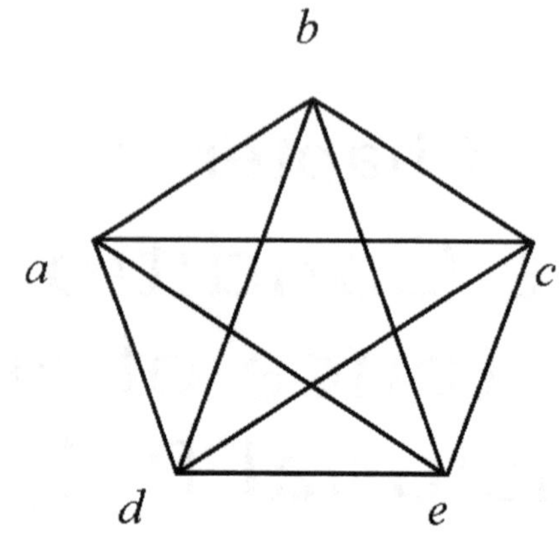

Figure 2. Cube

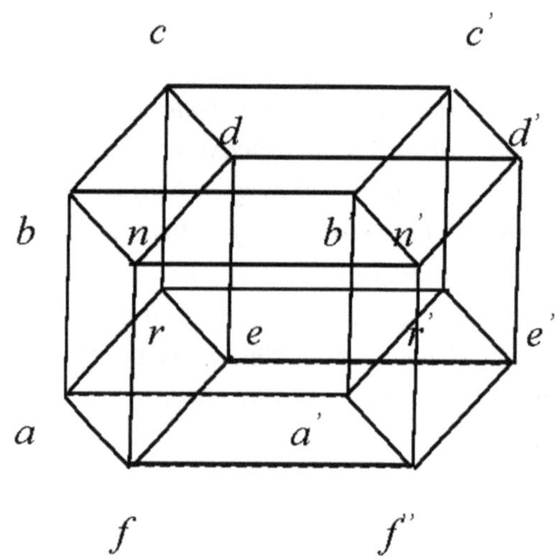

Figure 3. Cross - polytope

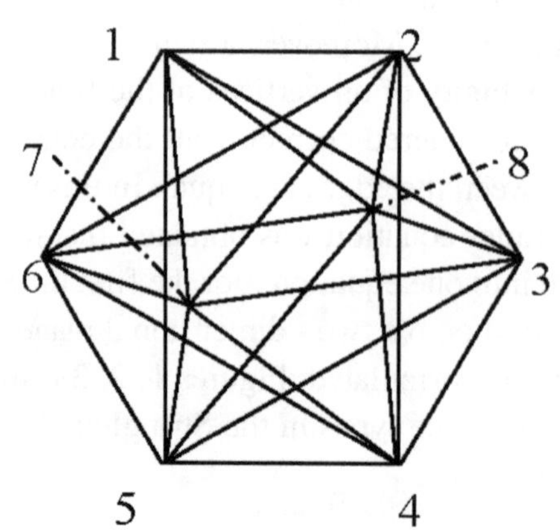

It is clearly seen that there are images the three - dimensional figures included in the four -dimensional figures and adjacent to each other along two - dimensional faces, as if pass through each other, being on one side of the two - dimensional face. This is a fundamental feature of multidimensional figures. Five years after the appearance of Stringham's work, Poincare's work (Poincaré, 1885) was published, in which, using topological analysis, a generalized Euler equation was obtained for arbitrary closed convex and non - convex multidimensional figures without determining specific numbers of faces of different dimensions of these figures. Therefore, hereinafter, the generalized Euler equation will be called the Euler - Poincaré equation. In 1967, Grünbaum also obtained the Euler - Poincaré equation without determining specific numbers of faces of various dimensions in convex n - dimensional figures (Grünbaum, 1967). We can say that the main condition for the existence of a multidimensional figure is the fulfillment of the Euler - Poincaré equation for it for some value of dimension n. In the monograph Zhizhin (Zhizhin, 2018), using the Gillespie data on the geometry of molecules (Gillespie, 1972; Gillespie & Hargittai, 1991), the dimensions of the molecules of most elements of the periodic system were determined and it was proved that almost all of them have a higher dimension. Then, the geometry of polytopes of molecules of chemical compounds was studied in detail (Zhizhin, 2019 a). It is noteworthy that among these polytopes, virtually no polytope is correct. Deviations from the correctness of polytopes can consist in the inequality of the angles between the edges in the vicinity of the same vertex, the inequality of the lengths of the edges, the inequality in the number of edges emanating from each vertex. Therefore, in this chapter, when determining the conditions for the existence of polytopes of higher dimension, the condition for the correctness of polytopes is not used. In this regard, the Euler - Poincaré equation is re - formulated with the determination of the number of faces of various dimensions in irregular polytopes of higher dimension. A specific determination of the numbers of elements of different dimensions for different types of polytopes of higher dimension allows us to establish previously unknown other conditions for the existence of polytopes of higher dimension, based on the patterns of incidence of elements of one dimension with respect to elements of another dimension. In the conclusion of the chapter, basing on previous studies of polytopes of higher dimensionality of chemical compounds (Zhizhin, 2018, 2019 a) and biomolecule compounds (Zhizhin, 2019 b), possible types of polytopes of higher dimensionality of chemical compounds are listed.

THE EQUATION OF EULER - POINCARÉ

The Euler theorem is known according to which, if S, A, and F are the numbers of vertices, edges, and faces of a convex three - dimensional polyhedron, then equality must hold for the latter

$$S - A + F = 2$$

This theorem was extended to Poincaré for polytopes of any dimension, both convex and nonconvex (Poincaré, 1885). In this monograph, only convex polytopes of dimension n are considered. For them, the Euler - Poincaré equation has the form

$$\sum_{i=0}^{n-1}(-1)^i f_i(n) = 1 + (-1)^n, \tag{1.1}$$

in (1.1) $f_i(n)$ is the number of elements of a polytope P of dimension n having dimension i.

Equality (1.1) is a necessary condition for the existence of a convex polytope of dimension n. In the general topological proof of equation (1.1), Poincaré was not given a concrete form of functions $f_i(n)$. The fulfillment of equation (1.1) for the correct n - dimensional figures (the correct simplex, cube, and cross - polytope) was previously confirmed (Zhizhin, 2013, 2014). However, the condition of correctness of figures of the highest dimension, often used in geometry, is rarely satisfied in the structures of substances. This was shown when studying the structures of molecules of chemical compounds (Zhizhin, 2018, 2019a). Therefore, in this monograph we abandon the limiting condition for the correctness of figures of higher dimension. We define the number of elements of different dimensions in the n - simplex, n – cross - polytope and n - prismahedron in the general case (the n - cube is a special case of the n - prismahedron in the case of equal lengths of all edges). Let us prove that equation (1.1) holds in these general cases with specific values of the functions $f_i(n)$. The determination of the numbers of elements of different dimensions that are part of the n - simplex, n - cube and n – cross - polytope, in addition, will allow us to study the geometric properties of these figures of higher dimension, which cannot be established on the basis of equation (1.1) in a general way. It turns out that these properties are also essential and necessary for the existence of figures of higher dimension, as the Euler - Poincaré condition.

THE NUMBER OF FACES OF VARIOUS DIMENSIONS IN THE N - SIMPLEX

The simplest n - dimensional polytope is the n - simplex, i.e. a figure with the number of vertices n + 1, in which each vertex is connected by an edge to any other vertex. In accordance with this definition in this polytope

$$f_0(n) = n+1 \tag{1.2}$$

The number of edges in this polytope is equal to the number of combinations of the total number of vertices in 2, i.e.

$$f_1(n) = C_{n+1}^2 = \frac{(n+1)n}{2}. \tag{1.3}$$

All two - dimensional faces in an n - simplex are 2 - simplexes, i.e. triangles. The number of two - dimensional faces is

$$f_2(n) = C_{n+1}^3 = \frac{(n+1)n(n-1)}{3!}. \tag{1.4}$$

Three - dimensional faces are composed of four two - dimensional faces (i.e., they have four vertices - tetrahedrons). Therefore, the number of three - dimensional faces is

$$f_3(n) = C_{n+1}^4 = \frac{(n+1)n(n-1)(n-2)}{4!}. \tag{1.5}$$

And go on. Thus, for the number of elements with dimension j in the n - simplex we have

$$f_j(n) = C_{n+1}^{j+1}. \tag{1.6}$$

We calculate the amount

$$\sum_{j=0}^{n-1}(-1)^j f_j(n) = \sum_{j=0}^{n-1}(-1)^j C_{n+1}^{j+1} = (-1)^0 C_{n+1}^1 + (-1)^1 C_{n+1}^2 + (-1)^2 C_{n+1}^3 + \ldots + (-1)^{n-1} C_{n+1}^n. \tag{1.7}$$

Taking into account that $C_{n+1}^k = C_{n+1}^{n-k}$, we find that on the right - hand side of (1.7) with even n, all terms are pairwise reduced, i.e. the amount $\sum_{j=0}^{n-1}(-1)^j f_j(n)$ is zero. If n is odd, then all terms on the right - hand side of (1.7) are added up in pairs with the exception of the central term located in place $\frac{n+1}{2}$. In this case, the right - hand side after the transformation takes the form $2[1+\frac{n!}{n!}(1-1)]$, i.e. equals 2. This proves the Euler – Poincaré formula for a simplex of dimension n, regardless of the general Poincaré proof for unknown functions $f_j(n)$.

It can also be noted that the right - hand side of equation (1.7) is the term in the expansion of Newton's binomial

$$(1-1)^{n+1} = 1-(-1)^n - \sum_{j=0}^{n-1}(-1)^j C_{n+1}^{j+1}.$$

Since in this case Newton's binomial is obviously equal to zero, this proves that the right - hand side of (1.7) is equal to zero for even n and equal to 2 for odd n.

THE NUMBER OF FACES OF VARIOUS DIMENSIONS IN THE N - PRISMAHEDRON

Multiply a segment (one - dimensional segment) by another segment (not necessarily equal to this). The rules for multiplying polytopes by each other can be found in the monograph (Zhizhin, 2019 a). As a result, we will have a rectangle. Multiply this rectangle by another segment. We get a parallelepiped, i.e. three - dimensional polyhedron. Multiply this box again by a certain segment. We get a four - dimensional prismahedron. If all segments have the same length, then this 4 - prismahedron is a 4 - cube. At each step of multiplication by a segment, the number of vertices in a figure multiply on 2. When multiplying m times, the number of vertices is equal 2^{m+1} and the dimension of the figure $n = 1 + m$, or $f_0(n)=2^n$. At each step of the multiplication, each vertex generates an edge, each edge generates a rectangle, each rectangle generates a parallelepiped, etc. The process of calculating the numbers of elements of different dimensions can be described by recurrence relations

$$f_{0(n)} = 2f_{0(n-1)},$$
$$f_{1(n)} = 2f_{1(n-1)} + 2f_{0(n-1)},$$
$$f_{2(n)} = 2f_{2(n-1)} + 2f_{1(n-1)}, \quad (1.8)$$
$$\vdots$$
$$f_{0(n)} = 1$$

Here, the first lower index indicates the dimension of the face, and the second lower index indicates the dimension of the space. Substituting successively recurrence relations into each other, taking into account that $f_{0(2)} = 4, f_{1(2)} = 4, f_{2(2)} = 1$, we obtain formulas more convenient for calculations for any finite figure dimension

$$f_0 = 2^n C_n^0, f_1 = 2^{n-1} C_n^1, f_2 = 2^{n-2} C_n^2, ..., f_n = 2^{n-n} C_n^n = 1.$$

Thus, the number of elements of dimension i in the n - prismahedron is expressed by the formula

$$f_i(n) = 2^{n-i} C_n^i, i = 0,1,...,n. \quad (1.9)$$

It is possible to precisely prove the fulfillment of the Euler - Poincaré condition for the n - prismahedron if we represent equation (1.1) in the form

$$\sum_{i=0}^{n} (-1)^i f_{i(n)} = 1. \quad (1.10)$$

Using recurrence relations (1.8), we will successively lower the dimension of polytopes on the left side of (1.1). Then we get

$$\sum_{i=0}^{n} (-1)^i f_{i(n)} = (-1)^0 f_{0(2)} + (-1)^2 f_{1(2)} + (-1)^2 = 4 - 4 + 1 = 1. \text{Q.E.D.}$$

THE NUMBER OF FACES OF VARIOUS DIMENSIONS IN THE N - POLYTOPES WITH ANTI - PARALLEL EDGES (N – CROSS - POLYTOPES)

Let we have some segment AB and some point O on the plane outside the segment AB. Rotate the segment AB around the point O on 180 degrees. We get the segment $B'A'$ antiparallel to the segment AB. Connect the edges vertices indicated different letters: B with A', A with B'. We get a quadrangle with

Figure 4. A quadrangle with antiparallel edges

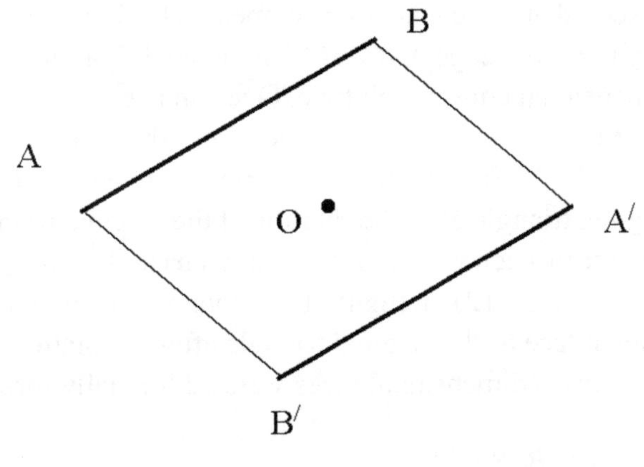

antiparallel sides. In this figure, four vertices formed from two vertices of the initial segment, i.e. vertices doubled (Figure 1.4).

Four edges were formed from one edge of the initial segment: the original *AB*, new $B'A'$, and one edge were formed from each vertex of the original segment *AB*. In addition, a two -dimensional region was formed.

The movement of triangles along a helix leads to the formation of polytopes with antiparallel edges. Consider an arbitrary triangle *ABC* on the plane. Choose some point *O*/ on the plane outside the triangle to his left. Let this point be the base of the axis of the helix passing through the triangle. Rotate the *ABC* triangle 180 degrees to the right, moving it up the helix, parallel to the initial plane. In the projection on the plane, both triangles *ABC* and the displaced triangle $A'B'C'$ will be located as shown in Figure 1.5.

Figure 5. Polytopes of dimension 3 with anti – parallel edges

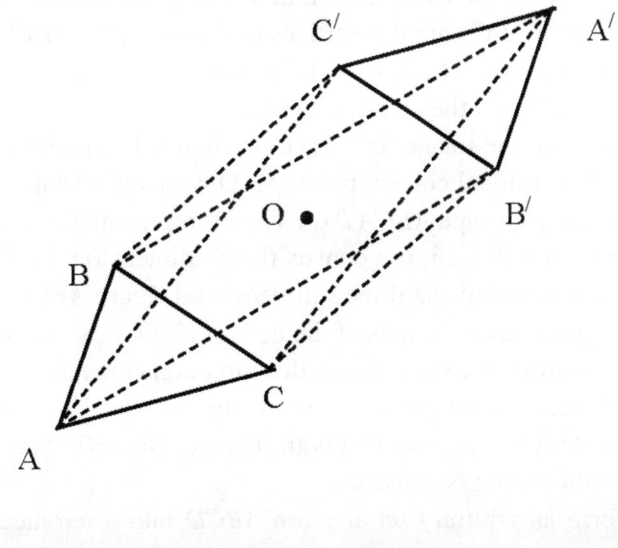

It is easy to see that the edges of the triangle *ABC* and *A/B/C/* are antiparallel. It can now connect in space the vertices of the triangle *ABC* with the vertices of the triangle *A/B/C/* so that there are no connections of the vertices with the same letters. In a projection on the plane the connection is represented by dotted segments. It can be seen that the connecting segments also break up into pairs of anti - parallel segments. Let us now verify that the image *ABCA′B′C′* in Figure 1.5, along with the dotted segments, is a projection of a three - dimensional convex polytope. One can use the Euler – Poincaré equation (1.1) for this aim. In Figure 1.5 *ABCA′B′C′* six vertices formed from three vertices of the initial triangle, i.e. vertices doubled (f_0= 3+3=6). In the figure *ABCA′B′C′* three more edges of the new triangle are added to the three edges of the original triangle after the rotation of the original triangle. In addition, two edges began to additionally proceed from each vertex of the original triangle. Thus, the number of edges in the figure *ABCA′B′C′* is 12 (f_1= 3+3+3•2=12). In Figure 1.5, another two - dimensional face of the triangle is added to one two - dimensional face of the original triangle after its rotation. In addition, at each vertex of the original triangle, two a two - dimensional faces were additionally formed

$$ABC', ACB', BCA', A'B'C, A'C'B, B'C'A.$$

Thus, the total number of two - dimensional faces is 8 (f_2= 2+6=8). Substituting these values of elements of different dimensions into equation (1.1), can to find 6 – 12 +8 = 2, i.e. the Euler – Poincaré equation holds in this case for $n = 3$. This proves that the resulting figure is a convex polytope of dimension 3 (if the figure were not convex, the Euler – Poincaré equation would be violated).

The point *O′* in Figure 1.5 coincides with the center of the three - dimensional figure *ABCA′B′C′* as diagonal figures pass through it and point *O′* located on the axis of the helix. Point *O′* can be considered as the origin of three - dimensional space. Coordinate axes *x, y, z* in this direction emanate from directions *AA′BB′CC′* respectively. Three pairs of these axes define the coordinate planes of the space of this shape.

Choose some point *O″* on the plane outside the triangle below it. Let this point be the base of the axis of the helix passing through the triangle. Rotate the *ABC* triangle 180 degrees to the left, moving it up the helix, parallel to the initial plane. In the projection on the plane, both triangles *ABC* and the displaced triangle *A″B″C″* will be located as shown in Figure 1.5.

It is easy to see that the edges of the triangle *ABC* and *A″B″C″* are antiparallel. It can now connect in space the vertices of the triangle *ABC* with the vertices of the triangle *A″B″C″* so that there are no connections of the vertices with the same letters. In a projection on the plane the connection is represented by dotted segments. It can be seen that the connecting segments also break up into pairs of anti - parallel segments. Let us now verify that the image *ABCA′B′C′* in Figure 1.5, along with the dotted segments, is a projection of a three - dimensional convex polytope. Obviously, a shape *ABCA″B″C″* has as many vertices, edges, and flat faces as a shape *ABCA′B′C′*. Therefore, it satisfies the Euler – Poincaré equation (1.1) with dimensionality $n = 3$, i.e. it is a convex three - dimensional polyhedron. The point *O″* in Figure 1.5 coincides with the center of the three - dimensional figure *ABCA″B″C″* as diagonal figures pass through it and point *O″* located on the axis of the helix. Point *O″* can be considered as the origin of three - dimensional space. Coordinate axes *x, y, z* in this direction emanate from directions *AA″BB″CC″* respectively. The order of the coordinate axes *x, y, z* in the figures *ABCA′B′C′* and *ABCA″B″C″*, as can be seen from Figure 1.5, coincide. This suggests that both figures *ABCA′B′C′* and *ABCA″B″C″* are topologically one and same figure - the wrong octahedron.

Interestingly, to transform an arbitrary tetrahedron *ABCD* into a tetrahedron *A′B′C′D′* with anti - parallel edges, it is not enough to rotate it along a helix by 180 degrees. To do this, you must turn the helix together with the tetrahedron and move the tetrahedron along the helix in the opposite direction,

Figure 6. Polytope of dimension 4 with anti - parallel edges

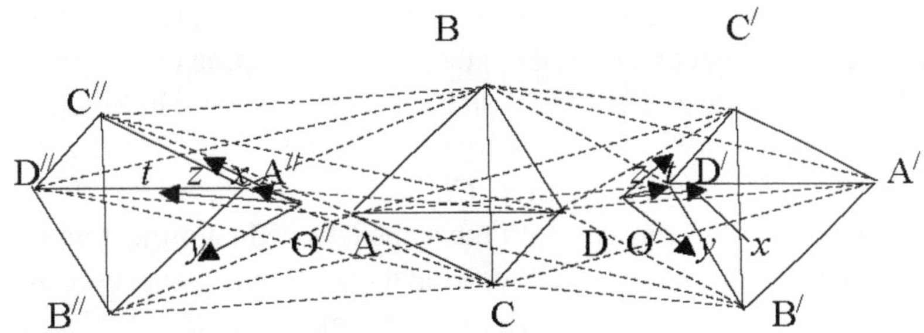

also rotating it 180 degrees. In the initial state, the tetrahedron on the initial helix and the tetrahedron on the reversed helix, after its rotation by 180 degrees, will have anti -parallel edges. Both tetrahedrons can $ABCD$ and $A'B'C'D'$ be present by the Figure 6 for rotation to the right. On Figure 1.6 the point O' is the point of rotation.

Now connect the vertices of the tetrahedrons so that the connecting edges (dotted segments) do not have the same letters. In the resulting figure (along with dotted edges) the number of vertices doubled ($f_0 = 8$). The number of the edges completed six edges initial tetrahedron and six edges final tetrahedron. In addition, three more edges are additionally formed at each vertex of the initial tetrahedron. Thus, the total number of edges is 24 (f_1= 6+6+3•4=24). The flat faces of the figure include 4 faces of the original tetrahedron and 4 faces of the final tetrahedron. In addition, at each vertex of the initial tetrahedron, six flat faces are additionally formed. Thus, the total number of flat faces of the figure is 32 (f_2= 4+4+4•6=32). The initial and final tetrahedra are included in the figure as three - dimensional faces. All tetrahedra do not have vertices of the same name by construction. Therefore, there are still 4 tetrahedra with three vertices of the original tetrahedron and one vertex of the final tetrahedron $ABDC', ACDB', BCDA', ABCD'$. In addition, there are 4 tetrahedra with one of the vertices of the initial tetrahedron and three vertices of the final tetrahedron $A'B'D'C, A'C'D'B, B'C'D'A, A'B'C'D$. There are also tetrahedrons of a tetrahedron with two vertices of the initial tetrahedron and with two vertices of the final tetrahedron $ABDC', ACDB', DCDA', ABCD'$. Thus, the total number of tetrahedrons in the figure is 16 (f_3= 2+4+4+6=16). Substituting these values into equation (1.1), can find 8 - 24 + 24 - 8 = 0.

Consequently, the Euler – Poincaré equation is satisfied in this case for $n = 4$. Thus, the polytope $ABCDA'B'C'D'$ in Figure 1.6 has dimension 4. Comparing figure $ABCDA'B'C'D'$ with Figure 1.3 from Stringham's work, it is easy to see that these figures are topologically equivalent. Thus, figure $ABCDA'B'C'D'$ is a 4 - cross - polytope. Unlike Stringham's work, the resulting figure is not necessarily correct. Thus, it was found that a 4 – cross - polytope is obtained as a result of rotation of the tetrahedron (3 - simplex) around the helicoid axis. Note that, as a result of the analysis, the octahedron can be considered as a 3 – cross - polytope.

The point O' in Figure 1.6 coincides with the center of the fourth - dimensional figure $ABCDA'B'C'D'$ as diagonal figures pass through it and point O' located on the axis of the helix. Point O' can be considered as the origin of forth - dimensional space. Coordinate axes *x, y, z, t* in this direction emanate from directions $AA'BB'CC'DD'$ respectively. Six pairs of these axes define the coordinate planes of the space of this shape.

Choose some point $O^{//}$ on the plane outside the tetrahedron $ABCD$ below it. Let this point be the base of the axis of the helix passing through the tetrahedron. Rotate the $ABCD$ tetrahedron 180 degrees to the left, moving it up the helix, parallel to the initial plane. In the projection on the plane, both tetrahedrons $ABCD$ and the displaced tetrahedron $A^{//}B^{//}C^{//}D^{//}$ will be located as shown in Figure 1.6.

It is easy to see that the edges of the tetrahedrons $ABCD$ and $A^{//}B^{//}C^{//}D^{//}$ are antiparallel. It can now connect in space the vertices of the tetrahedron $ABCD$ with the vertices of the tetrahedron $A^{//}B^{//}C^{//}D^{//}$ so that there are no connections of the vertices with the same letters. In a projection on the plane the connection is represented by dotted segments. It can be seen that the connecting segments also break up into pairs of anti - parallel segments. Let us now verify that the image $ABCDA^{//}B^{//}C^{//}D^{//}$ in Figure 1.6, along with the dotted segments, is a projection of a fourth - dimensional convex polytope. Obviously, a shape $ABCDA^{//}B^{//}C^{//}D^{//}$ has as many vertices, edges, and flat faces as a shape $ABCDA^{/}B^{/}C^{/}D^{/}$. Therefore, it satisfies the Euler – Poincaré equation (1.1) with dimensionality $n = 4$. Thus, the polytope $ABCDA^{//}B^{//}C^{//}D^{//}$ in Figure 1.6 has dimension 4. It is easy to see (Zizhin, 2018) that this too is 4 - cross - polytope.

The point $O^{//}$ in Figure 1.6 coincides with the center of the 4 – cross - polytope as diagonal figures pass through it and point $O^{//}$ located on the axis of the helix. Point $O^{//}$ can be considered as the origin of forth - dimensional space. Coordinate axes x, y, z, t in this direction emanate from directions $AA^{//}BB^{//}CC^{//}DD^{//}$ respectively. From Figure 1.6 it follows that the sequence of alternation of coordinate axes x, y, z, t in a 4 - cross – polytope $ABCDA^{/}B^{/}C^{/}D^{/}$ differs from the sequence of alternation of coordinate axes x, y, z, t in a 4 - cross – polytope $ABCDA^{//}B^{//}C^{//}D^{//}$. Thus, a surprising fact emerged: the figures $ABCDA^{/}B^{/}C^{/}D^{/}$ and $ABCDA^{//}B^{//}C^{//}D^{//}$, being 4 - cross - polytopes, are topologically different from each other. It is impossible to move from one of them to another by continuous transformation, since they have a different order of alternation of vertices.

Interestingly, to transform an arbitrary 4 – simplex $ABCDE$ into a 4 - simplex $A^{/}B^{/}C^{/}D^{/}E^{/}$ with anti - parallel edges, it is not enough to rotate it along a helix by 180 degrees. To do this, you must turn the helix together with the 4 - simplex and move the 4 - simplex along the helix in the opposite direction, also rotating it 180 degrees. In the initial state, the 4 - simplex on the initial helix and the 4 - simplex on the reversed helix, after its rotation by 180 degrees, will have anti -parallel edges. Both 4 - simplexes can $ABCDE$ and $A^{/}B^{/}C^{/}D^{/}E^{/}$ be present by the Figure 7 for rotation to the right. On Figure 1.7 the point O is the point of rotation.

Now connect the vertices of the 4 - simplexes so that the connecting edges (dotted segments) do not have the same letters. In the resulting figure (along with dotted edges) the number of vertices doubled ($f_0 = 10$). The number of the edges completed ten edges initial 4 - simplex and ten edges final 4 - simplex. In addition, four more edges are additionally formed at each vertex of the initial 4 - simplex. Thus, the total number of edges is 40 (f_1= 10+10+5•4=40). The flat faces of the figure include 10 faces of the original 4 - simplex and 10 faces of the final 4 - simplex. In addition, from the connections of each of the vertices of a finite 4 - simplex with some two vertices of the original 4 - simplex, 30 flat faces in the form of triangles are formed additionally:

$$\begin{array}{l} ABC', ABD', ABE', ACB', ACD', ACE', ADB', ADC', ADE', AEB', \\ AEC', AED', BCA', BCD', BCE', BDA', BDC', BDE', BEA', BEC', \\ BED', CDA', CDB', CDE', CEA', CEB', CED', DEA', DEB', DEC'. \end{array}$$

Figure 7. Polytope of dimension 5 with anti - parallel edges

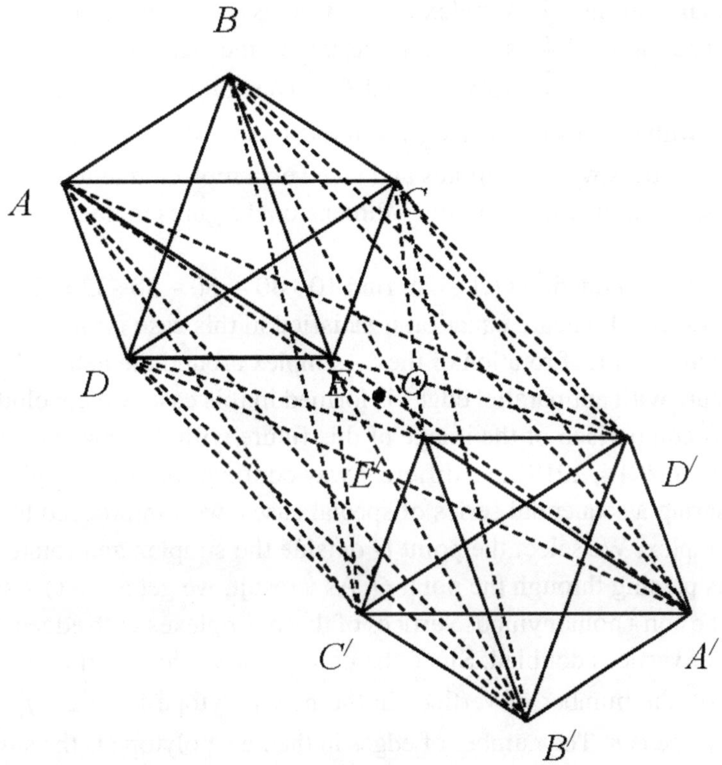

Thus, the total number of flat faces of the Figure 1.7 is 80 (f_2= 10+10+30+30=80). As three -dimensional faces, Figure 1.7 includes five tetrahedrons of the initial 4 - simplex and five tetrahedrons of the final 4 - simplex. In addition, tetrahedrons are formed from the connection of vertices from the initial 4 - simplex and the final 4 - simplex. Since there are 5 vertices in the 4 -simplex and 4 vertices in the tetrahedron, the following combinations are possible for the formation of tetrahedra with vertices from the initial 4 - simplex (letters without strokes without repetition) and from the final 4 - simplex (other letters with strokes). If there is only one vertex with a stroke, then the number of such combinations is equal to the number of combinations of five by three (C_5^3) multiplied by 2. This is because when choosing three letters from five for vertices without strokes, there are two more options for choosing letters with strokes for vertices of the final 4 - simplex ($2C_5^3$). This number must also be multiplied by 2, taking into account the possibility of completely replacing letters with strokes with letters without strokes and vice versa, replacing letters without strokes with letters with strokes ($2 \cdot 2C_5^3$). If there are two vertices with strokes in the tetrahedron, and, therefore, there are also two vertices without strokes, then the number of such letter combinations equals the product $C_5^2 C_{5-2}^2$. This is because after choosing two letters without strokes from five letters, it remains to choose two letter options from the remaining letters (5-2) for vertices with strokes. Since in this case the number of letters with strokes is equal to the number of letters without strokes, replacing letters with strokes with letters without strokes and vice versa does not give new options. Thus, the total number of three - dimensional faces (tetrahedrons) from the vertices of the initial and final 4 -simplexes is equal $f_3 = 5 + 5 + 4C_5^3 + C_5^2 C_3^2 = 10 + 40 + 30 = 80$.

When calculating the four -dimensional faces included in Figure 1.7, it is necessary to take into account the initial 4 - simplex and the final 4 - simplex (2), as well as 4 - simplexes formed from the vertices of the initial 4 -simplex and the final 4 - simplex at the same time. The number of variants of four letters without strokes with one letter with a stroke is equal C_5^4. The number of variants of three letters without strokes and two letters with strokes is equal C_5^2. You must add options to these options when replacing letters with strokes with letters without strokes and vice versa replacing letters without strokes with letters with strokes. Thus, the total number of four - dimensional figures included in the general Figure 1.7 is equal f_4= 2+5+10+5+10=32.

Substituting values f_i into equation (1.1), can find 10 - 40 + 80 – 80+32 = 2.

Consequently, the Euler – Poincaré equation is satisfied in this case for $n = 5$. Thus, the polytope in Figure 1.7 is obtained as a result of rotation of the 4 - simplex around the helicoid axis has dimension 5. Considering that a figure with antiparallel edges is formed in this case, we conclude that this figure is a 5 - cross - polytope. A comparison of the image of this figure with the image of a 5 – cross - polytope obtained earlier (Zhizhin, 2014, 2019) in a different way confirms their topological equivalence.

Now, from considering a sequential series of special cases, we can proceed to the general case. Let there be an (n -1) - simplex. We select the point O outside the simplex and rotate it 180 degrees along the helix with the axis passing through the point O. As a result, we get a (n -1) – simplex$'$ with antiparallel edges. Connect the non - homonymous vertices of these simplexes with edges. We obtain a polytope in which the number of vertices doubled. Since the number of vertices in the (n - 1) - simplex is equal C_n^1 (see formula (1.6)), the number of vertices in the new polytope is equal $f_0 = 2C_n^1 = 2C_n^{n-1}$. The dimension of this polytope is n. The number of edges in the new polytope is the sum of the edges in two (n -1) – simplexes $2C_n^2$ (see formula (1.6)). In addition, (n – 1) edges additionally emanate from each vertex of the original (n -1) - simplex. Thus, the total number of edges in the new polytope is equal

$$f_1 = 2C_n^2 + n(n-1) = 2\frac{n!}{2!(n-2)!} + n(n-1) = 2(n-1)n = 2^2 C_n^{n-2}.$$

The number of planar faces in the new polytope is formed from the sum of the planar faces in the initial and final polytopes (n -1) – simplexes $2C_n^3$. In addition, planar faces are formed due to the union of non - homonymous vertices of both (n - 1) - simplexes into triangles $2C_n^2(n-2)$. Thus, the total number of flat faces in the new polytope is equal

$$f_2 = 2C_n^3 + 2C_n^2(n-2) = 2\frac{n!}{(n-3)!3!} + 2\frac{n!(n-2)}{(n-2)!2!} = \frac{n!(2+3!)}{(n-3)!3!} = 2^3 C_n^{n-3}.$$

The output of the values of the numbers of elements of dimension 1, 2, 3 in the new polytope already shows a general formula for calculating these numbers for any dimension j

$$f_j(n) = 2^{j+1} C_n^{n-1-j}. \tag{1.11}$$

Calculations by the formula (1.11) of the concrete examples of the numbers of elements with dimensions $j=1, 2, 3, 4$ already considered confirm the values of these numbers obtained by direct determination. Comparing formulas (1.9) and (1.11), we conclude that the values $f_i(n)$ in formula (1.9) coincide with the values $f_j(n)$ in formula (1.11) for $j=n-i-1$. This means that the n – cross - polytope and the n - prismahedron are in relation to each other with respect to inclusion inversion. In this case, the number of faces of dimension $i = n-1$ in the n - prismahedron is equal to the number of faces of dimension 0 in the n – cross - polytope, the number of faces of dimension $i = n - 2$ in the n - prismahedron is equal to the number of faces of dimension 1 in the n – cross - polytope, etc. The number of faces of dimension 0 in the n - prismahedron is equal to the number of faces of dimension n - 1 in the n - prismahedron.

Thus, it is proved that, as well as for the n - prismahedron, the Euler - Poincaré equation is also valid for the n – cross - polytope.

It should be noted that the obtained formulas for the numbers of faces of different dimensions for the n - simplex, n - prismahedron and n – cross - polytope, respectively, coincide with the previously obtained formulas for the numbers of faces of different dimensions in the correct simplex, cube, and correct cross - polytope. It is also interesting to note the result obtained, that any cross - polytope is the result of a simplex rival along the helicoid axis.

INCIDENCE IN THE HIGHER DIMENSIONAL POLYTOPES

Let P is an n – polytope, i.e. a convex polytope of dimension n. We denote by B (P) the boundary complex P, that is, a complex consisting of all faces of P with dimension n - 1 or less (Grunbaum, 1967). The points not belonging to B (P) will be called interior points of P. The set of interior points of the polytope P will be denoted by V. Thus, we have equality $P = B(P)\bigcup V$. Since all the faces of a polytope from vertices (which in chemical compounds are images of atoms, molecules or functional groups) to facets enter the boundary complex, it is of interest to establish relationships between faces of different dimensions in the boundary complex B (P).

For each element b_j of the boundary complex B (P) (here j is the serial number of the element in the boundary complex) can to introduce of the notation its dimension $i_{b_j}, 0 \leq i_{b_j} < n$.

There is denoted $k_{b(i_1)b(i_2)}$ the incidence value of the element with dimension i_1 of the boundary complex of polytope P with dimension n to the elements of the boundary complex having dimension $i_2 (0 \leq i_1 < i_2 \leq n)$. This value is equal to the number of elements of the boundary complex of this polytope having dimension i_2 which the given element with dimension i_1 belongs.

Consider simplex polytopes of various dimensions. Let a polytope of simplex type have dimension 2 (triangle). For this polytope, the incidence coefficients have the following values

$$k_{b(0)b(1)} = C_2^1 = 2;\ k_{b(0)b(2)} = C_2^2 = 1;$$
$$k_{b(1)b(2)} = C_1^1 = 1.$$

Let a polytope of simplex type have dimension 3 (tetrahedron). For this polytope, the incidence coefficients have the following values

$k_{b(0)b(1)} = C_3^1 = 3;\ k_{b(0)b(2)} = C_3^2 = 3;\ k_{b(0)b(3)} = C_3^3 = 1;$
$k_{b(1)b(2)} = C_2^1 = 2;\ k_{b(1)b(3)} = C_2^2 = 1;$
$k_{b(2)b(3)} = C_1^1 = 1.$

Let a polytope of simplex type have dimension 4 (4 - simplex, Figure 1.1). For this polytope, the incidence coefficients have the following values

$k_{b(0)b(1)} = C_4^1 = 4;\ k_{b(0)b(2)} = C_4^2 = 6;\ k_{b(0)b(3)} = C_4^3 = 4;\ k_{b(0)b(4)} = C_4^4 = 1;$
$k_{b(1)b(2)} = C_3^1 = 3;\ k_{b(1)b(3)} = C_3^2 = 3;\ k_{b(1)b(4)} = C_3^3 = 1;$
$k_{b(2)b(3)} = C_2^1 = 2;\ k_{b(2)b(4)} = C_2^2 = 1;$
$k_{b(3)b(4)} = C_1^1 = 1.$

Let a polytope of simplex type have dimension 5 (5 - simplex, Figure 8).

Figure 8. The polytope 5 - simplex

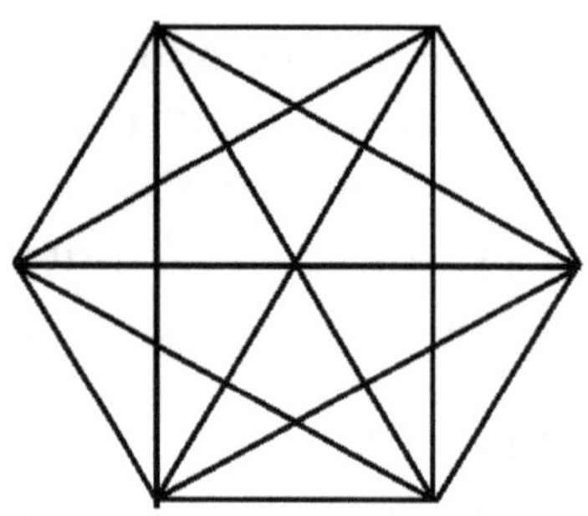

For this polytope, the incidence coefficients have the following values

$k_{b(0)b(1)} = C_5^1 = 5;\ k_{b(0)b(2)} = C_5^2 = 10;\ k_{b(0)b(3)} = C_5^3 = 10;\ k_{b(0)b(4)} = C_5^4 = 5;\ k_{b(0)b(5)} = C_5^5 = 1;$
$k_{b(1)b(2)} = C_4^1 = 4;\ k_{b(1)b(3)} = C_4^2 = 6;\ k_{b(1)b(4)} = C_4^3 = 4;\ k_{b(1)b(5)} = C_4^4 = 1;$
$k_{b(2)b(3)} = C_3^1 = 3;\ k_{b(2)b(4)} = C_3^2 = 3;\ k_{b(2)b(5)} = C_3^3 = 1;$
$k_{b(3)b(4)} = C_2^1 = 2;\ k_{b(3)b(5)} = C_2^2 = 1;$
$k_{b(4)b(5)} = C_1^1 = 1.$

Thus, it follows from the analysis that the general formula for calculating the incidence coefficient of an element of dimension i_1 with respect to elements of dimension i_2 in a polytope of type simplex of dimension n has the form

$$k_{b(i_1)b(i_2)}(n) = C_{n-i_1}^{i_2-i_1}, n \geq i_2 > i_1. \tag{1.12}$$

Consider n – prismahedron (or n - cube) polytopes of various dimensions. Let a polytope of prismahedron type have dimension 2 (square). For this polytope, the incidence coefficients have the following values

$$k_{b(0)b(1)} = 2;\ k_{b(0)b(2)} = 1;$$
$$k_{b(1)b(2)} = 1.$$

Let a polytope of prismahedron type have dimension 3 (parallelepiped, cube). For this polytope, the incidence coefficients have the following values

$$k_{b(0)b(1)} = 3;\ k_{b(0)b(2)} = 3;\ k_{b(0)b(3)} = 1;$$
$$k_{b(1)b(2)} = 2;\ k_{b(1)b(3)} = 1;$$
$$k_{b(2)b(3)} = 1.$$

Let a polytope of prismahedron type have dimension 4 (4 - prismahedron, 4 - cube, Figure 1.2). For this polytope, the incidence coefficients have the following values

$$k_{b(0)b(1)} = 4;\ k_{b(0)b(2)} = 6;\ k_{b(0)b(3)} = 4;\ k_{b(0)b(4)} = 1;$$
$$k_{b(1)b(2)} = 3;\ k_{b(1)b(3)} = 3;\ k_{b(1)b(4)} = 1;$$
$$k_{b(2)b(3)} = 2;\ k_{b(2)b(4)} = 1;$$
$$k_{b(3)b(4)} = 1.$$

From this analysis it follows that the values of the incidence coefficients of polytopes of the n -prismahedron type (n - cube) coincide with the values of the incidence coefficient of the n -simplex for the same dimension of polytopes n. Therefore, formula (1.12) obtained for incidence coefficients in the case of an n - simplex simultaneously refers to the incidence coefficients of elements of different dimensions in n - prismahedron. It can be shown that the right-hand side of formula (1.12) is expressed in a uniform manner through the number of elements of different dimensions of the polytopes of the n - simplex and n - prismahedron. It is argued that the n -simplexes and n - prismahedrons are valid the following formula

$$k_{b(i)b(j)}(n) = \frac{f_j(n)}{f_i(n)} f_i(j), n \geq j > i. \tag{1.13}$$

To prove this assertion in the case of an n - simplex, we express the functions on the right - hand side using equality (1.6)

$$k_{b(i)b(j)}(n) = \frac{!\,{}^{j+1}_{n+1}}{C^{i+1}_{n+1}} C^{i+1}_{j+1} = \frac{(n-i)!}{(n-j)!(j-i)!} = C^{j-i}_{n-i}.$$

In the case of a n – prismahedron we express the functions on the right – hand side of (1.13) using equality (1.9)

$$k_{b(i)b(j)}(n) = \frac{2^{n-j}\,!\,{}^{j}_{n}}{2^{n-i} C^{i}_{n}} 2^{j-i} C^{i}_{j} = \frac{(n-i)!}{(n-j)!(j-i)!} = C^{j-i}_{n-i}.\text{ Q.E.D.}$$

It follows directly from the universal formula (1.13) (the law of incidence) that the values of the incidence coefficients between elements of different dimensions in n - simplexes and in n - prismahedrons increase significantly with increasing dimension of polytopes. if we recall that the tops of polytopes contain material bodies (atoms, molecules, functional groups), then the incidence coefficients (belonging) characterize the exchange of information between material bodies. Consequently, the result indicates that with an increase in the dimension of polytopes, the intensity of the exchange of information between elements of polytopes sharply increases. This result does not follow from the Euler – Poincaré equation. It must be taken into account when constructing polytopes of higher dimension (in particular, parallelohedra). If there is no necessary increase in the incidence coefficients in the considered polytopes, then they cannot exist even if the Euler – Poincaré equation is satisfied. This newly discovered fact was not taken into account, for example, in the construction of parallelohedra of higher dimension earlier (Delaunay, 1937, 1961; Delaunay & Sandakova, 1961; Venkov, 1954, 1959; Alexandrov, 1934, 1954; Voronoi, 1952).

In particular, it follows from formula (1.13) that in n - simplices and in n – prismahedrons, the number of three - dimensional bodies based on one two - dimensional face increases linearly with increasing dimension n

$$k_{b(2)b(3)}(n) = C^{1}_{n-2} = n-2.$$

Consider n – cross - polytopes of various dimensions. Let a cross - polytope have dimension 3 (octahedron, Figure 1.5). For this polytope, the incidence coefficients have the following values

$$k_{b(0)b(1)} = 4;\ k_{b(0)b(2)} = 4;\ k_{b(0)b(3)} = 1;$$
$$k_{b(1)b(2)} = 2;\ k_{b(1)b(3)} = 1;$$
$$k_{b(2)b(3)} = 1.$$

Let a cross - polytope have dimension 4 (4 – cross - polytope, Figure 1.3, Figure 1.6). For this polytope, the incidence coefficients have the following values

$$k_{b(0)b(1)} = 6;\ k_{b(0)b(2)} = 12;\ k_{b(0)b(3)} = 6;\ k_{b(0)b(4)} = 1;$$
$$k_{b(1)b(2)} = 4;\ k_{b(1)b(3)} = 4;\ k_{b(1)b(4)} = 1;$$
$$k_{b(2)b(3)} = 2;\ k_{b(2)b(4)} = 1;$$
$$k_{b(3)b(4)} = 1.$$

Let a cross - polytope have dimension 5 (5 – cross - polytope, Figure 1.7). For this polytope, the incidence coefficients have the following values

$$k_{b(0)b(1)} = 8;\ k_{b(0)b(2)} = 20;\ k_{b(0)b(3)} = 20;\ k_{b(0)b(4)} = 8;\ k_{b(0)b(5)} = 1;$$
$$k_{b(1)b(2)} = 6;\ k_{b(1)b(3)} = 12;\ k_{b(1)b(4)} = 6;\ k_{b(1)b(5)} = 1;$$
$$k_{b(2)b(3)} = 4;\ k_{b(2)b(4)} = 4;\ k_{b(2)b(5)} = 1;$$
$$k_{b(3)b(4)} = 2;\ k_{b(3)b(5)} = 1;$$
$$k_{b(4)b(5)} = 1.$$

From the analysis of n – cross - polytopes, it follows that the values of incidence coefficients in cross - polytopes are significantly higher than the values of incidence coefficients in n -simplexes and n - prismahedrons with the same dimension of polytopes. In addition, the incidence coefficients in n – cross - polytopes increase more sharply with increasing polytope dimensions compared to the incidence coefficients in n - simplices and n - prismahedrons.

SOME TYPES OF POLYTOPES OF HIGHER DIMENSIONS OF CHEMICAL COMPOUNDS

In this section, examples of higher - dimensional polytopes of relatively simple chemical compounds of inorganic chemistry studied in the author's previous works (Zhizhin, 2018, 2019 a, b) will be given and shown. More complex chemical compounds and their corresponding polytopes will be partially considered in subsequent chapters.

Many chemical elements exhibit tetrahedral coordination (Gillespie, 1972; Gillespie & Hargittai, 1991). For example, element Cr have one valence electron on the 4 s - orbital and five of the electrons on 3d - orbital. This allows have of chromium a valence equal 6 in many compounds. Since crystalline chromium oxide CrO_3 consists of chains of tetrahedrons CrO_4, united in two vertices. Each tetrahedron has located in the center of an atom of chromium associated by double bond with each of the four oxygen atoms at the vertices of a tetrahedron. All molecules have the form of a tetrahedron with the center, as shown in (Zhizhin & Diudea, 2016; Zhizhin, 2018), have a dimension of 4, i.e., crystalline chromium oxide is a chain of polytopes of dimension 4, united in two vertices (Figure 9).

If instead of double bonds are one - time connection with the chromium atom monovalent groups (such as hydroxyl groups), the molecule will be the center of the octahedron, which also would have dimension 4 (Zhizhin, 2018). If the anomalous chemical elements has one electron in the outer orbital s and subshell d completely (or almost completely) filled, then the element at the expense of s electron

Figure 9. The chain of molecules CrO_3

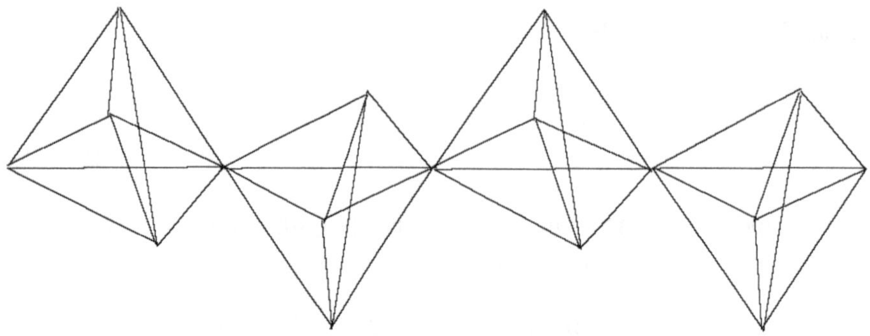

forms a linear molecule, such as a linear molecule oxide X_2 O, where X is the anomalous element (Cu, Pd, Ag, Pt, Au, Rg). However, due to the donor - acceptor bond linear molecule can form complex structures in the space. We choose element Cu from second group of anomalous elements. Figure 10 shows an exemplary structure formed by linear molecules Cu_2 O

Each oxygen atom in the structure of Figure 1.10 bonded to four metal atoms (Cu). Two covalent bonds due to the formation of electron pairs divided: one *s* - electron metal atom and a *p* - electron atom of oxygen. In addition, there are two more donor - acceptor chemical bond due to the transfer of two electrons from the *s* - orbital and two electrons from the *p* - orbitals of the oxygen atom to vacant quantum cell of orbital metal. Thus, oxygen atom acquires valence equal four. In addition, each metal atom is linearly between two oxygen atoms.

Figure 10. The structure of the compound Cu_2O
A black small circle is oxygen atom. A brown circle is copper atom.

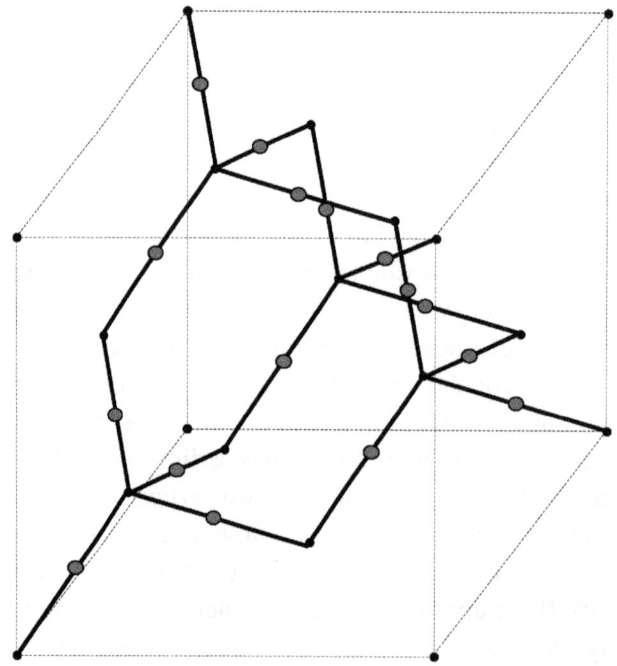

In this structure, oxygen atoms (except oxygen atoms located at the vertices of a cube) form structure is topologically equivalent to the structure of carbon atoms in the molecule of adamantane. The structure of the adamantane molecule is characteristic of many compounds (Ag_2O, Cu_2O, ZnS, CuCl, Pb_4O_6, As_4O_6, Sb_4O_6, P_4O_{10}, P_4O_4, $SiCl_4$, BeF_2, HgS). The adamantane molecule is a convex polytope in the 4*D* space (Figure 11).

Figure 11. The structure of molecule adamantane

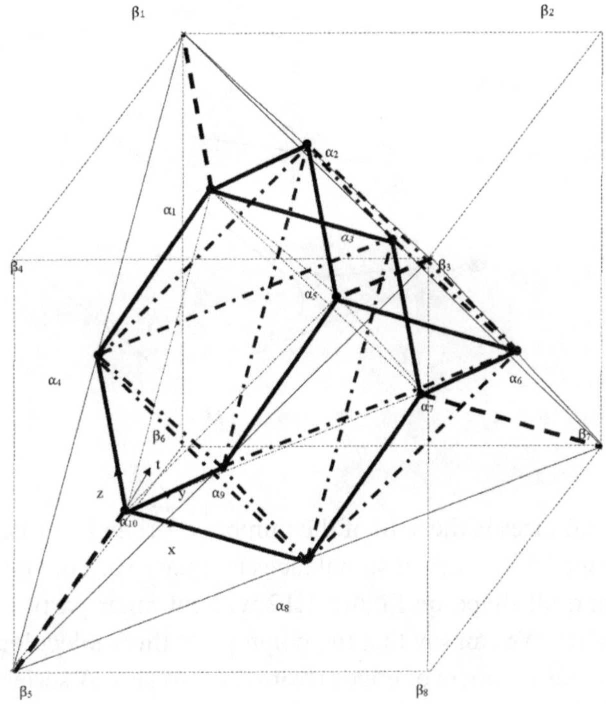

This polytope included 5 tetrahedrons, 6 prisms, 14 pyramids (Zhizhin, 2014 a, b; Zhizhin, Khalaj & Diudea, 2016).

A series of binary compound have a structure of the mineral rutile TiO_2. In this compound each titanium atom is surrounded by six the oxygen atoms in the octahedral coordination. To compounds with such structure to concern for example fluorides of copper, zinc, magnesium, manganese, cobalt, nickel. In work of Zhizhin & Diudea (2016) it is shown that octahedron with centrum have dimension 4. Therefore, all these structures have dimension 4. A series of binary compounds have structure of wurtzite - mineral ZnS, in which from compound ZnS with the structure of the adamantane zinc atom and sulfur atom have the tetrahedron coordination. The centrum each tetrahedron is vertex of another tetrahedron. The wurtzite structure has compounds ZnO, CdS, ZnS, BeO, AlN. The unit cell the lattice is a convex shape it is shown in Figure 12.

This figure is the unit cell structure of the wurtzite. In Figure 1.12, solid lines represent chemical bonds of the atoms, and the dotted lines are only geometric sense outlining contours of the figure. The number of vertices of this figure is equal to 14, i.e. f_0=14. The number of edges is equal to 29, i.e. f_1=29. The

Figure 12. The unit cell of the wurtzite

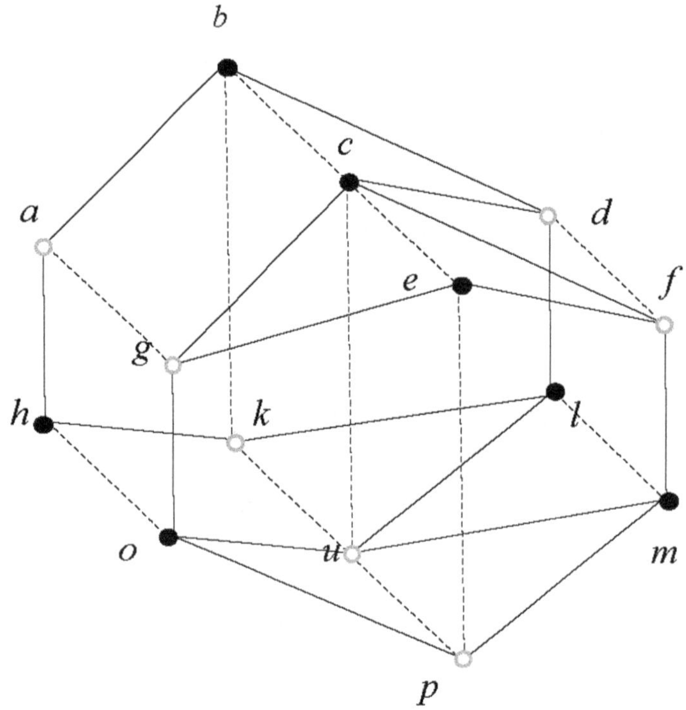

number of two - dimensional faces is the sum of the number of triangles (8) and number of quadrangles (13), i.e. f_2=21. The number of three - dimensional faces is equal to 6. The figures are *abcghkon, gceruo, cefump, cdfulm, bcdklu,* and all shape on Figure 1.12 without inner partitions. The dimension of this structure is 4 (Zhizhin, 2018). We can say that the polytope of this molecule is poly - incident, since its vertices are incident to various numbers of edges (from three to six). A series of binary compound have a structure of the mineral fluorite CdF_2, $MnCl_2$, $FeCl_2$, $CoCl_2$, $NiCl_2$, $ZnCl_2$, $CdCl_2$, Cu_2S, CaF_2, SrF_2, BaF_2, RbF_2, AlF_8, $MgCl_2$, $CaCl_2$, $SrCl_2$, $BaCl_2$, $PbCl_2$, $AlCl_3$, $SiCl_4$, Li_2O, Na_2O, K_2O, Rb_2O, Cs_2O, Li_2S, Na_2S, K_2S, Rb_2S. On example of compound $MnCl_2$ we look at the structure of fluorite. Isolate magnesium atoms lying at the centers of the cube faces (), and chlorine atoms (), forming a smaller cube inside the bigger cube, which are located at the vertices of magnesium atoms (Figure 13).

From Figure 1.13 it follows that the number of vertices is 14, i.e. f_0=14, the number of edges is 36, i.e. f_1=36, the number of flat faces is sum from number of triangles (24) and number of rectangles (6) (smaller cube faces), i.e. f_2=30. The number three - dimension shape is f_3=8. The dimension of unit cell of the fluorite is four (Zhizhin, 2018). This is also a poly - incident polytope, since six of its peaks are incident to four edges, and eight peaks are incident to six edges.

Oganov and his co - workers have found that under high pressure the structure of many compounds, including alkali metal compounds, acquires new unexpected properties (Zhung et al., 2013; Zhou et al., 2012; Zhu, Oganov & Lyakhov, 2013). In particular, it is shown that the structure of the sodium - chlorine compounds varies significantly (Zhung et al., 2013). The elementary cell of this structure is a cube with sodium atoms at its vertices, and an icosahedron with a center with chlorine atoms at its vertices is located inside the cube. This compound is denoted Pm3 – $NaCl_x$. The dimension of the unit cell Pm3

Figure 13. The unit cell of the fluorite
● *magnesium atom,* ▲ *chlorine atom*

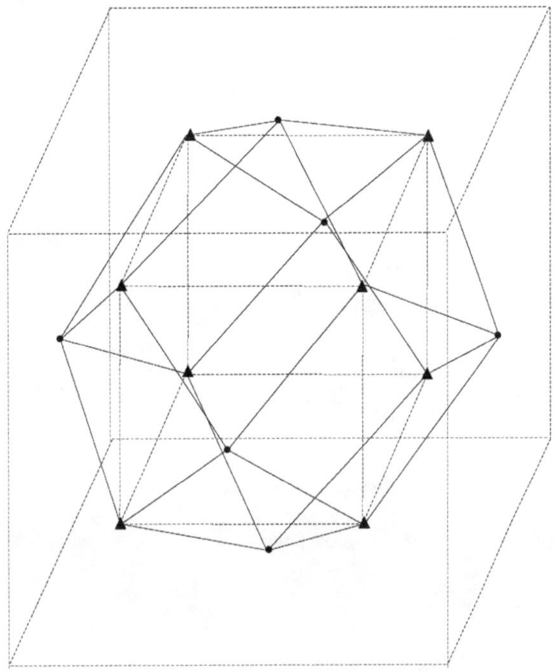

– $NaCl_x$ is 5 (Zhizhin, 2018). Figure 14 shows the structure of this compound, where sodium atoms are located at the vertices of the cube 13 - 19, and chlorine atoms are located at the vertices 1 - 12, 21.

The polytope in Figure 14 has 21 vertices, 48 different three - dimensional bodies, its edges have different incidence values in relation to the vertices.

Cyclic compounds have a dimension of more than four. For example, a compound $[(CH_3)_2AlF]_4$ has a dimension of five. Polytope (Figure 15) corresponding to these compounds contain many different three - dimensional bodies and are poly - incident (Zhizhin, 2018).

The bis(neopentyl) magnesium molecule has a dimension of 6 and is a 5 – cross - polytope with a center. It has 11vertices, it includes 108 different three - dimensional bodies. Its edges have different incidence values with respect to the peaks (Figure 16).

In general, it can be said that almost all chemical compounds form polytopes of higher dimension, consistent with the Euler - Poincaré equation. All of them are not correct. Deviations from the correctness of polytopes of all dimensions can be different. The main deviation from the correctness is the difference in the lengths of the edges within each of these polytopes. Many polytopes are poly - incidental. They simultaneously involve various three - dimensional figures (usually irregular), ribs in different numbers come from the vertices in the same polytope. The deviation of the real forms of chemical compounds from the correct polytopes is so significant that the scholastic constructions of polytopes of higher dimension are perplexing in the conditions of their complete ideality (correctness).

Figure 14. The structure of compound sodium and chlorine at high pressure

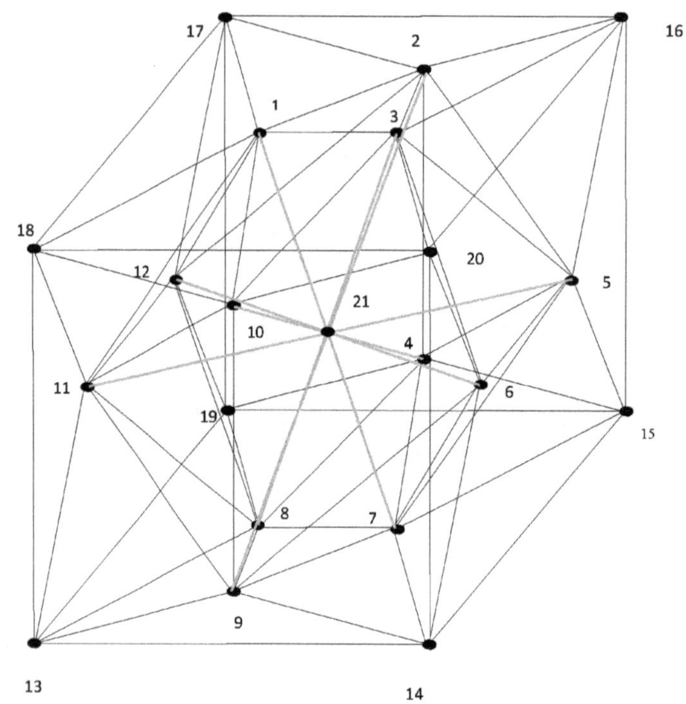

Figure 15. The convex polytope of cyclic compound $[(CH_3)_2AlF]_4$.

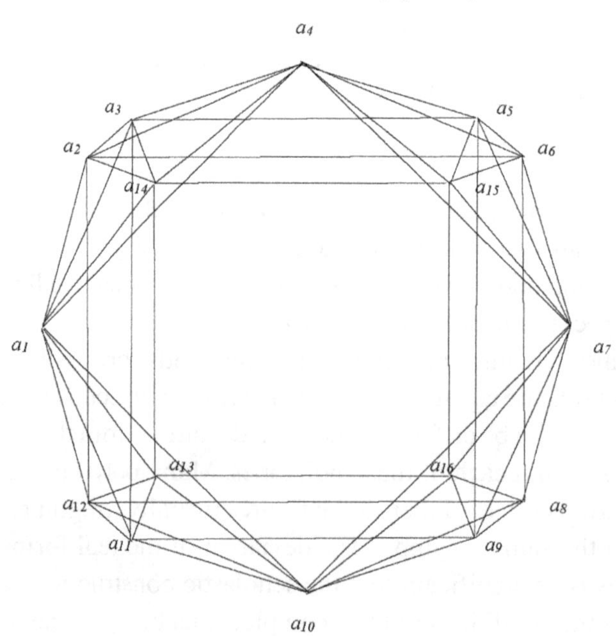

Figure 16. The structure of bis(neopentyl) magnesium molecule.

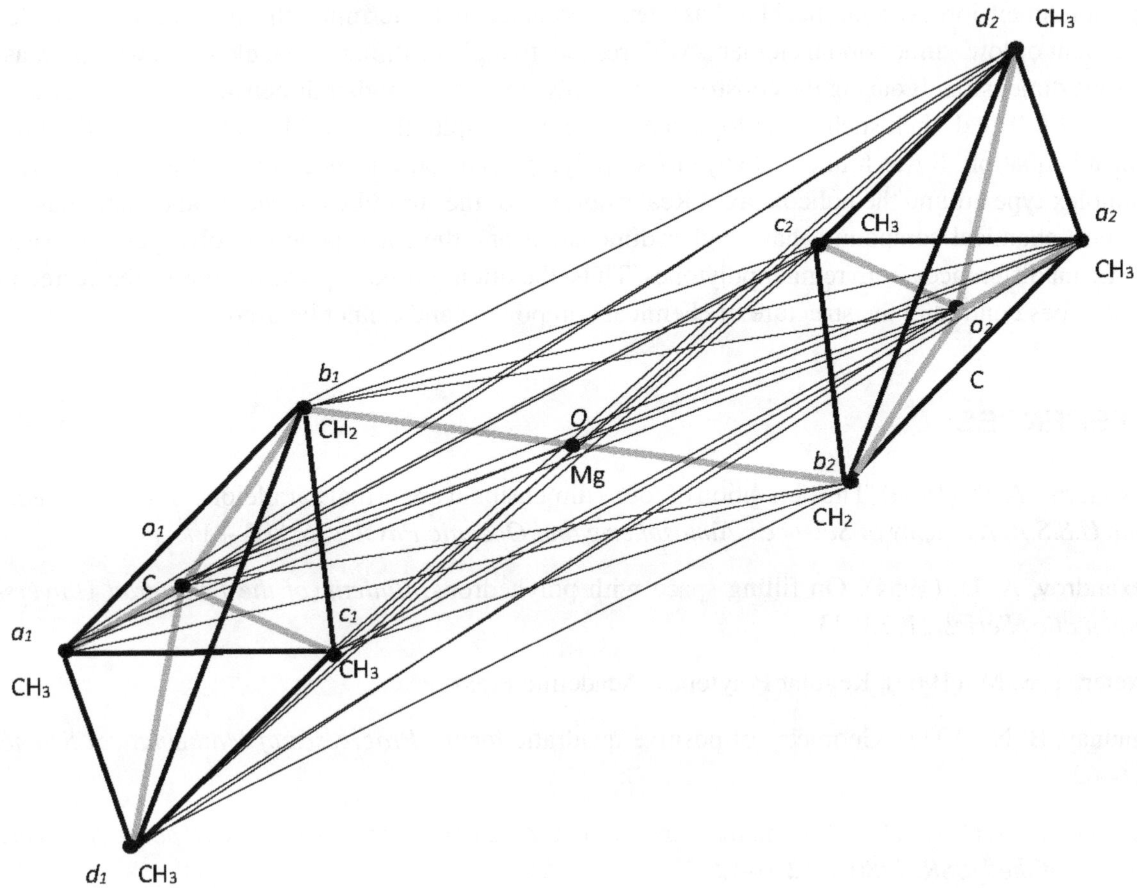

CONCLUSION

In 1880, Stringham for the first time demonstrated images of regular convex polytopes of dimension 4 and showed that between the numbers of faces of different dimensions of these polytopes a relation should be satisfied, which is a generalization of the Euler equation known for three - dimensional polyhedrons. In 1885, Poincaré derived topological methods to derive this generalized equation for both convex and non - convex polytopes, without determining the values of the numbers of elements of different dimensions that are part of the polytope. This chapter shows that this generalized equation of the relationship between the numbers of elements of different dimensions can be obtained by first determining the analytical expressions for the numbers of elements of different dimensions that are part of the polytope, from which the Euler -Poincaré equation automatically follows. This is illustrated by three types of irregular polytopes of dimension n that exist for any finite n. The fulfillment of the Euler - Poincaré equation is a necessary condition for the existence of a polytope of higher dimension. Without checking the fulfillment of the Euler - Poincaré equation, it cannot be argued that a polytope of higher dimension exists. In this chapter, analytical expressions for the numbers of geometric elements of various dimensions that are part of a polytope are defined for the first time without accepting the limiting conditions for the correctness of a polytope. The obtained analytical expressions for the numbers of elements of various dimensions that are part of the polytope allowed us to establish additional conditions for the existence of polytopes of higher dimension. An analytical law has been established that determines the increase in the incidence coefficient of low -dimensional elements with respect to higher - dimensional elements with increasing polytope dimension. If during the construction of polytopes of the higher dimension it turns out that this law is not fulfilled, then such a polytope does not exist, despite the possible fulfillment of the Euler - Poincaré equation. It was found that any cross - polytope is a consequence of the rotation of a polytope of simplex type around the helicoid axis. Real examples of the structures of chemical compounds show that most chemical compounds have higher dimensions and the corresponding polytopes have various significant differences from regular polytopes. Thus, the often is used approximations of the correctness of polytopes contradict the structure of chemical compounds and cannot be used.

REFERENCES

Alexandrov, A. D. (1934). The derivation of four - dimensional abnormal parallelohedrons. *Proceedings of the U.S.S.R. Academy of Sciences. Atmospheric and Oceanic Physics*, *6*, 803–817.

Alexandrov, A. D. (1954). On filling space with polyhedrons. *Bulletin of the Leningrad University. Mathematics Series.*, *1*, 33–43.

Coxeter, Y. S. M. (1963). Regular Polytopes. Academic Press.

Delaunay, B. N. (1937). Geometry of positive quadratic forms. *Progresses of Mathematical Sciences*, *3*, 16–62.

Delaunay, B. N. (1961). Proof of the main theorem of the theory of stereohedrons. *Reports. Academy of Sciences of the USSR*, *138*(6), 1270–1272.

Delaunay, B. N., & Sandakova, N. N. (1961). Theory of stereohedrons. *Proceeding Mathematical In. name V.A. Steklov of Academy Sciences of the USSR*, *64*, 28–51.

Gillespie, R. J. (1972). *Molecular Geometry*. Van Nostrand Reinhold Company.

Gillespie, R. J., & Hargittai, I. (1991). *The VSEPR Model of Molecular Geometry*. Allyn & Bacon.

Grünboum, B. (1967). *Convex Polytopes*. Springer.

Poincare A. (1895). Analysis situs. *J. de é Ecole Polytechnique, 1,* 1 – 121.

Stringham, W. I. (1880). Regular figures in *n* - dimensional space. *American Journal of Mathematics*, *3*(1), 1–14. doi:10.2307/2369441

Venkov, B. A. (1954). On a class of Euclidean polyhedrons. *Bulletin of the Leningrad University. Mathematics.*, *2*, 11–31.

Venkov, B.A. (1959). On the design of parallelohedrons. *Mathematical Collection, 49*(2), 207-224.

Voronoi, G. F. (1952). *Studies on primitive parallelohedrons. In Collected Works, 2*. Publishing House of the Academy of Sciences of the Ukrainian SSR.

Zhang, W., Oganov, A. R., Goncharov, A. F., Zhu, Q., Boulfelfel, S. E., Lyakhov, A. O., Stavrou, E., Somayazulu, M., Prakapenka, V. B., & Konopkova, Z. (2013). Unexpected Stable Stereochemical of Sodium Chlorides. *Science*, *342*(6165), 1502–1505. doi:10.1126cience.1244989 PMID:24357316

Zhizhin, G. V. (2013). Images of convex regular and semiregular *n* - dimensional polytopes. In *Proceedings of the 9th All - Russian Scientific School "Mathematical Research in the Natural Sciences"*. Apatity: Geological Institute, KSC RAS.

Zhizhin, G. V. (2014a). On the higher dimension in nature. *Biosphere*, *6*(4), 313–318.

Zhizhin, G. V. (2014b). *World – 4D*. Polytechnic Service.

Zhizhin, G. V. (2018). *Chemical Compound Structures and the higher Dimension of Molecules: Emerging research and Opportunities*. IGI Global. doi:10.4018/978-1-5225-4108-0

Zhizhin, G. V. (2019 a). *The Geometry of Higher-Dimensional polytopes*. IGI Global. doi:10.4018/978-1-5225-6968-8

Zhizhin, G. V. (2019 b). *Attractors and Higher Dimensions in Population and Molecular Biology: Emerging Research and Opportunities*. IGI Global. doi:10.4018/978-1-5225-9651-6

Zhizhin, G. V., & Diudea, M. V. (2016). Spaces of Nanoworld. In M. V. Putz & M. C. Mirica (Eds.), *Sustainable nanosystems, development, properties and applications* (pp. 214–236). IGI Global.

Zhizhin, G. V., Khalaj, Z., & Diudea, M. V. (2016). Geometrical and topology dimensions of the diamond. In A. R. Ashrafi & M. V. Diudea (Eds.), *Distance, symmetry and topology in carbon nanomaterials* (pp. 167–188). Springer. doi:10.1007/978-3-319-31584-3_12

Zhou, X. F. (2012). First Principles Determination of the Structure of Magnesium Borohydride. American Physical Society, 109, 245503-1 – 245503-5.

Zhu, Q., Oganov, A. R., & Lyakhov, A. O. (2013). Novel stable compounds in the Mg – O system under high pressure. *Phys. Chem.*, *15*(20), 7696–7700. doi:10.1039/c3cp50678a PMID:23595296

KEY TERMS AND DEFINITIONS

Divided Electron Pair: The binding electron pair, which simultaneously belongs to two atoms in the molecule.

Incidence in Polytopes: Incidence in polytopes define the number of elements of higher dimension the given element of lower dimension belongs.

N-Cross-Polytope: The convex polytope of dimension n in which opposite related of centrum edges not have connection of edge.

N-Simplex: The convex polytope of dimension n in which each vertex is joined by edges with all remain vertices of polytope.

Poly-Incident Polytopes: Polytopes in which elements of lower dimension have different incidence values for elements of higher dimension. Polytopes that are dual to polytopes products are poly-incident polytopes.

S- and P-Elements: The chemical elements in which is filling with electrons s- and p-orbitals of atoms.

Tetrahedral Coordination of Electron Pairs: The location of the electronic pairs of the outer and the pre-outer electron layer at the vertices of the tetrahedron.

Undivided Electron Pair: A non-bonding electron pair belonging to one atom in a molecule.

Chapter 2
Clusters of Chemical Compounds as Polytopes of the Highest Dimension

ABSTRACT

General analytical expressions are obtained for calculating the dimension of multi-shell clusters with a common center of shells in those cases when there is a metal atom in the center of the shells and when it is not. The shells can be in the shape of any body of Plato. It has been established that the gamma-copper cluster has the form of a cross-polytope of high dimension. The forms of clusters with ligands of the core of which is a chain of metal atoms or a metal polyhedron are geometrically investigated. It is shown that if the skeleton is a chain of metal atoms, then the cluster is polytope composed of two polytopes of higher dimension adjacent to each other along a flat section containing a chain. If the skeleton is a metal polyhedron, then a cluster of higher dimension has several ligand shells.

INTRODUCTION

In the previous chapter, the main types of polytopes of the highest dimension that form the molecules of chemical compounds in inorganic chemistry were listed. At present, the question of determining the structural principles of nature, in which clusters of atoms and molecules appear, is important. The literature presents known concrete clusters of chemical compounds (see, for example, Lord, Mackay & Ranganathan, 2006). However, the dimension of the clusters in accordance with the Euler – Poincaré equation (Poincaré, 1895) was not determined, but a priori was assumed to equal three. In this chapter, we will consider the main types of polytopes of higher dimension, which form clusters of chemical compounds. In it, starting with the study of clusters of intermetallic compounds (ligand - free clusters) they are sequentially considered. After that, clusters linear chains of metal atoms are considered with ligands attached to them and after that are considered clusters of metals with ligands at the polyhedral core of metal atoms. Analytical expressions are obtained for calculating the dimension of multi - shell clusters depending on the number of shells.

DOI: 10.4018/978-1-7998-6768-5.ch002

THE DIMENSION OF METALLIC CLUSTERS OF SEVERAL SHELLS IN THE FORM OF PLATO`S BODIES

The icosahedron is only one of Plato's bodies found in the skeleton of metal clusters (Gubin, 2019). Suppose there is a cluster, each shell of which is a convex regular three - dimensional polyhedron (Plato's body) with some possible number of vertices *n* and the same for all shells of this cluster. The flat sides of the shells are regular *m* - corner. All shells of this cluster differ only in size and have a common center. Assume that the number of flat edges in the shell is *j*. One denoted S_j a shell satisfying the indicated conditions. Then the following statement is true.

Theorem 2.1

The dimension *d* of a cluster of *N* shells with a common center is $N + 2$, if there is no atom in the common center, and is equal to $N + 3$, if there is an atom in the common center.

Proof

Denote by the symbol *t* the number of edges emanating from each vertex of the shell S_j. Then the total number of edges of the shell is $nt/2$, and the number of faces of the shell is $j=nt/m$. The number *j* in Plato's body also gives the form of the shell, i.e. sets the numbers *n, m, t* (Table 1).

Table 1. Defining Plato bodies by the number of two - dimensional faces of the outer shell

J	*n*	*m*	*T*	**Polyhedron**
4	4	3	3	Tetrahedron
6	8	4	3	Cube
8	6	3	4	Octahedron
12	20	5	3	Dodecahedron
20	12	3	5	Icosahedron

Let a shell with the number of faces *j* and be given. Therefore, the number of vertices n_j, the number of sides m_j at the flat face, the number of edges t_j emanating from each vertex are given. Consider two shells of different sizes with a common center, arranged so that every two corresponding vertices of both shells are on the same straight line connecting them with a common center (there is no atom in the center). Then the space between the shells is filled with prisms, the bases of which are flat faces of the larger and smaller shells. The number of these prisms is equal to the number of faces of the shell, i.e. *j*. The number of vertices in a polytope of two shells is $2n_j=f_0$. The number of edges of this polytope, taking into account the edges connecting the corresponding vertices of the shells, is $n_jt_j+n_j=f_1$. The number of flat edges in a polytope is equal to twice the number of flat edges in each shell and the number of edges in one of the shells, i.e. $2j+\frac{n_jt_j}{2}=f_2$. The number of three - dimensional figures in the polytope is equal to $j+2=f_3$. Substitute these numbers in the equation (2.1) of the Euler - Poincaré (Poincaré, 1895)

$$\sum_{i=0}^{d-1}(-1)^i f_i(P) = 1 + (-1)^{d-1} . \tag{2.1}$$

In (2.1) $f_i(P)$ is the number of faces with dimension i in polytope P with dimension d. Thus, we can get

$$2n_j - n_j(t_j + 1) + 2j + \frac{n_j t_j}{2} - j - 2 = (n_j - \frac{n_j t_j}{2} + j) - 2 = 0.$$

There $(n_j - \frac{n_j t_j}{2} + j) = 2$ on the equation (1) for the polytope of dimension 3.

This proves that a polytope composed of two shells S_j with a common center (in the absence of an atom in the center) has dimension 4 for any possible j.

If the cluster consists of three shells S_j with a common center, then $f_0=3n_j$. The number of edges in this cluster, taking into account the edges connecting of the corresponding vertices in shells S_j, is equal to $f_1 = 3\frac{n_j t_j}{2} + 2n_j$. The number of two - dimensional faces is $3j+n_j t_j = f_2$. The number of three - dimensional faces is $2j+3=f_3$. The number of four - dimensional faces is $f_4 = C_3^2 = 3$. Substitute these numbers in the equation (2.1) Euler - Poincaré

$$3n_j - n_j(\frac{3}{2}t_j + 2) + 3j + n_j t_j - (2\,\mathrm{j} + 3) + 3 = n_j - \frac{n_j t_j}{2} + j = 2.$$

This proves that dimension of the cluster of tree shells S_j is equal 5.

One can write a general expression for the numbers of elements of different dimensions in a cluster with an arbitrary number N of shells S_j and get a formula for calculating its dimension. So, in the N - shell cluster of shells S_j, we have

$$\begin{aligned} &f_0 = Nn_j, f_1 = N\frac{n_j t_j}{2} + (N-1)n_j, f_2 = Nj + (N-1)\frac{n_j t_j}{2}, f_3 = (N-1)\,\mathrm{j} + N, \\ &f_4 = C_N^2, f_5 = C_N^3, \ldots, f_{N+1} = C_N^{N-1}. \end{aligned} \tag{2.2}$$

Substituting the values (2.2) in the equation (2.1) and opening the brackets, one can see that in this case the left side of the Euler - Poincaré equation (2.1) takes the form

$$\sum_{i=0}^{d-1}(-1)^i f_i(P) = 2 - N + \sum_{k=2}^{N-1} C_N^k (-1)^k . \tag{2.3}$$

To calculate the sum on the right side of equation (2.3), we use the expression for the alternating series of combinations (Vilenkin, 1969)

$$C_N^0 - C_N^1 + C_N^2 - ... + (-1)^N C_N^N = 0. \tag{2.4}$$

From the series (2.4), taking into account the equalities $C_N^0 = C_N^N = 1, C_N^1 = N$, it follows that the sum $\sum_{k=2}^{N-1} C_N^k (-1)^k$ in the right side of equation (2.3) is N - 2 for N even and N for N odd. Therefore, the right side of equation (2.3) coincides with the right side of equation (2.1). This proves that the figures in question are closed convex polytopes and they satisfy the Euler - Poincaré equation.

At the same time, from equation (2.2), since $f_{N+2} = C_N^N = 1$, it follows that the dimension of a cluster of N shells S_j with a common center if at center not atoms is equal $d = N + 2$.

Now consider a cluster of several shells S_j with a common center in which the atom is located. The shell S_j with a central atom has $n_j+1=f_0$ vertices, $\frac{n_j t_j}{2} + n_j = f_1$ edges, $\frac{n_j t_j}{2} + j = f_2$ flat faces, $j+1=f_3$ three – dimensional faces. Substituting the values in the equation (2.1) one can get

$$n_j + 1 - (n_j + \frac{n_j t_j}{2}) + j + \frac{n_j t_j}{2} - (j-1) = 0.$$

This proves that the shell S_j with its center has a dimension of 4.

If one continue the edges going from the center of the shell S_j to its vertices and at the appropriate distance arrange more atoms forming the second a shell S_j of a larger size. Such a construction of two the shells S_j with a common atom in center will have $2n_j+1=f_0$ vertices, $n_j t_j+2n_j=f_1$ edges, $2j+n_j t_j=f_2$ flat faces, $2j+2=f_3$ three - dimensional faces, $f_4 = C_3^2 = 3$ four - dimensional faces. Substituting these values in the equation (2.1), one can get

$$2n_j + 1 - 2n_j(1 + t_j / 2) + 2j + n_j t_j - (2j + 2) + 3 = 2.$$

This proves that the two shells S_j with common atoms in the center has a dimension 5.

If one continue again the edges coming from the center of the shells S_j, and on these edges one construct the third shell S_j of a still larger size, then such a construction will have $f_0=3n_j+1$ vertices, $f_1 = 3\frac{n_j t_j}{2} + 3n_j$ edges, $f_2 = 3j + 3\frac{n_j t_j}{2}$ flat faces, $f_3=3j+3$ three - dimensional faces, $f_4 = C_4^2 = 6$ four - dimensional faces, $f_5 = C_4^3 = 4$ five - dimensional faces. Substituting these values into the equation (2.1), one can get

$$3n_j + 1 - 3(n_j + \frac{n_j t_j}{2}) + 3(j + \frac{n_j t_j}{2}) + 3(j+1) + 6 - 4 = 0.$$

This proves that cluster of the tree shells S_j with common atom at center has a dimension 6.

Can write a general expression for the numbers of elements of different dimensions in a cluster with a center for arbitrary number N of shells S_j and give a formula for calculating its dimension. Then in the cluster of N shell S_j with atom in a common center, we have

$$f_0 = Nn_j + 1, f_1 = N\frac{n_j t_j}{2} + Nn_j, f_2 = Nj + N\frac{n_j t_j}{2}, f_3 = Nj + N,$$
$$f_4 = C_{N+1}^2, f_5 = C_{N+1}^3, \ldots, f_{N+2} = C_{N+1}^N. \quad (2.5)$$

Substituting these values into equation (2.1), one can see that in this case the left side of the Euler - Poincaré equation (2.1) takes the form

$$\sum_{i=0}^{d-1}(-1)^i f_i(P) = 1 - N + \sum_{k=2}^{N} C_{N+1}^k (-1)^k. \quad (2.6)$$

To calculate the sum on the right side of equation (6), we use the expression (2.2). From the series (2.2), taking into account the equalities $C_N^0 = C_N^N = 1, C_N^1 = N$, it follows that the sum $\sum_{k=2}^{N} C_{N+1}^k (-1)^k$ in the right side of equation (2.4) is $N + 1$ for N even and N -1for N odd. Therefore, the right side of equation (2.4) coincides with the right side of equation (2.1). This proves that the figures in question are closed convex polytopes and they satisfy the Euler - Poincaré equation.

At the same time, from equation (2.3), since $f_{N+3} = C_{N+1}^{N+1} = 1$, it follows that the dimension of a cluster of N shells S_j with an atom in a common center is equal $d = N + 3$.

Theorem 2.1 allows one to easily and quickly determine the dimensions of many clusters. In particular, the monograph (Zhizhin, 2019 b, 2020) showed that the Bergman, Mackay, and Samson (Bergman et al., 1952, 1957; Mackay, 1962; Samson, 1972; Lord, Mackay, & Ranganathan, 2006) clusters can be considered as consisting of a number of icosahedral shells with a common center. In this case, an atom may be absent in the common center or an atom may be present. The dimension of these clusters as proved in the theorem depends on the number of shells in the cluster.

From Theorem 2.1, for example, it follows that the dimension of a giant palladium cluster (Vargaftik, et al., 1985), containing 561 atoms in five shells, is 8. The claims of some authors that the palladium cluster in this case is an E_8 lattice (Shevchenko, 2011) are groundless. The lattice E_8, as you know (Conway, & Sloane, 1988), is a collection of points in an eight -dimensional space with coordinates (±1,±1,0,0,0,0,0,0), where units can stand anywhere on the line with arbitrary signs, as well as points with coordinates

$$(\pm\frac{1}{2},\pm\frac{1}{2},\pm\frac{1}{2},\pm\frac{1}{2},\pm\frac{1}{2},\pm\frac{1}{2},\pm\frac{1}{2},\pm\frac{1}{2}).$$

Obviously, this lattice has nothing to do with the structure of a cluster consisting of five icosahedral shells, although it is for such a cluster that the number of atoms gives 561 (Lord, et al., 2006). It is assumed (Coxeter, 1963) that the lattice E_8 corresponds to the polytope of Gosset (Gosset, 1900), which draws from simplexes and cross - polytopes. But from the previous it follows that the polytope corresponding to the giant palladium cluster does not include either simplexes or cross - polytopes.

The structure of γ - brass was described as early as 1926 in the concept of a lattice of cubic cells (Bradley, &Thewlis, 1926). Pauling, considering the structure of γ - brass, noticed that this structure is icosahedral (Pauling, 1960). Alloys of γ - brass, as well as other intermetallic alloys, are conveniently considered using tetrahedrons (Lord, Mackay & Ranganathan, 2006). As part of the last presentation, it was shown (Zhizhin, 2019 b, 2019 c) that a γ - brass cluster has the dimension $d = 4 + 3n$, where n is the number of cluster shells ($n = 0, 1, 2,...$). The cluster is a d - cross – polytope.

LINEAR HOMO: ELEMENT METAL CHAINS

An example of a linear homo - element metal chain is shown in Figure 1 (Smart, Cook, & Woodward, 1977; Evans, Okrasinski, Pribula, & Norton, 1976; Gochin, & Moss, 1980).

Figure 1. Schematic structure of cluster connection $Os_3(CO)_{12}R_2$

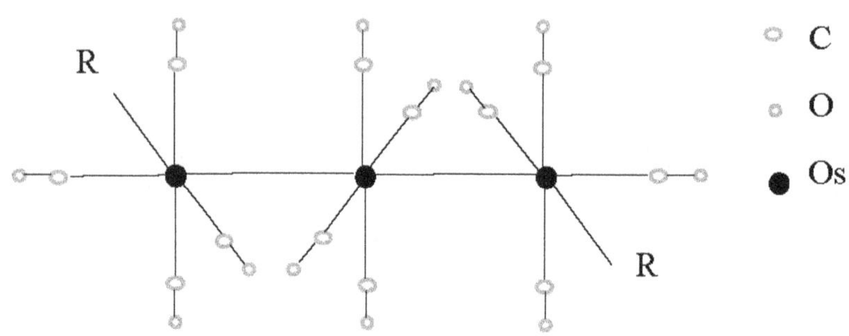

In this example, we illustrate the sequence of actions to calculate the dimension of the metal chain. When determining the dimension of metal chains, atoms that are linearly located and do not have branching of the chain at their location do not matter (Zhizhin, 2018). At the same time, not only metal atoms, but also ligand atoms are of importance for determining the dimension of a compound. Therefore, Figure 1 can be represented as Figure 2.

In Figure 2.2, the red edges correspond to chemical bonds, and the black edges serve only to create a convex closed figure. It can be seen that each link of the metal chain creates an octahedron with the center as a convex figure. Moreover, the center of each octahedron is simultaneously the vertex of another neighboring octahedron or two neighboring octahedrons. In Figure 2.2 at the centers of octahedrons are disposed Os atoms (vertices 2, 3, 4). At the vertices 1, 6, 7, 8, 9, 10, 11, 12, 13, 14, 15, 17 are disposed the functional group CO, at the vertices 5, 16 are disposed the functional group R (Zhizhin, 2016). It is easy to prove that an octahedron with a center has a dimension of 4. To do this, refer to Figure 3, which shows separately the first of the chain octahedron.

Figure 2. The geometric structure of cluster connection $Os_3(CO)_{12}R_2$

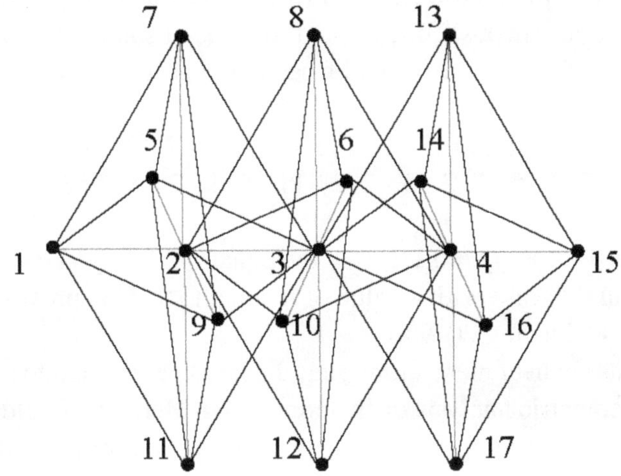

Figure 3. Octahedron with a center

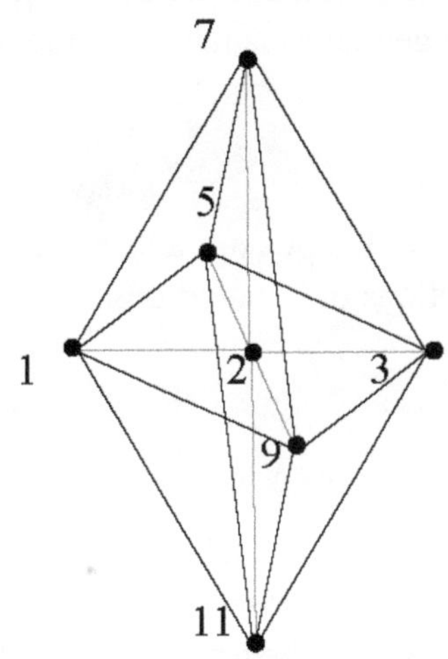

Polytope on Figure 2.3 has 7 vertices (f_0=7), 18 edges (f_1=18)

1 - 2, 2 - 3, 2 - 5, 2 - 9, 1 - 5, 5 - 3, 3 - 9, 1 - 9, 2 - 4, 2 - 11, 1 - 7, 5 - 7, 3 - 7, 7 - 9, 1 - 11, 5 - 11, 9 - 11, 3 - 11.

Polytope composed of eight tetrahedrons

1 - 2 - 9 - 11, 2 - 9 - 3 - 11, 2 - 5 - 3 - 11, 1 - 2 - 5 - 11, 1 - 2 - 5 - 7, 2 - 5 - 3 - 7, 2 - 9 - 3 - 7, 1 - 2 - 9 - 7.

To these three - dimensional figures you need to add an octahedron without a center 1 - 5 - 3 - 7 - 9 - 11. Thus, the total number of three - dimensional figures is equal to 9 (f_3=9). Each flat face of any of these nine polyhedrons is simultaneously one of the flat faces of some other of these nine polyhedrons. Therefore, the total number of two - dimensional faces in the polytope in Figure 2.3 is 16 (f_2=16):

1 - 9 - 11, 9 - 3 - 11, 5 - 3 - 11, 1 - 5 - 11, 5 - 3 - 7, 1 - 5 - 7, 1 - 7 - 9, 7 - 3 - 9, 2 - 9 - 11, 2 - 3 - 11, 1 - 2 - 11, 5 - 2 - 11, 1 - 2 - 7, 2 - 5 - 7, 2 - 3 - 7, 2 - 11 - 9, 1 - 2 - 5, 2 - 5 - 3, 2 - 9 - 3, 1 - 2 - 9.

Substituting the obtained values f_i(i=0,1,2,3) in the equation (2.1) you can see that the Euler – Poincaré equation is satisfied in this case with a value of n equal to 4. This proves that the dimension of an octahedron with center is 4 (Zhizhin, 2019 a).

In a linear homogeneous chain of metal atoms, tetrahedrons penetrate into each other (Figure 2). This greatly complicates the geometric analysis of the compound. Here one could use the same method of forming a convex figure, corresponding to a chain, as in the analysis of individual molecules. However, the construction of an additional system of edges, taking into account the possible growth of the chain, greatly complicates the convex figure. In this case, the dimension of the whole figure, the corresponding chain, will increase dramatically with the growth of the chain. This approach seems to be irrational. It would be logical to present these chains in the form of a convex figure consisting of a sequence of convex polytopes adjacent to each other in the geometric analysis of the chains of metal atoms.

Theorem 2.2

Linear homo - element metal chains with ligands are a sequence of convex polytopes of dimension 4 adjacent to each other along whole sections of octahedrons passing through the center and the vertices of the octahedrons.

Proof

Taking into account that metal chains with two chain links exist (Cotton, & Walton, 1982), consider two adjacent chain links in Figure 2.2. To create a closed shape from two octahedrons interconnected, it is necessary to fill the free space between the octahedrons. This can be done by connecting the vertices of two octahedrons, which were not yet connected to each other (Figure 4). These compounds (7 - 8, 5 - 6, 9 - 10, 11 - 12) in Figure 2.4 are indicated in green.

In Figure 2.4, the region in which the octahedrons enter each other

5 - 7 - 8 - 6 - 10 - 12 - 9 - 11 - 2 - 3

should be distinguished. As well as two regions in which only of the octahedron is present

1 - 5 - 7 - 9 - 11 - 2, 4 - 8 - 6 - 10 - 12 - 3

should be distinguished. An area in which only one the octahedron is present, for example,

1 - 5 - 7 - 9 - 11 - 2 (Figure 5)

Figure 4. Convex figure built on a metal chain of two metal atoms with ligands

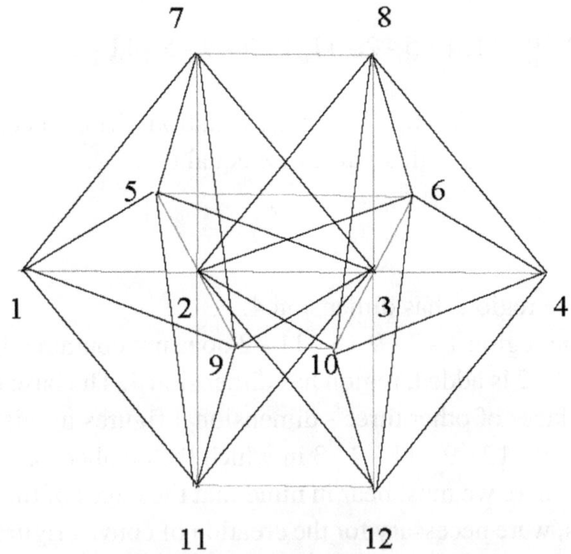

Figure 5. An area in which only one the octahedron is present

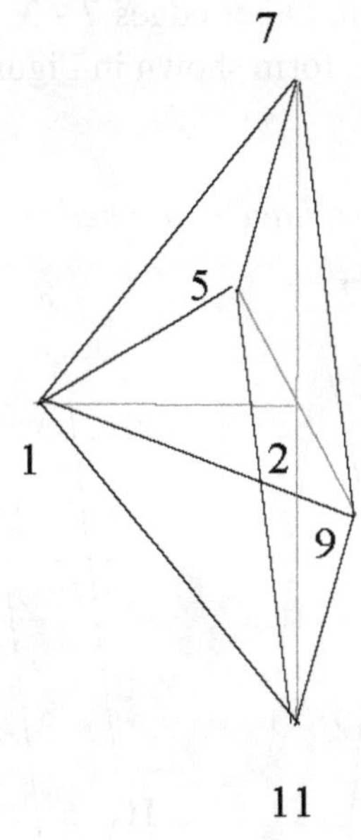

has 6 vertices (f_0=6), 13 edges (f_1=13)

1 - 5, 1 - 7, 1 - 2, 1 - 9, 1 - 11, 5 - 7, 7 - 9, 9 - 11, 5 - 11, 2 - 5, 2 - 7, 2 - 9, 2 - 11,

12 flat two - dimensional faces (f_2=12)

1 - 7 - 5, 1 - 5 - 2, 1 - 7 - 2, 1 - 9 - 7, 1 - 2 - 9, 1 - 9 - 11, 1 - 2 - 11, 1 - 5 - 11, 2 - 5 - 7, 2 - 7 - 9, 2 - 11 - 9, 5 - 2 - 11, 5

three - dimensional areas (f_3=5)1

- 5 - 2 - 7, 1 - 2 - 7 - 9, 1 - 2 - 9 - 11, 1 - 5 - 2 - 11, 1 - 7 - 9 - 5 - 11 - 2.

Substituting the obtained values f_i(i=0,1,2,3) in the equation (2.1), you can see that the Euler - Poincaré equation is satisfied in this case with a value of n equal to 4

$$6 - 13 + 12 - 5 = 0.$$

This proves that each of these regions has dimension 4.

It should be noted that the region 1 - 7 - 9 - 5 - 11 - 2 does not contain edge 1 - 2, therefore it has dimension 3. As soon as edge 1 - 2 is added, region has dimension 4. The base of this figure is constructed from four triangles that are faces of other three - dimensional figures in this polytope.

The region 5 - 7 - 8 - 6 - 10 - 12 - 9 - 11 - 2 - 3 in which the octahedrons penetrate into each other is quite complicated. However, here we must bear in mind that the edges of the octahedrons in this region, which are not valence bonds, were necessary for the creation of convex figures (octahedrons) when considering these figures separately. When the general region, which includes octahedrons, is considered, their purpose is lost, due to the introduction of other edges that create a unifying closed convex figure. Therefore, in this region one should leave only the edges corresponding to the valence bonds and the edges forming the boundary of the region. Other edges 7 - 3, 8 - 2, 5 - 3, 2 - 6, 2 - 10, 3 - 9 must be removed. In this case, the region takes the form shown in Figure 6.

Figure 6. Simplified region penetrated by adjacent octahedrons in chains

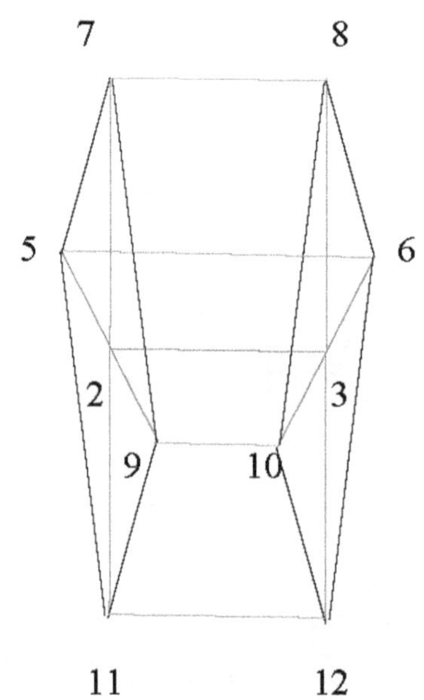

The polytope on Figure 2.6 has 10 vertices (f_0=10), 21 edges (f_1=21)

7 - 8, 6 - 5, 2 - 3, 9 - 10, 11 - 12, 5 - 7, 7 - 9, 9 - 11, 11 - 5, 2 - 5, 2 - 7, 2 - 9, 2 - 11, 8 - 6, 6 - 12, 12 - 10, 10 - 8, 3 - 8, 3 - 6, 3 - 12, 3 - 10,

16 two - dimensional faces (f_2=16)

5 - 7 - 8 - 6, 5 - 2 - 6 - 3, 7 - 8 - 2 - 3, 7 - 8 - 9 - 10, 9 - 10 - 11 - 12, 2 - 3 - 11 - 12, 2 - 3 - 9 - 10, 2 - 3 - 5 - 6, 2 - 5 - 7, 2 - 7 - 9, 2 - 9 - 11, 2 - 5 - 11, 3 - 6 - 8, 3 - 6 - 12, 3 - 10 - 12, 3 - 8 - 10,

5 tree - dimensional faces (f_3=5)

2 - 5 - 7 - 3 - 8 - 6, 2 - 7 - 9 - 3 - 8 - 10, 2 - 9 - 11- 3 - 10 - 12, 2 - 5 - 11 - 3 - 6 - 12, 5 - 7 - 8 - 6 - 9 - 10 - 11 - 12 - 2 - 3 (not contain edge 2 - 3) .

Substituting the obtained values f_i(i=0,1,2,3) in the equation (2.1) you can see that the Euler - Poincaré equation is satisfied in this case with a value of n equal to 4

$10 - 21 + 16 - 5 = 0.$

This proves that polytope on Figure 6 has dimension 4.

Let us return to the chain containing three metal atoms (Figure 2). Its image using spaces of higher dimension, according to the developed description, has the form shown in Figure 7.

Figure 7. Image of a chain with three metal atoms in a higher dimension space

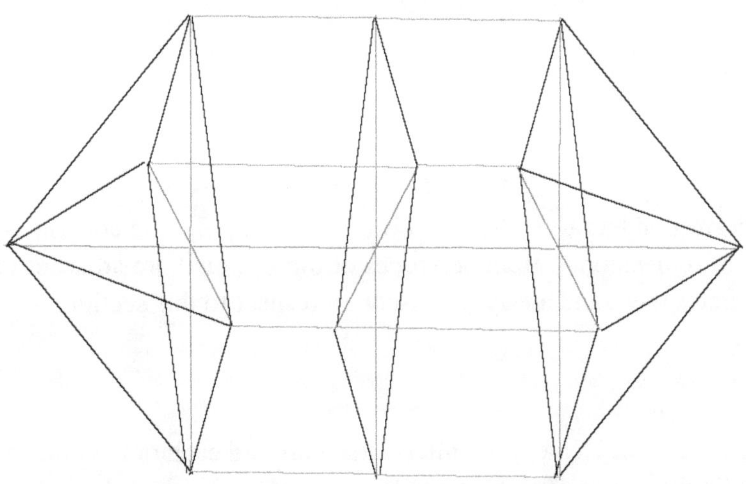

It is a chain of convex closed polytopes of dimension 4 adjoining each other along vertical sections of octahedrons passing through the center of the octahedrons and their vertices. It is obvious that the chain of polytopes can be continued. It will include a chain of polytopes in Figure 2.6, in alternating positions. At the ends of the chain are the polytopes in Figure 2.5, also in reversed positions. This is the spatial implementation of a polymetallic linear chain using a higher dimension space. Obviously, other linear metal chains with similar attachments of ligands will have the same appearance. For example, metal chains $[Mn_3(CO)_{14}]^-$ (Bau, Kirtley, Sorrel, & Winarko, 1974), in which the atoms of the metal are manganese atoms, and all ligands are functional groups CO.

CLUSTER CONNECTIONS WITH A BACKBONE IN THE FORM OF NONLINEAR HOMO-ELEMENT METAL CHAIN

The existence of cluster compounds with a skeleton in the form of nonlinear (bent) homo - element metal chains was experimentally established. An example of such a compound can be hydride $H_2Os_3(CO)_{12}$ (Aleksandrov, Zolnikova, Kritskaya, & Pods, 1980). A diagram of it is presented in Figure 8.

Figure 8. One of the possible of hydride $H_2Os_3(CO)_{12}$ structures

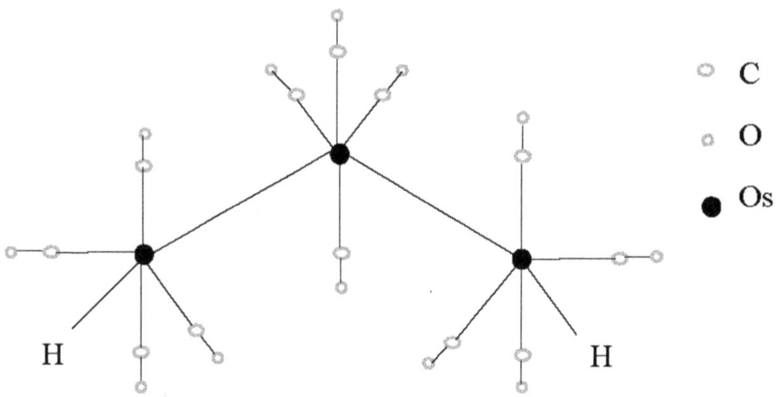

Theorem 2.3

The spatial figure correspondent to hydride $H_2Os_3(CO)_{12}$ is a polytope consists of two parts, each of which is a polytope of dimension 4. Both polytopes of dimension 4 are adjacent to each other along a flat section that separates them and are symmetric with respect to this section.

Proof

The octahedral coordination of elements manifests itself around each of the three atoms of the osmium and the octahedrons (in the center of each octahedron, an osmium atom is located) penetrate each other. The spatial image of this compound is shown in Figure 9.

Figure 9. The spatial image of hydride $H_2Os_3(CO)_{12}$

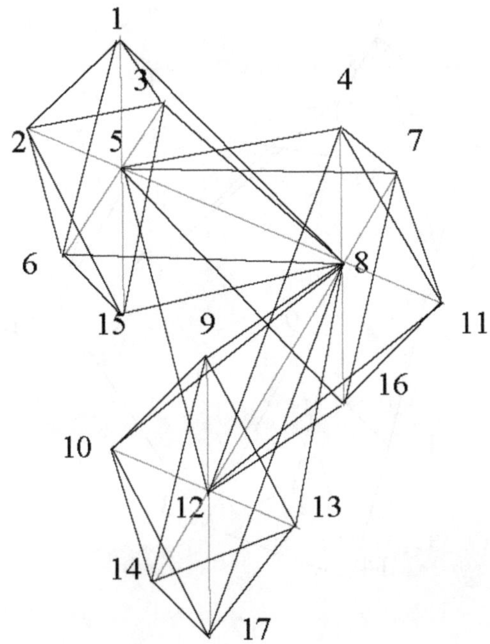

On Figure 2.9 the valence bonds are marked in red, it is taken into account that the metal - metal bond length is substantially greater than the bond length of the metal with the ligands. The edges of the octahedrons are marked in black. At the vertices 5, 8, 12 (centers of octahedrons), atoms of the osmium are located; hydrogen atoms are located at the vertices 2, 14; at vertices 3, 4, 6, 7, 9, 10, 11, 13, 14, 15, 16, 17 functional groups CO are located. Each octahedron with a center has a dimension of 4. However, the common connection of octahedrons is a non - convex and closed figure. To bring the image of this compound to the form of a figure closed and convex, one should fill the space between the octahedrons with additional figures. The result of such a construction will give a complex figure for analysis. However, here we must recall the procedure applied in the previous section. It is necessary to connect the vertices cut off from each other by the edges and remove the edges of the octahedrons, which have lost their significance for the creation of a convex closed figure around an Os atom and do not solve this problem in a combined figure. At same time, all the valence bonds are preserved and the edges delineating the outer contour of the figure are preserved. So, let is connect vertices of octahedrons 1, 4, 9 edges with each other (blue color), connect vertices of octahedrons 15, 16, 17 edges with each other (blue color). Connect vertex 3 with vertex 7, connect vertex 6 with vertex 10, connect vertex 11 with vertex 13 (all edges are black). Remove the edges 5 - 7, 1 - 8, 3 - 8, 12 - 11, 13 - 8, 5 - 16, 6 - 8, 15 - 8, 12 - 16. The number of remote edges turned out to be equal to the number of newly entered edges. Then all valence compounds will be inside the new closed convex figure. Moreover, the figure turned out to be a folded of two convex closed polytopes, applied to each other along a two - dimensional (horizontal) flat section 2 - 3 - 7 - 11 - 13 - 14 - 10 - 6 - 2 made up of several sections. The image of the upper polytope (above the horizontal section) is shown in Figure 10.

Figure 10. The top of the polytope corresponding to the compound $H_2Os_3(CO)_{12}$

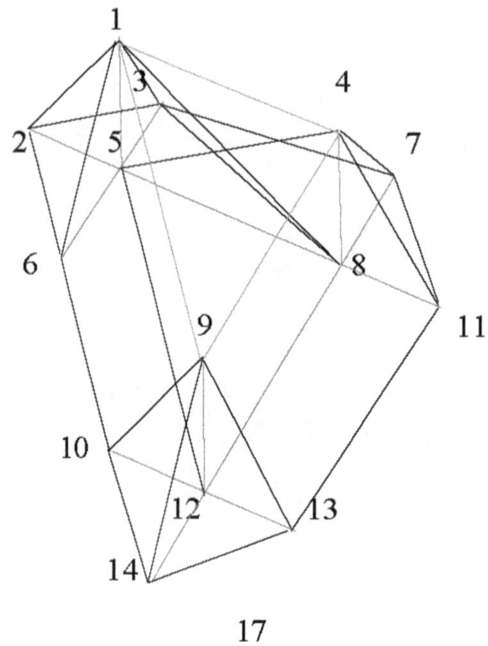

The image of the lower polytope (below the horizontal section) is shown in Figure 11.

Figure 11. The bottom of the polytope corresponding to the compound $H_2Os_3(CO)_{12}$

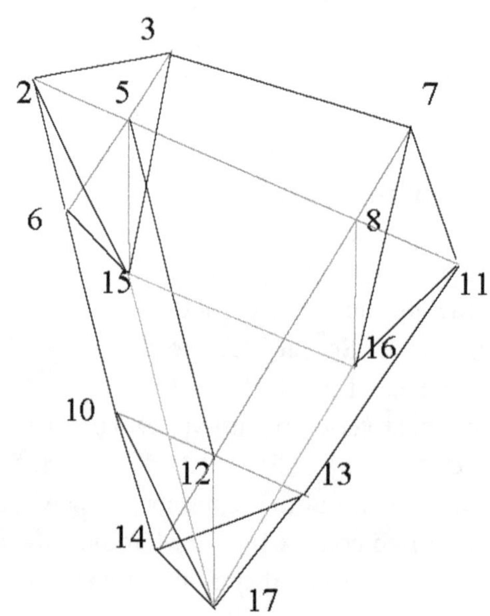

One define the dimension of the polytope in Figure 2.10.The polytope on Figure 2.10 has 14 vertices (f_0=14), 32 edges (f_1=32)

1 - 2, 1 - 3,1 - 4, 1 - 5, 1 - 6, 1 - 9, 2 - 5, 2 - 6, 3 - 5, 3 - 7, 4 - 9, 4 - 7, 4 - 11, 4 - 8, 5 - 8, 5 - 6, 6 - 10, 7 - 8, 7 - 11, 8 - 11, 8 - 12, 9 - 10, 9 - 12, 9 - 13, 9 - 14, 10 - 12, 10 - 14, 11 - 13, 12 - 13, 12 - 14, 13 - 14, 5 - 12,

28 flat two - dimension sections (f_2=28)

1 - 4 - 9, 1 - 2 - 6, 1 - 2 - 3, 1 - 2 - 5, 1 - 5 - 3, 1 - 4 - 5 - 8, 1 - 5 - 6, 1 - 3 - 4 - 7, 1 - 5 - 9 - 12, 1 - 6 - 9 - 10, 2 - 3 - 5, 2 - 5 - 6, 3 - 5 - 7 - 8, 4 - 7 - 8, 4 - 7 - 11, 4 - 8 - 11, 4 - 9 - 11 - 13, 4 - 8 - 9 - 12, 5 - 6 - 10 - 12, 5 - 8 - 12, 7 - 8 - 11, 8 - 11 - 12 - 13, 9 - 10 - 12, 9 - 12 - 13, 9 - 10 - 14, 9 - 13 - 14, 10 - 12 - 14, 12 - 13 - 14,

10 tree – dimension area (f_3=10)

1 - 2 - 3 - 5, 1 - 2 - 5 - 6, 1 - 3 - 4 - 7 - 5 - 8, 1 - 4 - 9 - 5 - 8 - 12, 1 - 9 - 5 - 12 - 6 - 10, 4 - 7 - 8 - 11, 4 - 8 - 11 - 9 - 12 - 13, 9 - 12 - 13 - 14, 1 - 2 - 3 - 4 - 5 - 6 - 7 - 8 - 9 - 10 - 11 - 12 - 13 - 14 (without edges 1 - 5, 4 - 8, 9 - 12).

Substituting the obtained values f_i(i=0,1,2,3) in the equation (2.1) you can see that the Euler - Poincaré equation is satisfied in this case with a value of n equal to 4

$$14 - 32 + 28 - 10 = 0.$$

This proves that polytope on Figure 2.10 has dimension 4.

One define the dimension of the polytope in Figure 2.11.The polytope on Figure 2.11 has 14 vertices (f_0=14), 32 edges (f_1=32)

15 - 2, 15 - 3,15 - 16, 15 - 5, 15 - 6, 15 - 17, 2 - 5, 2 - 6, 3 - 5, 3 - 7, 16 - 17, 16 - 7, 16 - 11, 16 - 8, 5 - 8, 5 - 6, 6 - 10, 7 - 8, 7 - 11, 8 - 11, 8 - 12, 17 - 10, 17 - 12, 17 - 13, 17 - 14, 10 - 12, 10 - 14, 11 - 13, 12 - 13, 12 - 14, 13 - 14, 5 - 12,

28 flat two - dimension sections (f_2=28)

15 - 16 - 17, 15 - 2 - 6, 15 - 2 - 3, 15 - 2 - 5, 15 - 5 - 3, 15 - 16 - 5 - 8, 15 - 5 - 6, 15 - 3 - 16 - 7, 15 - 5 - 17 - 12, 15 - 6 - 17 - 10, 2 - 3 - 5, 2 - 5 - 6, 3 - 5 - 7 - 8, 16 - 7 - 8, 16 - 7 - 11, 16 - 8 - 11, 16 - 17 - 11 - 13, 16 - 8 - 17 - 12, 5 - 6 - 10 - 12, 5 - 8 - 12, 7 - 8 - 11, 8 - 11 - 12 - 13, 17 - 10 - 12, 17 - 12 - 13, 17 - 10 - 14, 17 - 13 - 14, 10 - 12 - 14, 12 - 13 - 14,

10 tree - dimension area (f_3=10)

15 - 2 - 3 - 5, 15 - 2 - 5 - 6, 15 - 3 - 16 - 7 - 5 - 8, 15 - 16 - 17 - 5 - 8 - 12, 15 - 17 - 5 - 12 - 6 - 10, 16 - 7 - 8 - 11, 16 - 8 - 11 - 17 - 12 - 13, 17 - 12 - 13 - 14, 15 - 2 - 3 - 16 - 5 - 6 - 7 - 8 - 17 - 10 - 11 - 12 - 13 - 14 (without edges 15 - 5, 16 - 8, 17 - 12).

Substituting the obtained values $f_i(i0,1,2,3)$ in the equation (2.1) you can see that the Euler - Poincaré equation is satisfied in this case with a value of n equal to 4

$14 - 32 + 28 - 10 = 0.$

This proves that polytope on Figure 2.11 has dimension 4.

It can be noted that the upper and lower parts of the polytope corresponding to the compound are symmetrical to each other with respect to the flat section separating them.

COMPOUNDS HAVING A METAL OCTAHEDRON FRAME

A cluster of phosphine gold, discussed in the previous section, is a special case of numerous compounds in which the octahedron participates as a structural unit. If a complex functional group is opened in gold phosphine, the dimension of the compound in this case will become much higher. Clusters with a structural unit octahedron exist with different metals in the core: Pt, Pd, Ni, Cu, Ti, Fe, Ru and other metals (Gubin, 2019). They are distinguished by sufficient structural complexity, leading to the formation of polytopes of a higher dimension of new types. So far, none of them have been analyzed in the space of higher dimension. As an illustration of such an analysis, we consider the structure of the cobalt tetra - anion cluster $[Co_6(\mu-Co)8_{(}CO)6_{]}4^{-}$ The scheme of this compound is shown in Figure 12 (Johnson, & Benfield, 1981).

The desire to create a convex model of this compound leads to the appearance of various three -dimensional polyhedrons and four - dimensional polytopes entering into each other.

Figure 12. Scheme of cobalt tetra - anion cluster $[Co_6(\mu-Co)8_{(}CO)6_{]}4^{-}$

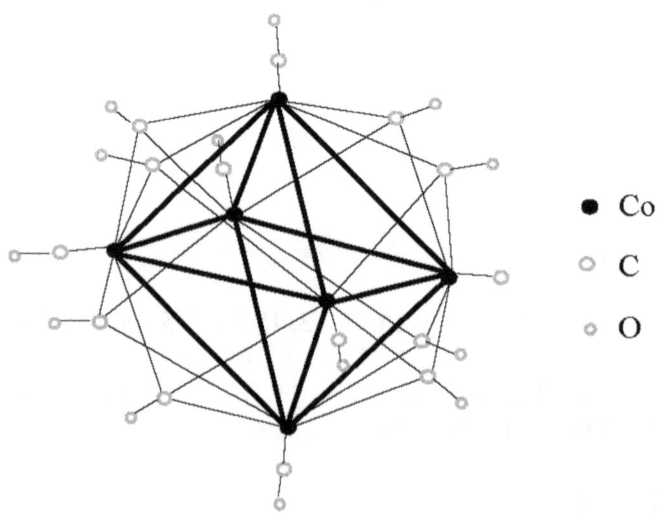

Theorem 2.4

The dimension of cobalt tetra - anion cluster $[Co_6(\mu{-}Co)8_{(}CO)6_{]}4^{-e}$qual 5.

Proof

To build a spatial model of cobalt tetra - anion cluster $[Co_6(\mu{-}Co)8_{(}CO)6_{]}4^{-}$ let us turn to its scheme in Figure 2.12. From this figure, it can be seen that the ligands (μ - CO) that bind two metal atoms form a cube. In addition, each vertex of the metal core in the form of an octahedron is formed as a result of constructing a pyramid on each face of the ligand cube. The free vertices of these pyramids give the vertices of the octahedron of atoms cobalt. The six faces of the cube thus give six vertices of the octahedron. Consider first the figure that forms as a result of this construction. The spatial image of this figure is shown in Figure 13.

Figure 13. The spatial image of compound $Co_6(\mu{-}Co)8$

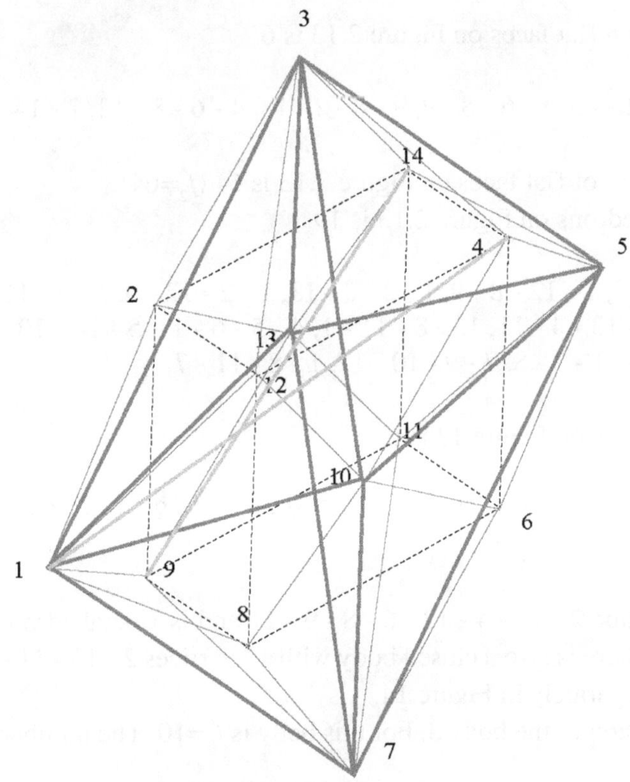

The valence bonds are indicated in this figure with red edges. The edges of the ligand cube are marked with dotted black lines. The edges of the pyramids, with the exception of the grounds, denoted by solid lines in black. Thus, the ligands (μ - CO) are located at the vertices 2, 4, 6, 8, 9, 11, 12, 14. Metal atoms are located at the vertices 1, 3, 5, 7, 10, 13. To create a closed convex figure in the figure, two edges are added blue 1 - 4, 9 - 14. Determine the dimension of this compound. There are 14 vertices here (f_0=14). The numbers of edges on Figure 2.10 is 50 (f_1=50).

1 - 2, 1 - 3, 1 - 13, 1 - 12, 1 - 4, 1 - 10, 1 - 9, 1 - 8, 1 - 7, 2 - 3, 2 - 13, 2 - 14, 2 - 12, 2 - 9, 3 – 4, 3 - 5, 3 - 14, 3 - 13, 3 - 12, 3 - 10, 4 - 5, 4 - 6, 4 - 10, 4 - 12, 4 - 14, 5 - 14, 5 - 11, 5 - 10, 5 - 6, 5 - 7, 5 - 13, 6 - 7, 6 - 10, 6 - 11, 6 - 8, 7 - 10, 7 - 11, 7 - 13, 7 - 8, 7 - 9, 8 - 9, 8 - 10, 8 - 12, 9 - 11, 9 - 13, 9 - 14, 10 - 12, 11 - 13, 11 - 14, 13 - 14.

The number of triangles on Figure 2.13 is 58

1 - 2 - 3, 1 - 3 - 13, 1 - 3 - 12, 1 - 2 - 13, 1 - 2 - 12, 1 - 2 - 9, 1 - 9 - 8, 1 - 9 - 7, 1 - 12 - 8, 1 - 13 - 7, 1 - 10 - 7, 1 - 10 - 8, 1 - 10 - 12, 1 - 10 - 3, 1 - 9 - 13, 1 - 8 - 7, 2 - 12 - 3, 2 - 13 - 3, 2 - 14 - 3, 2 - 14 - 13, 2 - 9 - 13, 3 - 12 - 4, 3 - 13 - 14, 3 - 13 - 5, 3 - 4 - 5, 3 - 14 - 5, 3 - 14 - 4, 3 - 10 - 12, 3 - 10 - 4, 3 - 10 - 5, 4 - 5 - 14, 4 - 10 - 12, 4 - 5 - 10, 4 - 6 - 10, 4 - 5 - 6, 5 - 13 - 11, 5 - 13 - 14, 5 - 14 - 11, 5 - 11 - 6, 5 - 11 - 7, 5 - 10 - 6, 5 - 7 - 6, 5 - 10 - 7, 6 - 10 - 7, 6 - 11 - 7, 6 - 8 - 7, 6 - 8 - 10, 7 - 11 - 13, 7 - 8 - 10, 7 - 8 - 9, 7 - 9 - 11, 7 - 9 - 13, 8 - 10 - 12, 9 - 11 - 13, 9 - 13 - 14, 9 - 14 - 11, 1 - 12 - 4, 1 - 10 - 4.

The number of tetragon flat faces on Figure 2.13 is 6

2 - 14 - 4 - 13, 14 - 4 - 11 - 6, 11 - 6 - 8 - 9, 9 - 8 - 2 - 13, 4 - 6 - 8 - 12, 2 - 14 - 11 - 9.

Thus, the common number of flat faces on Figure 2.13 is 64 (f_2=64).

The number of tetrahedrons on Figure 2.13 is 19

1 - 2 - 3 - 12, 1 - 12 - 9 - 7, 1 - 12 - 8 - 9, 1 - 2 - 3 - 13, 1 - 2 - 12 - 9, 1 - 3 - 11 - 13, 1 - 9 - 8 - 7, 2 - 12 - 14 - 3, 3 - 4 - 5 - 14, 3 - 13 - 4 - 11, 13 - 8 - 11 - 1, 4 - 5 - 6 - 11, 5 - 14 - 10 - 12, 5 - 11 - 7 - 6, 5 - 10 - 7 - 6, 5 - 12 - 7 - 10, 6 - 11 - 7 - 8, 7 - 9 - 10 - 12, 1 - 8 - 11 - 7.

The number of pyramids on Figure 13 is 6

2 - 12 - 4 - 6 - 3, 4 - 14 - 6 - 11 - 5, 6 - 8 - 9 - 11 - 7, 9 - 8 - 12 - 2 - 1, 4 - 6 - 8 - 12 - 10, 2 - 14 - 11 - 9 - 13.

On Figure 2.13 is 1 cube 2 - 14 - 4 - 12 - 6 - 8 - 9 - 11 and is 1 octahedron 1 - 3 - 5 - 7 - 10 - 13. In addition, in Figure 2.13 there is also a closed body with 12 vertices 2 - 13 - 11 - 10 - 6 - 12 - 3 - 5 - 1 - 7. His image is presented separately in Figure 14.

Determine the dimension of the body S. For this body is f_0=10. The number of edges on Figure 2.14 is 22 (f_1=22)

1 - 2, 1 - 13, 1 - 12, 1 - 10, 1 - 7, 2 - 3, 2 - 13, 2 - 12, 3 - 5, 3 - 10, 3 - 13, 3 - 12, 5 - 13, 5 - 11, 5 - 10, 6 - 7, 6 - 10, 6 - 11, 7 - 10, 7 - 11, 10 - 12, 11 - 13.

The number of triangles on Figure 2.14 is 8

2 - 3 - 12, 2 - 3 - 13, 1 - 2 - 13, 6 - 5 - 10, 6 - 5 - 11, 1 - 2 - 12, 7 - 6 - 10, 7 - 6 - 11.

Figure 14. The image a closed body S = 2 - 13 - 11 - 10 - 6 - 12 - 3 - 5 - 1 - 7

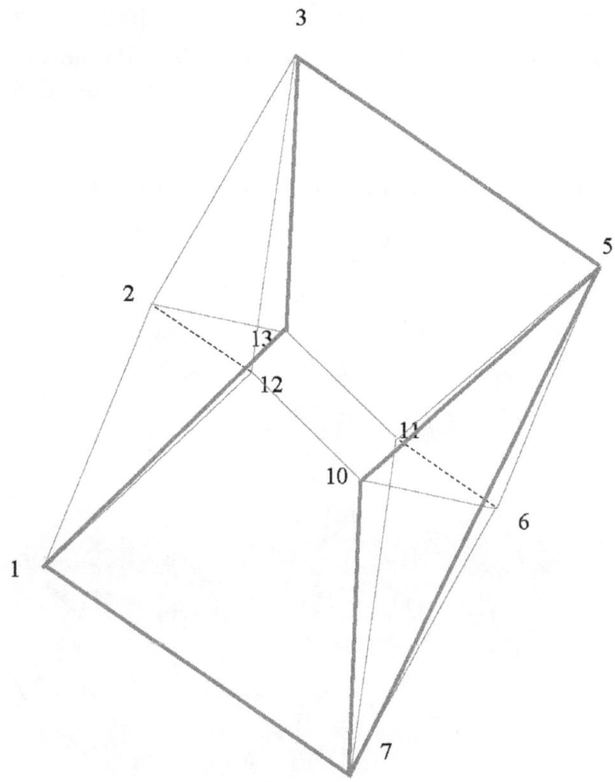

The number of tetragon flat faces on Figure 2.14 is 4

13 - 3 - 5 - 11, 12 - 3 - 5 - 10, 1 - 7 - 13 - 11, 1 - 12 - 10 - 7.

Thus, the common number of flat faces on Figure 2.14 is 12 (f_2=12).

Substituting the obtained values f_i(i=0,1,2) in the equation (2.1) you can see that the Euler - Poincaré equation is satisfied in his case for $n = 3$

$$10 - 22 + 12 = 2.$$

This proves that Figure 2.14 is polyhedron with dimension 3. Consequently, the total number of three - dimensional figures included in the polytope in Figure 2.13 is 19 + 9 = 28, i.e. for this figure f_3=28.

Substituting the obtained for Figure 2.13 values f_i(i=0,1,2,3) in the equation (2.1) you can see that the Euler - Poincaré equation is satisfied in his case for $n = 4$

$$14 - 50 + 64 - 28 = 0.$$

This proves that Figure 2.13 is polytope with dimension 4. Now back to the review of cobalt tetra - anion cluster $[Co_6(\mu{-}Co)8_{(}CO)6_{]}4^{-}$ in Figure 12.

Figure 13 shows the spatial image of only part of this compound. To create the spatial image of this compound, it is necessary to release the edges into the outer part from the vertices of the octahedron in Figure 2.13 and build a larger octahedron 15 - 16 - 17 - 18 - 19 - 20 on the free vertices of these edges. This are red edges 1 - 18, 3 - 15, 5 - 16, 7 - 17, 13 - 19, 10 – 20 on Figure 15.

Figure 15. Spatial image of cobalt tetra - anion cluster $[Co_6(\mu{-}Co)_8(CO)_6]^{4-}$

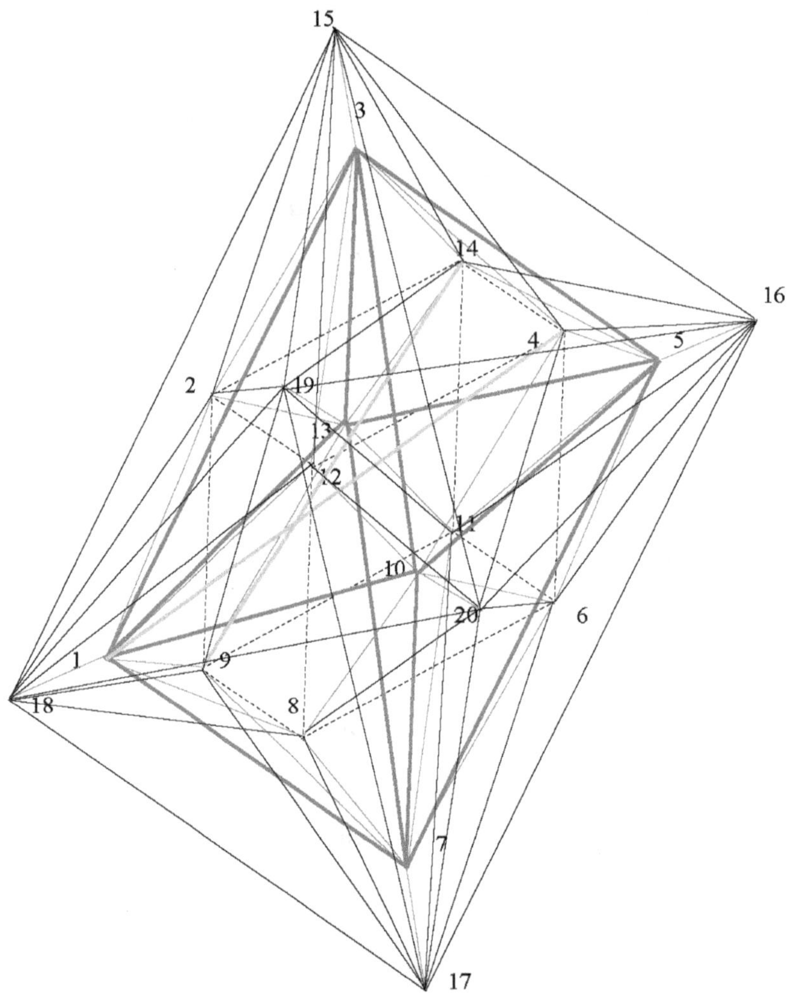

Thus, the number of vertices on Figure 15 equal 20 (f_0=20). The construction of a large octahedron taking into account the edges connecting two octahedrons adds 16 edges to the body in Figure 2.13. A three - dimensional figure is formed between each face of a large octahedron and the corresponding face of a small octahedron. For example, a three - dimensional figure is formed between the face of 1 - 10 - 7 small octahedron and the face 18 - 17 - 20 of the large octahedron (see Figure 2.12) 1 - 10 - 7 - 18 - 17 - 20. The number of such three - dimensional figures is equal to the number of faces of the octahedron - 8. In addition, to fill the space between the large octahedron and other convex bodies that make up the polytope in Figure 2.13, it is necessary to connect the vertices of the large octahedron with the vertices of the cube 2 - 14 - 4 - 12 - 6 - 8 - 9 - 11. This adds to the body in Figure 2.13 another 24 edges

2 - 15, 12 - 15, 14 - 15, 4 - 15, 14 - 16, 4 - 16, 6 - 16, 11 - 16, 6 - 17, 8 - 17, 9 - 17, 11 - 17, 8 - 18, 9 - 18, 2 - 18, 12 - 18, 2 - 19, 14 - 19, 9 - 19, 11 - 19, 4 - 20, 12 - 20, 8 - 20, 6 - 20.

Thus, a common number of the edges on Figure 15 is 50 + 16 + 24 = 90 (f_1=90).

The construction of a large octahedron taking into account the edges connecting two octahedrons adds 20 flat faces to body on Figure 2.13 (see previous section). Connection the vertices of the large octahedron with the vertices of the cube adds 48 flat faces to Figure 2.13. For example, the face of the cube 2 - 14 - 4 - 12 when building a large pyramid on it gives an additional 8 flat faces 2 - 12 - 15, 4 - 12 - 15, 4 - 14 - 15, 2 - 14 - 15, 12 - 3 - 15, 2 - 3 - 15, 4 - 3 - 15, 14 - 3 - 15. Since the faces at cube 6 one gets an additional 48 flat faces. In this way, the total number of flat edges in Figure 15 is 64 + 20 + 48 = 132, (f_2=132).

The construction of a large octahedron taking into account the edges connecting two octahedrons adds 9 three - dimensional faces to body on Figure 2.13. Connection the vertices of the large octahedron with the vertices of the cube adds 30 three - dimensional faces to Figure 2.13. For example, the face of the cube 2 - 14 - 4 - 12 when building a large pyramid on it gives an additional 5 three - dimensional faces 2 - 12 - 15 - 3, 4 - 12 - 15 - 3, 4 - 14 - 15 - 3, 2 - 14 - 15 - 3, 12 - 2 - 15 - 4 - 14. Since the faces at cube 6 we get an additional 30 three - dimensional faces. In addition, 4 tetrahedrons are additionally formed between the cube and the large octahedron, each of which has an edge of one of the edges of a rectangular cross section of 15 - 16 - 17 - 18 large octahedron. For example, a tetrahedron 18 - 17 - 8 - 9 is formed on the edge 17 - 18. Thus, the total number of three - dimensional areas in the body in Figure 2.15 is 28 + 9 + 30 + 4 = 71 (f_3=71).

Connection the vertices of the large octahedron with the vertices of the cube adds 2 four - dimensional faces to Figure 2.13, one of which has an edge 18 - 17 and second has edges 15 - 16 of a rectangular cross section of 15 - 16 - 17 - 18 large octahedron. For example, on the edge 17 - 18 formed area 18 - 1 - 9 - 8 - 7 - 17 (see Figure 2.15). This area has 6 vertices (f_0=6). It has14 edges (f_1=14) 1 - 9, 1 - 8, 1 - 7, 1 - 18, 7 - 8, 7 - 9, 7 - 17, 8 - 9, 8 - 18, 8 - 17, 9 - 18, 9 - 17, 9 - 8, 18 - 17. It has 14 two - dimensional faces (f_2=14) 1 - 9 - 18, 1 - 9 - 8, 1 - 8 - 7, 1 - 7 - 17 - 18, 7 - 9 - 17, 7 - 8 - 9, 7 - 9 - 17, 8 - 17 - 18, 8 - 9 - 17, 8 - 9 - 18, 9 - 18 - 17. It has 6 three - dimensional faces (f_3=6) 1 - 9 - 8 - 18, 1 - 9 - 8 - 7, 18 - 9 - 8 - 17, 1 - 18 - 9 - 7 - 17, 9 - 8 - 7 - 17, 1 - 8 - 18 - 7 - 17. Substituting the obtained for Figure 2.13 values f_i(i=0,1,2,3) in the equation (2.1) you can see that the Euler - Poincaré equation is satisfied in his case for $n = 4$

$$14 - 50 + 64 - 28 = 0.$$

This proves that body 18 - 1 - 9 - 8 - 7 - 17 is polytope with dimension 4. It is also proved that the body 15 - 16 - 5 - 4 - 14 - 3 has the dimension 4. It can be shown that, due to the orientation of the cube with respect to the large octahedron, similar four - dimensional polytopes with the edges of the large octahedron 18 - 15 and 17 - 16 do not form.

Connection the vertices of the large octahedron with the vertices of the cube adds yet 6 four - dimensional faces to Figure 13, associated with the formation of structures of the pyramid in the pyramid on the faces of the cube. Consider for example two pyramids on the face of a cube 2 - 12 - 14 - 4. This body has 6 vertices (f_0=6), 13 edges (f_1=13)

2 - 3, 2 - 15, 12 - 3, 12 - 15, 4 - 3, 14 - 3, 4 - 15, 14 - 15, 3 - 15, 2 - 12, 2 - 14, 4 - 12, 4 - 14,

13 two - dimensional faces (f_2=13)

2 - 15 - 12, 2 - 3 - 12, 4 - 3 - 14, 4 - 15 - 14, 3 - 4 - 12, 15 - 4 - 12, 2 - 14 - 3, 2 - 14 - 15, 12 - 3 - 15, 2 - 3 - 15, 4 - 3 - 15, 14 - 3 - 15, 2 - 14 - 12 - 4,

and 6 three - dimensional faces (f_3=6) 2 - 14 - 4 - 12 - 3, 2 - 14 - 4 - 12 - 15, 4 - 3 - 15 - 14, 12 - 4 - 3 - 15, 2 - 12 - 3 - 15, 2 - 3 - 14 - 15.

Substituting the obtained for two pyramids values $f_i(i$=0,1,2,3) in the equation (1) you can see that the Euler - Poincaré equation is satisfied in his case for $n = 4$

$$14 - 50 + 64 - 28 = 0.$$

This proves that construction from two pyramids is polytope with dimension 4.

The four - dimensional structure of two pyramids on the remaining five faces of the cube is proved in a similar way. In addition, the polytope in Figure 2.15 has two polytopes formed by inscribing a cube into an octahedron (large and small). Each of these polytopes has dimension 4 (this has already been proved). Finally, the polytope in Figure 2.15 has a polytope octahedron in an octahedron. This polytope also has dimension 4 (see previous section). Therefore, the total number of polytopes of dimension 4 that make up a polytope in Figure 2.15 is 11 (f_4=11). Substituting the obtained for Figure 15 values $f_i(i$=0,1,2,3,4) in the equation (2.1) you can see that the Euler - Poincaré equation is satisfied in his case for $n = 5$

$$20 - 90 + 132 - 71 + 11 = 2.$$

This proves that dimension of cobalt tetra - anion cluster $[Co_6(\mu{-}Co)8_(CO)6_]4^-$ equal 5.

CONCLUSION

Analytical expressions are obtained for calculating the dimension of multi - shell clusters depending on the number of shells. These expressions are applicable to clusters consisting of shells in the form of a polyhedral of Plato with a common center, if an atom is located or dislocated in this center. From the obtained analytical formulas in it follows that in both cases the dimension of the clusters increases linearly with an increase in the number of shells. Thus, for the clusters of Mackay (Mackay, 1962), Bergman (Bergman et al.,1952, 1957, 1996), Samson (Samson, 1972) known in the literature, which can be approximately described by such multi - shell models, expressions are obtained for calculating their dimensions and proved that they have a higher dimension. It should be noted an interesting fact that the clusters of γ - brass studied for a long time turned out to have geometry of a high - dimensional cross – polytope.

Geometrically investigated the structure of clusters, the core of which represent the metal chains of both identical and different elements. Chains are linear or curved. For the first time it was shown that the dimension of the structures of these clusters is more than three. The octahedral coordination of ligands around metal atoms in chains leads to the penetration of octahedrons into each other in both linear and non - linear metal chains. To create a model of these chains in a higher dimension space, a new geometric approach has been developed, which allows us to construct a convex, closed polytopes of these chains. It consists in removing part of the octahedron edges necessary for constructing the octahedron and adding the same number of new edges necessary to build a closed polytope chain, while maintaining the number of metal atoms and ligands and their valence bonds. As a result, it was found that metal chain polytopes consist of polytopes of higher dimension, adjacent to each other along flat sections. This allowed us to simplify the geometry of the clusters and to construct their structure. Including it was possible to detect the existence of clusters with a closed cycle of polytopes in the absence of closed cycles of metal atoms.

The geometry of clusters with a polyhedral metal core with ligands was studied. The attachment of ligands to the skeleton of metal atoms forming a three - dimensional polyhedron should naturally lead to the formation of a cluster of the higher dimension. The analysis of the geometry of a tetra - anion cobalt cluster can serve as an effective example of the application of geometric analysis of clusters of higher dimensions. It is proved that the dimension of this cluster is 5. The structure of this cluster includes octahedrons, a cube, a set of twin pyramids, and different three – dimensional body, 11 four - dimensional bodies. In this cluster, the metal core was located between the two shells of the ligands. Binding ligands form a small ligand polyhedron. Terminal ligands form a large ligand polyhedron. Both ligand polyhedrons limit the metal core "bottom" and "top".

REFERENCES

Aleksandrov, G.G., Zolnikova, G.P., Kritskaya, I.I., & Struchkov, Yu.T. (1980). Molecular structure $Os_3(\mu$–Br$)2_(CO)1_0$ of the reaction product O*s3*$_($*CO)*2 with (3 -bromomethylnaphthalene). C*oordination Chemistry,* 6*(*4), 626 - 628.

Bau, R., Kirtley, S. W., Sorrell, T. N., & Winarko, S. (1974). Preparation and structure determination of the manganese carbonyl - anion. Staggered vs eclipsed carbonyl groups and linear vs bent M – M – M - [metal – metal – metal] and M – H – M [metal – hydrogen – metal] bonds. *Journal of the American Chemical Society*, *96*(4), 988–993. doi:10.1021/ja00811a006

Benfield, R. E., & Johnson, B. F. G. (1981). Structures and fluxional behaviour of transition metal cluster carbonyls. *Transition Metal Chemistry (Weinheim)*, *6*(3), 131–144. doi:10.1007/BF00624332

Bergman, G., Waugh, J. L. T., & Pauling, L. (1952). Crystal structure of the intermetallic compound Mg32(Al,Zn)49. *Nature*, *169*(4312), 1057–1058. doi:10.1038/1691057a0

Bergman, G., Waugh, J. L. T., & Pauling, L. (1957). The crystal structure of the metallic phase Mg32(Al,Zn)49. *Acta Crystallographica*, *10*(4), 254–259. doi:10.1107/S0365110X57000808

Bredley, A. J., & Thewlis, J. (1926). The structure of γ - brass. *Proceedings of the Royal Society of London. Series A, Containing Papers of a Mathematical and Physical Character*, *112*(762), 678–681. doi:10.1098/rspa.1926.0134

Conway, J. H., & Sloane, N. J. A. (1988). *Sphere Packings, Lattices, and Groups*. Springer – Verlag. doi:10.1007/978-1-4757-2016-7

Cotton, F. L., & Walton, R. A. (1982). *Multiple bonds between metal atoms*. Wiley.

Coxeter, H. S. M. (1963). *Regular Polytopes*. Dower.

Evans, J., Okrasunski, S. J., Pribula, A. J., & Norton, J. K. (1976). Synthesis of di- and trinuclear methyl osmium complexes via cis-hydridomethyltetracarbonylosmium. *Journal of the American Chemical Society*, *98*(13), 4000–4001. doi:10.1021/ja00429a045

Gochin, M., & Moss, Y. R. (1980). On the structure of H2Os3(CO)12. *Journal of Organometallic Chemistry*, *192*(3), 409–419. doi:10.1016/S0022-328X(00)81233-4

Gosset, T. (1900). On the regular and semi-regular figures in space of *n* dimensions. *Messenger of Mathematics.*, *29*, 43–48.

Gubin, S. P. (2019). *Chemistry clusters. Basics of classification and structure*. Editorial URSS.

Lord, E. A., Mackay, A. L., & Ranganathan, S. (2006). *New Geometry for New Materials*. Cambridge University Press.

Mackay, A. L. (1962). A dense non - crystalline packing of equal spheres. *Acta Crystallographica*, *15*(9), 1916–1918. doi:10.1107/S0365110X6200239X

Poincaré, A. (1895). Analysis situs. *J. de é Ecole Polytechnique, 1,* 1 - 121.

Pouling, L. (1960). *The Nature of the Chemical Bond and the Structure of Molecules and Crystals: An Introduction to Modern Structural Chemistry* (3rd ed.). Cornell University Press.

Samson, S. (1972). Complex cubic A_6B compounds. II. The crystal structure of Mg_6Pd. *Acta Crystallographica. Section B, Structural Crystallography and Crystal Chemistry*, *28*(3), 936–945. doi:10.1107/S0567740872003437

Shevchenko, V. Y. (2011). *Search in Chemistry, Biology and Physics of the Nanostate*. Lema.

Smart, L., Cook, N., & Woodward, P. (1977). Article. *Journal of the Chemical Society, Dalton Transactions: Inorganic Chemistry*, 4312.

Vargaftik, M. N. (1985). Article. *Reports of the Academy of Sciences of the USSR*, *284*(4), 896–899.

Vilenkin, N. V. (1969). *Combinatorics*. Science.

Zhizhin, G. V. (2016). The structure, topological and functional dimension of biomolecules. *J. Chemoinformatics and Chemical Engineering*, *5*(2), 44–58. doi:10.4018/IJCCE.2016070104

Zhizhin, G. V. (2018). *Chemical Compound Structures and the Higher Dimension of Molecules: Emerging Research and Opportunities*. IGI Global. doi:10.4018/978-1-5225-4108-0

Zhizhin, G. V. (2019a). *The Geometry of Higher - Dimensional Polytopes*. IGI Global. doi:10.4018/978-1-5225-6968-8

Zhizhin, G. V. (2019b). Higher Dimensions of Clusters of Intermetallic Compounds: Dimensions of Metallic Nanoclusters. *International Journal of Applied Nanotechnology Research*, *4*(1), 8–25. doi:10.4018/IJANR.2019010102

Zhizhin, G. V. (2019c). On the structure of parallelohedrons of higher dimension: Hilbert's 18th problem. *Int. Journal of Chemoinformatic and Chemistry Engireeng*, *8*(2), 69–96. doi:10.4018/IJCCE.20190701.oa2

KEY TERMS AND DEFINITIONS

Dimension of the Space: The member of independent parameters needed to describe the change in position of an object in space.

Hetero-Element Metal Chains: Clusters in which the core contains atoms of different elements.

Homo-Element Metal Chains: Clusters in which the core contains atoms of the same element.

Linear Metal Chains: Metal chains in which metal atoms are located in a straight line.

Metal Chains: Cluster compounds that have a skeleton in the form of metal chains, i.e. polymetallic chains formed by metal-metal localized covalent bonds.

Nanocluster: A nanometric set of connected atoms, stable either in isolation state or in building unit of condensed matter.

Nonlinear (Curved) Metal Chains: Clusters in which metal-metal valent bonds form an angle different from 180 degrees between them.

Polytope: Polyhedron in the space of higher dimension.

Chapter 3
The Law of Conservation of Incidents in the Space of Nanoworld

ABSTRACT

For the first time it was established that for any convex polytope of higher dimension there is an integral equality in the transfer of information from low-dimensional elements to higher-dimensional elements and vice versa. This integral equality is called the law of conservation of incidents. In previous works of the author, this law was established for some polytopes of a particular kind. There is the incidence interpreted as the transfer of information from one material body to another. It is shown that the law of conservation of incidents is valid for both regular bodies and irregular bodies, which can be clusters of chemical compounds. The incident conservation law can serve as a mathematical basis for the recently discovered epigenetic principle of the transmission of hereditary information without changing the sequence of genes in DNA and RNA molecules.

INTRODUCTION

In previous works of the author, it was convincingly shown that various objects of the nanoworld have a higher dimension (Zhizhin, 2012; 2013; 2014 a, b, c; 2015; 2016; 2018; 2019 a, b, c, d, e, f; Zhizhin & Diudea, 2016; Zhizhin, Khalaj, & Diudea, 2016). Geometrically, these objects are convex closed polytopes whose components are geometric elements with a monotonic change of the dimension from zero to the dimension of the polytope. Moreover, many of them are constructions with the main part in the form of an n – simplex or n – cross – polytopes. Strictly speaking, any polytope can be reduced to an n – simplex by adding a sufficient number of edges to it without changing the number of vertices. As it turned out, an important place among polytopes describing chemical compounds are occupied by the cross – polytopes of higher dimension. In the monograph "The Geometry of Higher - Dimensional Polytopes" (Zhizhin, 2019 a) has studied in detail the geometry of higher - dimensional polytopes, based on an analysis of the structures of chemical compounds (Zhizhin, 2018). The monograph (Zhizhin, 2019

DOI: 10.4018/978-1-7998-6768-5.ch003

a) introduced the concept of the incidence coefficients of elements of lower dimension with respect to elements of higher dimension and elements of higher dimension with respect to elements of lower dimension. The first characterizes the number of elements of a certain higher dimension to which the given element of a lower dimension belongs. The second characterizes the number of elements of a given lower dimension that are included in a particular element of a higher dimension. In monograph (Zhizhin, 2019 a) for some types of polytopes, a law was established according to which the sum of incidence coefficients from low -dimensional elements to high - dimensional elements is equal to the sum of incidence coefficients of high - dimensional elements to low - dimensional elements. This integral equality is called the law of conservation of incidents. In this chapter, based on previous studies for a particular type of polytopes, the law of conservation of incidents is established for any convex polytopes. Here we must remember that the vertices of geometric elements of various dimensions are atoms, molecules or functional groups. Therefore, the incidence of geometric elements to a friend means contact between particles of matter, including living matter. The contact between particles of matter can be interpreted as the transfer of information on material structures, including biological structures. The monograph "Attractors and higher dimensions in population and molecular biology: emerging research and opportunities" (Zhizhin, 2019 d) showed that the elements of nanostructures of living matter are objects of higher dimension. Information sharing is an inherent property of living beings, without which even primitive organisms would not be able to maintain an extremely delicate balance, which depends on their survival (Mancuso, &Viola, 2013; Mancuso, 2017). This provides the memory of living organisms, i.e. transfer of inherited traits without changing the sequence of genes in DNA (Mancuso, 2017; Lindquist et al., 2016; Sanbonmatsu et al., 2016). Therefore, it is of interest to study in detail the incidence of the entire polygamy of elements of different dimensions in higher dimension polytopes, considering it as a study of the possibility of transmitting information in biological molecules.

INCIDENT CONSERVATION LAW FOR N - SIMPLEX

In n - simplex (Zhizhin, 2013, 2014 c, 2019 a) the number of elements with dimension d is

$$f_d(n) = C_{n+1}^{d+1}. \tag{3.1}$$

For example, if $d = 0$, so

$$f_0(n) = C_{n+1}^{1} = \frac{(n+1)!}{n!} = n+1. \tag{3.2}$$

This is the number of vertices in n - simplexes. If $d = 1$, so

$$f_1(n) = C_{n+1}^{2} = \frac{(n+1)!}{(n-1)!2!} = \frac{(n+1)n}{2}. \tag{3.3}$$

This is the number of edges in n - simplexes. We introduce the notation: $k_{d_j^i d_u}$ is the number of elements of dimension u, which include an element of dimension j ($u > j$) with number i. Thus, $k_{d_j^i d_u}$ is the incidence factor of element i with dimension j relative to elements with dimension u. We introduce the notation also: $k_{d_j d_u^i}$ is the number of elements of dimension j, which included in element i with a dimension u ($u > j$). Thus, $k_{d_j d_u^i}$ is the incidence factor of element i with dimension u relative to elements with dimension j.

If $n = 2$, so $f_0(2)=3$, $f_1(2)=3$, see (3.2), (3.3). The factors of incidents (from smaller dimension to larger) are $k_{d_0^i d_1} = 2, i = 1,2,3;\ k_{d_1^i d_2} = 1, i = 1,2,3.$

Sum up the incidence coefficients for all vertices and edges of the triangle

$$\sum_{i=1}^{3} k_{d_0^i d_1} = 6, \sum_{i=1}^{3} k_{d_1^i d_2} = 3, \sum_{i=1}^{3} k_{d_0^i d_1} + \sum_{i=1}^{3} k_{d_1^i d_2} = 9. \tag{3.4}$$

The factors of incidents (from larger dimension to smaller) are $k_{d_0 d_1^i} = 2, i = 1,2,3; k_{d_1 d_2} = 3.$

Sum up the incidence coefficients for all elements of the triangle with dimension larger of zero

$$\sum_{i=1}^{3} k_{d_0 d_1^i} + k_{d_1 d_2} = 9. \tag{3.5}$$

Comparing (3.4) and (3.5) you can see that the sum of incidents in a triangle from elements with a lower dimension to elements with a higher dimension is equal to the sum of incidents from elements with a higher dimension to elements with a lower dimension. Thus, the sum of incidents retains its value when the direction of the relationship between the elements it is change.

If $n = 3$, so $f_0(3)=4$, $f_1(3)=6$, $f_2(3)=4$. The factors of incidents (from smaller dimension to larger) are

$$k_{d_0^i d_1} = 3, i = 1 \div 4;\ k_{d_0^i d_2} = 3, i = 1 \div 4;\ k_{d_0^i d_3} = 1, i = 1 \div 4;$$
$$k_{d_1^i d_2} = 2, i = 1 \div 6;\ k_{d_1^i d_3} = 1, i = 1 \div 6;\ k_{d_2^i d_3} = 1, i = 1 \div 4.$$

Sum up the incidence coefficients for all elements of the tetrahedron

$$\begin{gathered} \sum_{i=1}^{4} k_{d_0^i d_1} = 12, \sum_{i=1}^{4} k_{d_0^i d_2} = 12, \sum_{i=1}^{4} k_{d_0^i d_3} = 4, \sum_{i=1}^{6} k_{d_1^i d_2} = 12, \sum_{i=1}^{6} k_{d_1^i d_3} = 6, \\ \sum_{i=1}^{4} k_{d_2^i d_3} = 4, \sum_{i=1}^{4} k_{d_0^i d_1} + \sum_{i=1}^{4} k_{d_0^i d_2} + \sum_{i=1}^{4} k_{d_0^i d_3} + \sum_{i=1}^{6} k_{d_1^i d_2} + \sum_{i=1}^{6} k_{d_1^i d_3} + \sum_{i=1}^{4} k_{d_2^i d_3} = 50. \end{gathered} \tag{3.6}$$

The factors of incidents (from larger dimension to smaller) are

$$k_{d_0 d_1^i} = 2, i = 1 \div 6; k_{d_0 d_2^i} = 3, i = 1 \div 4; k_{d_1 d_2^i} = 3, i = 1 \div 4; k_{d_0 d_3} = 4, k_{d_1 d_3} = 6, k_{d_2 d_3} = 4.$$

Sum up the incidence coefficients for all elements of the tetrahedron

$$\sum_{i=1}^{6} k_{d_0 d_1^i} = 12, \sum_{i=1}^{4} k_{d_0 d_2^i} = 12, \sum_{i=1}^{4} k_{d_1 d_2^i} = 12, \sum_{i=1}^{6} k_{d_0 d_1^i} + \sum_{i=1}^{4} k_{d_0 d_2^i} + \sum_{i=1}^{4} k_{d_1 d_2^i} + k_{d_0 d_3} + k_{d_1 d_3} + k_{d_2 d_3} = 50. \tag{3.7}$$

Comparing (3.6) and (3.7) you can see that the sum of incidents in a tetrahedron from elements with a lower dimension to elements with a higher dimension is equal to the sum of incidents from elements with a higher dimension to elements with a lower dimension. Thus, the sum of incidents retains its value when the direction of the relationship between the elements it is change.

Let us prove the general statement for a simplex of arbitrary finite dimension *n*.

Theorem 3.1

In any simplex of dimension *n*, the sum of all incidents of elements of a lower dimension with respect to all elements of a higher dimension is equal to the sum of all incidents of elements of a higher dimension with respect to all elements of a lower dimension and equals the sum of the series

$$f_0(n)(2^n - 1) + f_1(n)(2^{n-1} - 1) + f_2(n)(2^{n-2} - 1) + ... + f_{n-1}(n), f_d(n) = C_{n+1}^{d+1}, d = 0 \div (n-1).$$

Proof

The number of edges (elements of dimension 1) that have a certain vertex d_0^i its end point (given that in a simplex each vertex is connected by edges with all other vertices) is determined by the incidence factor

$$k_{d_0^i d_1} = C_n^1 = f_0(n) - 1 = n, i = 1 \div f_0(n). \tag{3.8}$$

The number of elements of dimension 2 that have its vertex a certain vertex d_0^i is determined by the incident factor

$$k_{d_o^i d_2} = C_n^2, i = 1 \div f_0(n). \tag{3.9}$$

The number of elements of dimension 3 that have its vertex a certain vertex d_0^i is determined by the incident factor $k_{d_o^i d_3} = C_n^3, i = 1 \div f_0(n)$. And so on.

The number of elements of dimension (n - 1) that have its vertex a certain vertex d_0^i is determined by the incident factor $k_{d_o^i d_{n-1}} = C_n^{n-1}, i = 1 \div f_0(n)$. The number of elements of dimension n that have its vertex a certain vertex d_0^i is determined by the incident factor $k_{d_o^i d_n} = C_n^n = 1, i = 1 \div f_0(n)$.

The total number of incidents of vertices to elements of higher dimensions in a n - simplex will be

$$f_0(n)\sum_{j=1}^{n} k_{d_0^i d_j} = f_0(n)(C_n^1 + C_n^2 + ... + C_n^n) = f_0(n)(2^n - 1).$$

The number of elements of dimension 2 that have a certain edge d_1^i its edge is determined by the incidence factor

$$k_{d_1^i d_2} = C_{n-1}^1 = f_0(n) - 2 = n - 1, i = 1 \div f_1(n).$$

The number of elements of dimension 3 that have its edge an element d_1^i is determined by the incident factor $k_{d_1^i d_3} = C_{n-1}^2, i = 1 \div f_1(n)$. And so on.

The number of elements of dimension (n - 1) that have its edge a certain edge d_1^i is determined by the incident factor $k_{d_1^i d_{n-1}} = C_{n-1}^{n-2}, i = 1 \div f_1(n)$. The number of elements of dimension n that have its edge a certain edge d_1^i is determined by the incident factor $k_{d_1^i d_n} = C_{n-1}^{n-1} = 1, i = 1 \div f_1(n)$.

The total number of incidents of edges to elements of higher dimensions in n - simplex will be

$$f_1(n)\sum_{j=2}^{n} k_{d_1^i d_j} = f_1(n)(C_{n-1}^1 + C_{n-1}^2 + ... + C_{n-1}^{n-1}) = f_1(n)(2^{n-1} - 1). \quad (3.10)$$

In the same way, it is proved that the total number of incidences of elements of dimension 2 to elements of higher dimension in a n - simplex is determined by the expression

$$f_2(n)\sum_{j=3}^{n} k_{d_2^i d_j} = f_2(n)(C_{n-2}^1 + C_{n-2}^2 + ... + C_{n-2}^{n-2}) = f_2(n)(2^{n-2} - 1).$$

The total number of incidences of elements of dimension 3 to elements of higher dimension in a n - simplex is determined by the expression

$$f_3(n)\sum_{j=4}^{n} k_{d_3^i d_j} = f_3(n)(C_{n-3}^1 + C_{n-3}^2 + ... + C_{n-3}^{n-3}) = f_3(n)(2^{n-3} - 1).$$

And so on. Thus, the sum of all incidents of elements of a lower dimension with respect to all elements of a higher dimension in n - simplexes is equal

$$f_0(n)(2^n-1)+f_1(n)(2^{n-1}-1)+f_2(n)(2^{n-2}-1)+....+f_{n-1}(n). \tag{3.11}$$

To complete the proof of Theorem 3.1, it is necessary to show that the same expression is equal to the sum of all incidents of elements of a higher dimension with respect to all elements of a lower dimension in n - simplexes.

We pose the question: how many elements of a lower dimension belong to one element of a higher dimension? For example, how many vertices belong to a single edge $d_1^i, i=1\div f_1(n)$? Obviously, this number is 2. This is the value of the coefficient of incidence $k_{d_0 d_1^i}=C_2^1=2$. The product $C_2^1 f_1(n)$ gives the number of vertices belonging to all edges of an n - simplex. This number $2C_{n+1}^2=2\dfrac{(n+1)!}{(n-1)!2!}=(n+1)n$ is equal to the product $k_{d_0^i d_1}=C_n^1 C_{n+1}^1=n(n+1)$, i.e. the number of edges with a peak point $d_0^i, i=1\div f_0(n)$ (8). The number of vertices belonging to a single element of dimension 2 (triangle) is given by the incidence coefficient $k_{d_0 d_2^i}, i=1\div f_2(n)$. Obviously, this number is $3=C_3^1$. The product $C_3^1 f_2(n)$ corresponds to the number of vertices belonging to all elements of dimension 2 in the n - simplex. This number is $C_3^1 f_2(n)=\dfrac{3!}{2!}C_{n+1}^1=\dfrac{3!}{2!}\dfrac{(n+1)!}{(n-2)!3!}=\dfrac{(n+1)!}{(n-2)!2!}$.

Let us compare this number with the number of elements of dimension 2 with vertices (9), i.e. with the product

$$C_n^2 f_0(n)=\frac{n!}{(n-2)!2!}C_{n+1}^1=\frac{n!}{(n-2)!2!}\frac{(n+1)!}{n!}=\frac{(n+1)!}{(n-2)!2!}.$$

It is clear that $C_n^2 f_0(n)=C_3^1 f_2(n)$. This proves that the number of vertices belonging to all elements of dimension 2 in an n - simplex is equal to the number of elements of dimension 2 that have their vertices d_0^i.

Let $k_{d_0 d_j^i}$ be the number of vertices belonging to some one element of dimension j, that is d_j^i. Obviously, this number is equal to $C_{j+1}^1=j+1$, $i=1\div f_j(n)$. The product $C_{j+1}^1 f_j(n)$ corresponds to the number of vertices belonging to all elements of dimension j in the n - simplex. Compare this number with the number of elements of dimension j having vertices d_0^i at $i=f_0(n)$, i.e. with the product

$$C_n^j f_0(n)=(n+1)\frac{n!}{(n-j)!j!}.$$

On the other hand, we have

$$C_{j+1}^1 f_0(n)=\frac{(j+1)!}{j!}\frac{(n+1)!}{(n-j)!(j+1)!}=\frac{n!(n+1)}{j!(n-j)!}.$$

It can be seen that these values coincide. This proves that the number of vertices belonging to all elements of dimension j in the n - simplex is equal to the number of elements of dimension j that have their vertices d_0^i.

Let $k_{d_1 d_j^i}$ be the number of edges belonging to some one element of dimension j, that is $d_j^i, j>1, i=1\div f_j(n)$. Obviously, this number edges belonging all elements with dimension j is

$$f_1(j)f_j(n)=C_{j+1}^{2}C_{n+1}^{J+1}=\frac{(j+1)!}{(j-1)!2!}\frac{(n+1)!}{(n-j)!(j+1)!}=\frac{(n+1)!}{(j-1)!2!(n-j)!}.$$

Compare this number with the number of elements dimension j having element d_1 (see (3.10))

$$f_1(n)k_{d_1^i d_j}=f_1(n)C_{n-1}^{j-1}=C_{n+1}^{2}C_{n-1}^{j-1}=\frac{(n+1)n}{2}\frac{(n-1)!}{(n-j)!(j-1)!}.$$

It can be seen that these numbers are the same.

Let $k_{d_h d_j^i}$ be the number of elements with dimension h belonging to some one element of dimension j ($h<j$), that is $d_j^i, i=1\div f_j(n)$. Obviously, this number equal $f_h(j)$. The product $f_h(j)f_j(n)$ corresponds to the number of elements with dimension h belonging to all elements of dimension j in the n - simplex. This number is

$$f_h(j)f_j(n)=C_{j+1}^{h+1}C_{n+1}^{j+1}=\frac{(j+1)!}{(j-h)!(h+1)!}\frac{(n+1)!}{(n-j)!(j+1)!}=\frac{(n+1)!}{(j-h)!(h+1)!(n-j)!}.$$

Compare this number with the number of elements dimension j having elements with dimension h ($d_h^i, i=f_h(n)$)

$$f_h(n)k_{d_h^i d_j}=f_h(n)C_{n-h}^{j-h}=C_{n+1}^{h+1}C_{n-h}^{j-h}=\frac{(n+1)!}{(n-h)!(h+1)!}\frac{(n-h)!}{(n-j)!(j-h)!}.$$

It can be seen that these numbers are the same.

This proves that the number of elements of dimension h, belonging to all elements of dimension j in n - simplex, is equal to the number of elements of dimension j, which have elements of dimension h. Thus, the total number of incidents of elements of a smaller dimension to elements of a larger dimension is equal to the total number of incidents of elements of a higher dimension to elements of a smaller dimension. This exists in spite of the difference in the incidence coefficients of elements of a smaller dimension to elements of a higher dimension and the incidence coefficients of elements of a higher dimension to elements of a smaller dimension.

This fact can be considered the law of conservation of incidences in the n - simplex.

The total number of incidents in the n - simplex is described by equality (3.11).

It can be seen that the total number of incidents in the n - simplex depends only on the dimension n and increases sharply with increasing n. For example, it is 12 for $n = 2$, 50 for $n = 3$, 180 for $n = 4$, 602 for $n = 5$.

INCIDENT CONSERVATION LAW FOR N - CUBE

In n - cube (Zhizhin, 2013, 2014, 2019 a) the number of elements with dimension d is

$$f_d(n) = 2^{n-d} C_n^d, d = 1 \div n. \tag{3.12}$$

For example, if $d = 0$, so

$$f_0(n)=2^n. \tag{3.13}$$

This is the number of vertices in n - cube. If $d = 1$, so

$$f_1(n) = 2^{n-1} C_n^1 = 2^{n-1} n. \tag{3.14}$$

This is the number of edges in n - cube.

We introduce the notation: $k_{d_j^i d_u}$ is the number of elements of dimension u, which include an element of dimension j ($u > j$) with number i. Thus, $k_{d_j^i d_u}$ is the incidence factor of element i with dimension j relative to elements with dimension u. We introduce the notation also: $k_{d_j d_u^i}$ is the number of elements of dimension j, which included in element i with a dimension u ($u > j$). Thus, $k_{d_j d_u^i}$ is the incidence factor of element i with dimension u relative to elements of dimension j.

If $n = 2$, so $f_0(2)=4, f_1(2)=4$, see (3.13), (3.14). The factors of incidents (from smaller dimension to larger, Figure 1) are

$$k_{d_0^i d_1} = 2 = C_2^1, i = 1 \div 4; \; k_{d_0^i d_2} = 1, i = 1 \div 4; \; k_{d_1^i d_2} = 1, i = 1 \div 4.$$

On Figure 3.1 the edges are $d_1^1 = d_0^1 d_0^2, d_1^2 = d_0^2 d_0^3, d_1^3 = d_0^3 d_0^4, d_1^4 = d_0^4 d_0^1$.

Sum up the incidence coefficients for all vertices and edges of the triangle

$$\sum_{i=1}^{4} k_{d_0^i d_1} = 2 \cdot 4 = 8, \sum_{i=1}^{4} k_{d_0^i d_2} = 1 \cdot 4 = 4, \sum_{i=1}^{4} k_{d_1^i d_2} = 1 \cdot 4 = 4, \sum_{i=1}^{4} k_{d_0^i d_1} + \sum_{i=1}^{4} k_{d_0^i d_2} + \sum_{i=1}^{4} k_{d_1^i d_2} = 16. \tag{3.15}$$

The factors of incidents (from larger dimension to smaller, Figure 1) are

$$k_{d_0 d_1^i} = 2, i = 1 \div 4; k_{d_1 d_2} = 4, k_{d_0 d_2} = 4.$$

Figure 1. The 2 - cube

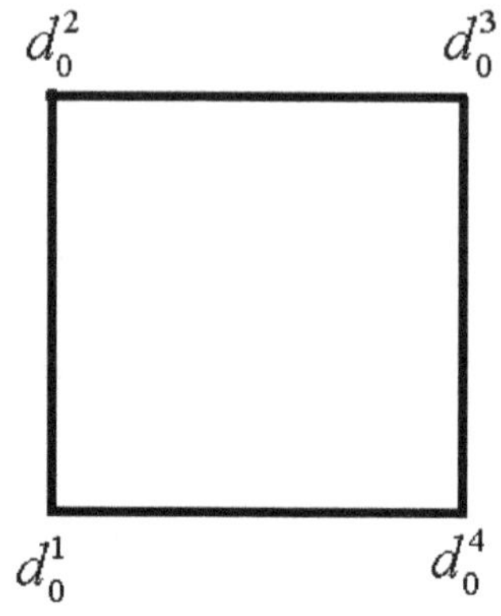

Sum up the incidence coefficients for all elements of the 2 - cube with dimension larger of zero

$$\sum_{i=1}^{4} k_{d_0 d_1^i} + k_{d_1 d_2} + k_{d_0 d_2} = 16. \tag{3.16}$$

Comparing (3.15) and (3.16) you can see that the sum of incidents in a 2 - cube from elements with a lower dimension to elements with a higher dimension is equal to the sum of incidents from elements with a higher dimension to elements with a lower dimension. Thus, the sum of incidents remains unchanged when the direction of the relationship between the elements it is change.

If $n = 3$, so $f_0(3)=8$, $f_1(3)=12$, $f_2(3)=6$, see (3.13), (3.14). The factors of incidents (from smaller dimension to larger, Figure 2) are

$$k_{d_0^i d_1} = 3, i = 1 \div 8;\ k_{d_0^i d_2} = 3, i = 1 \div 8;\ k_{d_0^i d_3} = 1, i = 1 \div 8;$$
$$k_{d_1^i d_2} = 2, i = 1 \div 12;\ k_{d_2^i d_3} = 1, i = 1 \div 6;\ k_{d_1^i d_3} = 1, i = 1 \div 12.$$

Figure 2. The 3 - cube

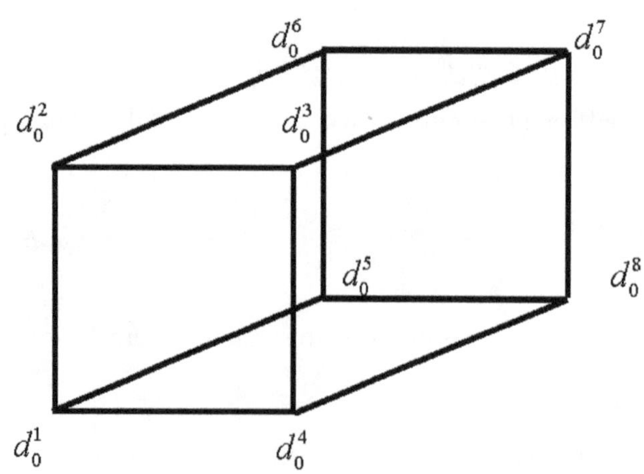

Sum up the incidence coefficients for all vertices, edges and two - dimension faces of the 3 - cube

$$\sum_{i=1}^{8} k_{d_0^i d_1} = 3\cdot 8 = 24, \sum_{i=1}^{8} k_{d_0^i d_2} = 3\cdot 8 = 24, \sum_{i=1}^{8} k_{d_0^i d_3} = 1\cdot 8 = 8, \sum_{i=1}^{12} k_{d_1^i d_2} = 2\cdot 12 = 24, \sum_{i=1}^{6} k_{d_2^i d_3} = 1\cdot 6 = 6,$$
$$\sum_{i=1}^{12} k_{d_1^i d_3} = 1\cdot 12 = 12, \sum_{i=1}^{8} k_{d_0^i d_1} + \sum_{i=1}^{8} k_{d_0^i d_2} + \sum_{i=1}^{12} k_{d_1^i d_2} + \sum_{i=1}^{8} k_{d_0^i d_3} + \sum_{i=1}^{6} k_{d_2^i d_3} + \sum_{i=1}^{12} k_{d_1^i d_3} = 98. \qquad (3.17)$$

The factors of incidents (from larger dimension to smaller, Figure 3.2) are

$$k_{d_0 d_1^i} = 2, i = 1 \div 12; k_{d_1 d_2^i} = 4, i = 1 \div 6; k_{d_0 d_2^i} = 4, i = 1 \div 6; k_{d_0 d_3} = 8; k_{d_1 d_3} = 12; k_{d_2 d_3} = 6.$$

Sum up the incidence coefficients for all elements of the 3 - cube with dimension larger of zero

$$\sum_{i=1}^{12} k_{d_0 d_1^i} + \sum_{i=1}^{6} k_{d_0 d_2^i} + \sum_{i=1}^{6} k_{d_1 d_2^i} + k_{d_1 d_3} + k_{d_0 d_3} + k_{d_2 d_3} = 98. \qquad (3.18)$$

Comparing (3.17) and (3.18) you can see that the sum of incidents in a 3 - cube from elements with a lower dimension to elements with a higher dimension is equal to the sum of incidents from elements with a higher dimension to elements with a lower dimension. Thus, the sum of incidents retains its value when the direction of the relationship between the elements it is changes.

If $n = 4$, so $f_0(4)=16$, $f_1(4)=32$, $f_2(4)=24$, $f_3(4)=8$, see (12) - (14). The factors of incidents (from smaller dimension to larger, Figure 3) are

$$k_{d_0^i d_1} = 4, i = 1 \div 16; k_{d_0^i d_2} = 6, i = 1 \div 16; k_{d_0^i d_3} = 4, i = 1 \div 16; k_{d_0^i d_4} = 1, i = 1 \div 16; k_{d_1^i d_2} = 3, i = 1 \div 32;$$
$$k_{d_1^i d_3} = 3, i = 1 \div 32; k_{d_1^i d_4} = 1, i = 1 \div 32; k_{d_2^i d_3} = 2, i = 1 \div 24; k_{d_2^i d_4} = 1, i = 1 \div 24; k_{d_3^i d_4} = 1, i = 1 \div 8.$$

Figure 3. The 4 - cube

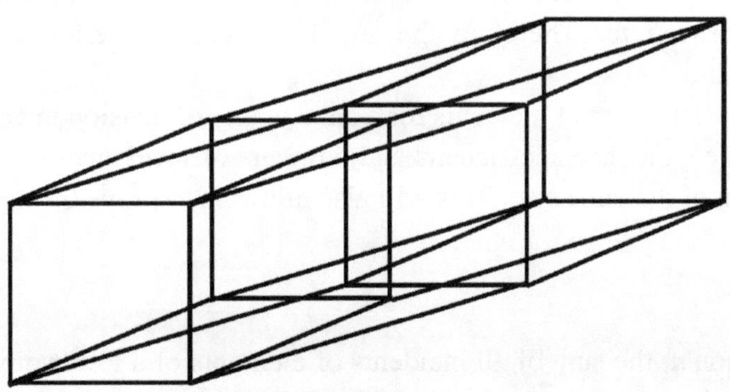

Sum up the incidence coefficients for all vertices, edges, two - dimension faces and three - dimension faces of the 4 - cube

$$\sum_{i=1}^{16} k_{d_0^i d_1} = 4\cdot 16 = 64, \sum_{i=1}^{16} k_{d_0^i d_2} = 6\cdot 16 = 96, \sum_{i=1}^{16} k_{d_0^i d_3} = 4\cdot 16 = 64, \sum_{i=1}^{16} k_{d_0^i d_4} = 16, \sum_{i=1}^{32} k_{d_1^i d_2} = 3\cdot 32 = 96,$$

$$\sum_{i=1}^{32} k_{d_1^i d_3} = 3\cdot 32 = 96, \sum_{i=1}^{32} k_{d_1^i d_4} = 32, \sum_{i=1}^{24} k_{d_2^i d_3} = 2\cdot 24 = 48, \sum_{i=1}^{24} k_{d_2^i d_4} = 24, \sum_{i=1}^{8} k_{d_3^i d_4} = 1\cdot 8 = 8.$$

$$\sum_{i=1}^{16} k_{d_0^i d_1} + \sum_{i=1}^{16} k_{d_0^i d_2} + \sum_{i=1}^{16} k_{d_0^i d_3} + \sum_{i=1}^{16} k_{d_0^i d_4} + \sum_{i=1}^{32} k_{d_1^i d_2} + \sum_{i=1}^{32} k_{d_1^i d_3} + \sum_{i=1}^{32} k_{d_1^i d_4} + \sum_{i=1}^{24} k_{d_2^i d_3} + \sum_{i=1}^{24} k_{d_2^i d_4} + \sum_{i=1}^{8} k_{d_3^i d_4} = 544. \quad (3.19)$$

The factors of incidents (from larger dimension to smaller, Figure 3.3) are

$$k_{d_0 d_1^i} = 2, i = 1 \div 32; k_{d_0 d_2^i} = 4, i = 1 \div 24; k_{d_0 d_3^i} = 8, i = 1 \div 8; k_{d_0 d_4} = 16; k_{d_1 d_2^i} = 4, i = 1 \div 24;$$

$$k_{d_1 d_3^i} = 12, i = 1 \div 8; k_{d_1 d_4} = 32; k_{d_2 d_3^i} = 6, \mathrm{i} = 1 \div 8; k_{d_2 d_4} = 24; k_{d_3 d_4} = 8.$$

Sum up the incidence coefficients for all elements of the 4 - cube with dimension larger of zero

$$\sum_{i=1}^{32} k_{d_0 d_1^i} \sum_{0\ i=1}^{24} k_{d_0 d_2^i} + \sum_{i=1}^{24} k_{d_1 d_2^i} + \sum_{i=1}^{8} k_{d_0 d_3^i} + \sum_{i=1}^{8} k_{d_1 d_3^i} + \sum_{i=1}^{8} k_{d_2 d_3^i} + k_{d_0 d_4} + k_{d_1 d_4} + k_{d_2 d_4} + k_{d_3 d_4} = 544. \quad (3.20)$$

Comparing (3.19) and (3.20) you can see that the sum of incidents in a 4 - cube from elements with a lower dimension to elements with a higher dimension is equal to the sum of incidents from elements with a higher dimension to elements with a lower dimension. Thus, the sum of incidents retains its value when the direction of the relationship between the elements it is changes.

It is easy to see that each incidence coefficient from a smaller dimension to a higher dimension corresponds to its inverse incidence factor from a larger dimension to a smaller dimension. For example, $k_{d_0^i d_2} = 6, k_{d_0 d_2^i} = 4$. Moreover, the number of such incidents from a smaller dimension to a higher dimension is 16, the number of corresponding inverse incidents is 24. So, the product of the incidence coefficient and the corresponding number retains its value, 6•6=4•24=96. It can be seen that this situation exists for all incidence rates. Therefore, the sum of incidents in the forward and reverse direction is preserved.

In order to write down the sum of incidents for a cube of any dimension in general, it is necessary to write expressions for the incidence coefficients and numbers of incidences in the forward and reverse directions, as a function of the cube size. This will also allow us to prove the next statement:

Theorem 3.2

In any cube of dimension n, the sum of all incidents of elements of a lower dimension with respect to all elements of a higher dimension is equal to the sum of all incidents of elements of a higher dimension with respect to all elements of a lower dimension and equals the sum of the series

$$f_0(n)(2^n-1)+f_1(n)(2^{n-1}-1)+f_2(n)(2^{n-2}-1)+....+f_{n-1}(n), f_d(n)=C_{n+1}^{d+1}, d=0\div(n-1).$$

Proof

The number of vertices in n - cube is equal $f_0=2^n$ (see (3.12)). The number of edges in n - cube incident to any of the vertices of a cube, by its definition, is equal to $k_{d_0^i d_1}=C_n^1, i=1\div f_0$.

The number of elements of two - dimension incident to any of the vertices of a cube is equal $k_{d_0^i d_2}=C_n^2, i=1\div f_0$. We can also record the number of elements of any dimension incident to any of the vertices of the cube up to the cube itself $k_{d_0^i d_n}=C_n^n, i=1\div f_0$.

Multiplying each of these incidence coefficients by the number of vertices in the n - cube and summing up the obtained products, we find that the total number of incidences of the vertices to the elements of higher dimensionality in the n - cube is determined by the expression

$$f_0(C_n^1+C_n^2+...+C_n^n)=f_0(2^n-1).$$

This expression coincides with the expression of the total number of incidences of vertices to elements of higher dimension in the n - simplex. The numerical difference between these expressions consists in a different value of the vertices in the n - cube and in the n - simplex, i.e. in the meaning of $f_0(n)$.

The number of edges in n - cube is equal $f_1(n)=2^{n-1}C_n^1$ (see (3.12)). The number of two - dimension elements in n - cube incident to any of the edges of a cube, by its definition, is equal to $k_{d_1^i d_2}=C_{n-1}^1, i=1\div f_1$.

The number of three - dimension elements incident to any of the edges of a cube is equal $k_{d_1^i d_3}=C_{n-1}^2, i=1\div f_1$. We can also record the number of elements of any dimension incident to any of the edges of the cube up to the cube itself $k_{d_1^i d_n}=C_{n-1}^{n-1}, i=1\div f_1$.

Multiplying each of these incidence coefficients by the number of edges in the n - cube and summing up the obtained products, we find that the total number of incidences of the edges to the elements of higher dimensionality in the n - cube is determined by the expression

$$f_1(C_{n-1}^1+C_{n-1}^2+...+C_{n-1}^{n-1})=f_1(2^{n-1}-1).$$

This expression coincides with the expression of the total number of incidences of edges to elements of higher dimension in the n - simplex. The numerical difference between these expressions consists in a different value of the edges in the n - cube and in the n - simplex, i.e. in the meaning of $f_1(n)$.

Continuing to write expressions for the total numbers of incidents of elements of different dimensions to elements of a higher dimension in a n - cube, summing up these numbers, we obtain a general expression for the number of incidences of elements of a lower dimension in relation to elements of a higher dimension in n - cube

$$f_0(n)(2^n-1)+f_1(n)(2^{n-1}-1)+f_2(n)(2^{n-2}-1)+....+f_{n-1}(n),$$
$$f_d(n)=C_{n+1}^{d+1}, d=0\div(n-1). \tag{3.21}$$

It is easy to make sure that the calculations of the incidence flow in the n - cube according to expression (3.21) at values n=2÷4 coincide with the values of these quantities obtained earlier by analyzing the images of n - cube at these values n.

Let $k_{d_h d_j^i}$ be the number of elements of dimension h in a n - cube belonging to some one element of dimension j ($h<j$), that is d_j^i. Obviously, this number is equal to $f_h(j)$, and $i=1\div f_j(n)$. The product $f_h(j)$ $f_j(n)$ corresponds to the number of elements of dimension h belonging to all elements of dimension j. For a n - cube, this product is equal to

$$f_h(j)f_j(n)=2^{j-h}C_j^h 2^{n-j}C_n^j=2^{n-h}\frac{n!}{(n-j)!(j-h)!h!}.$$

Let us compare this number with the number of elements of dimension j, which have elements of dimension h in the n - cube

$$k_{d_h^i d_j}f_h(n)=C_{n-h}^{j-h}2^{n-h}C_n^h=2^{n-h}\frac{n!}{(n-h)!h!}\frac{(n-h)!}{(n-j)!(j-h)!}.$$

Obviously, these numbers are equal to each other. This proves that the number of elements of dimension h in a n - cube belonging to all dimensions of j ($j>h$) in a n - cube is equal to the number of elements of dimension j, which have elements of dimension h as elements.

Thus, the total number of incidents of elements of a smaller dimension with respect to elements of a higher dimension is equal to the total number of incidents of elements of a higher dimension with respect to elements of a smaller dimension. This the total number define expression (3.21).

It is significant that equality of incidence (information) flows exists despite the difference in the incidence coefficients of elements of a lower dimension in relation to elements of a higher dimension and the incidence coefficients of elements of a higher dimension in relation to elements of a smaller dimension. Therefore, the incidence conservation law found in the previous section for the n - simplex is also valid for the n - cube. Moreover, the analytical form of this law is preserved, expressed through functions $f_i(n)$ that determine the values of the numbers of the elements of dimension i in the polytope of dimension n.

INCIDENT CONSERVATION LAW FOR N - CROSS - POLYTOPE

In n - cross - polytope (Zhizhin, 2013, 2014 c, 2019 a) the number of elements with dimension d is

$$f_d(n)=2^{d+1}C_n^{n-1-d}, d=0\div n. \tag{3.22}$$

For example, if $d = 0$, so

$$f_0(n) = 2C_n^{n-1} = 2n. \tag{3.23}$$

This is the number of vertices in n - cross - polytope. If $d = 1$, so

$$f_1(n) = 2^2 C_n^{n-2} = 2\frac{n!}{(n-2)!} = 2(n-1)n. \tag{3.24}$$

This is the number of edges in n - cross - polytope. If $d = 2$, so

$$f_2(n) = 2^3 C_n^{n-3} = \frac{4}{3}(n-2)(n-1)n. \tag{3.25}$$

This is the number of two - dimension faces in n - cross - polytope. If $d = 3$, so

$$f_3(n) = 2^4 C_n^{n-4} = \frac{2}{3}(n-3)(n-2)(n-1)n. \tag{3.26}$$

This is the number of three - dimension faces in n - cross - polytope.

The smallest dimension of the cross - polytope is 4. From (3.23) - (3.26) it follows that in this polytope there are 8 vertices, 24 edges, 32 two - dimensional faces, 16 three - dimensional faces (Figure 4).

Figure 4. The 4 - cross - polytope

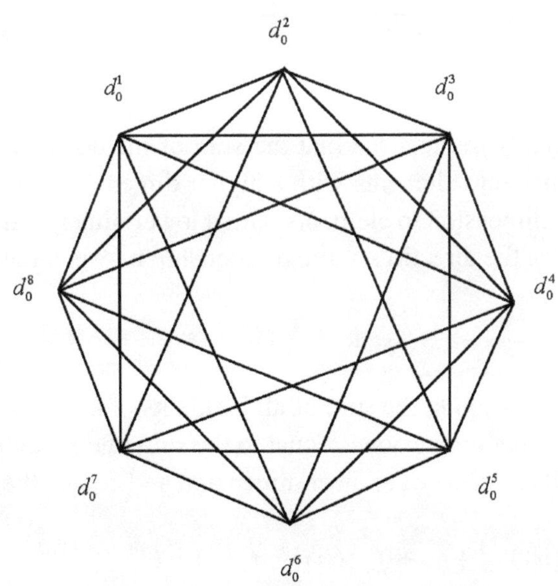

The factors of incidents (from smaller dimension to larger) are

$$k_{d_0^i d_1} = 6, i = 1 \div 8; k_{d_0^i d_2} = 12, i = 1 \div 8; k_{d_0^i d_3} = 8, i = 1 \div 8; k_{d_0^i d_4} = 1, i = 1 \div 8; k_{d_1^i d_2} = 4, i = 1 \div 24;$$
$$k_{d_1^i d_3} = 4, i = 1 \div 24; k_{d_1^i d_4} = 1, i = 1 \div 24; k_{d_2^i d_3} = 2, i = 1 \div 32; k_{d_2^i d_4} = 1, i = 1 \div 32; k_{d_3^i d_4} = 1, i = 1 \div 16.$$

Sum up the incidence coefficients for all vertices, edges, two - dimension faces and three -dimension faces of the 4 - cross - polytope

$$\sum_{i=1}^{8} k_{d_0^i d_1} = 6 \cdot 8 = 48, \sum_{i=1}^{8} k_{d_0^i d_2} = 12 \cdot 8 = 96, \sum_{i=1}^{8} k_{d_0^i d_3} = 8 \cdot 8 = 64, \sum_{i=1}^{8} k_{d_0^i d_4} = 8, \sum_{i=1}^{24} k_{d_1^i d_2} = 4 \cdot 24 = 96,$$
$$\sum_{i=1}^{24} k_{d_1^i d_3} = 4 \cdot 24 = 96, \sum_{i=1}^{24} k_{d_1^i d_4} = 24, \sum_{i=1}^{32} k_{d_2^i d_3} = 2 \cdot 32 = 64, \sum_{i=1}^{32} k_{d_2^i d_4} = 32, \sum_{i=1}^{16} k_{d_3^i d_4} = 16.$$
$$\sum_{i=1}^{8} k_{d_0^i d_1} + \sum_{i=1}^{8} k_{d_0^i d_2} + \sum_{i=1}^{8} k_{d_0^i d_3} + \sum_{i=1}^{8} k_{d_0^i d_4} + \sum_{i=1}^{24} k_{d_1^i d_2} + \sum_{i=1}^{24} k_{d_1^i d_3} + \sum_{i=1}^{24} k_{d_1^i d_4} + \sum_{i=1}^{32} k_{d_2^i d_3} + \sum_{i=1}^{16} k_{d_3^i d_4} + \sum_{i=1}^{32} k_{d_2^i d_4} = 544. \quad (3.27)$$

The factors of incidents (from larger dimension to smaller, Figure 4) are

$$k_{d_0 d_1^i} = 2, i = 1 \div 24; k_{d_0 d_2^i} = 3, i = 1 \div 32; k_{d_0 d_3^i} = 4, i = 1 \div 16; k_{d_0 d_4} = 8; k_{d_1 d_2^i} = 3, i = 1 \div 32;$$
$$k_{d_1 d_3^i} = 6, i = 1 \div 16; k_{d_1 d_4} = 24; k_{d_2 d_3^i} = 4, \mathrm{i} = 1 \div 8; k_{d_2 d_4} = 32; k_{d_3 d_4} = 16.$$

Sum up the incidence coefficients for all elements of the 4 - cube with dimension larger of zero

$$\sum_{i=1}^{24} k_{d_0 d_1^i} + \sum_{i=1}^{32} k_{d_0 d_2^i} + \sum_{i=1}^{32} k_{d_1 d_2^i} + \sum_{i=1}^{16} k_{d_0 d_3^i} + \sum_{i=1}^{16} k_{d_1 d_3^i} + \sum_{i=1}^{16} k_{d_2 d_3^i} + k_{d_0 d_4} + k_{d_1 d_4} + k_{d_2 d_4} + k_{d_3 d_4} = 544. \quad (328)$$

Comparing (3.28) and (3.27) you can see that the sum of incidents in a 4 - cross - polytope from elements with a lower dimension to elements with a higher dimension is equal to the sum of incidents from elements with a higher dimension to elements with a lower dimension. Thus, the sum of incidents retains its value when changes the direction of the relationship between the elements.

Theorem 3.3

In any cross - polytope of dimension n, the sum of all incidents of elements of a lower dimension with respect to all elements of a higher dimension is equal to the sum of all incidents of elements of a higher dimension with respect to all elements of a lower dimension and equals the sum of the series

$$f_0(n)3^{n-1} + f_1(n)3^{n-2} + f_2(n)3^{n-3} + ... + f_{n-1}(n)3^0, f_d(n) = C_n^{n-1-d} 2^{1+d}, d = 0 \div (n-1).$$

Proof

According to equation (3.23), each cross - polytope of dimension n has $2n = f_0(n)$ vertices. The peculiarity of a cross - polytope is that each of its vertices has an opposite vertex, with which it is not connected by an edge. Moreover, there is one edge between this vertex and all other vertices. We subtract from the total number of vertices two vertices (the selected vertex and its opposite) 2n - 2. This is the possible number of edges emanating from the selected vertex. Thus, the incidence coefficient of the edges of any vertex is $k_{d_0^i d_1} = 2(n-1) = 2\,\mathrm{C}_{n-1}^1, i = 1 \div f_0(n)$. A cross - polytope of any dimension can be depicted as a projection on a two - dimensional plane (Zhizhin, 2019 a). In this image, all its vertices are located on a circle, with the selected vertex and its opposite vertex located symmetrically relative to the center of the circle. A mentally drawn line through these two vertices halves the circle (Figure 3.4). Therefore, the number of variants the location of the edges from the selected vertex to the other vertices in one of the halves of the circle is n - 1, i.e. the number of combinations of n - 1 vertices one by one. Since there are two halves of a circle, then for the total number of vertex selection options for edge formation, this number of combinations should be multiplied by 2. This is the meaning of the expression for the incidence rate of any vertex in the n - cross - polytope with respect to the edge. To further prove the theorem 3.3 and clarify the nature of the formation of the coefficients of incidence in the n - cross - polytope, we will use this technique. We arrange the vertices of the 4 - cross - polytope in two lines, so that the vertices unlinked by an edge form vertical pairs: each vertex in the top line is not connected by an edge to the vertex in the bottom line located strictly under this vertex in the top line

$$\begin{matrix} d_0^1 & d_0^2 & d_0^3 & d_0^4 \\ d_0^5 & d_0^6 & d_0^7 & d_0^8 \end{matrix} . \tag{3.29}$$

We choose some vertex in the top line (3.29), for example d_0^1 . It cannot be connected by an edge with a vertex d_0^5 in the bottom line, but it can be connected by an edge with other vertices of the bottom line. The number of such options is 3, i.e. with n = 4 is $3 = C_{n-1}^1 = C_3^1$. However, a vertex d_0^1 may be connected by an edge with any of the remaining three vertices in the top line (3.29). Therefore, the incidence coefficient of a vertex with respect to an edge in 4 - cross - polytope $k_{d_0^i d_1} = 2\,\mathrm{C}_{n-1}^1 = 6, i = 1 \div 8$. When considering the belonging of a vertex d_0^1 to two - dimensional faces, it is necessary to determine the number of options for the participation of two vertices in the top line (3.29) (without a vertex d_0^1) and in the bottom line (3.29) (without a vertex d_0^5). This will be the number $2C_3^1$. In addition, as the vertices of the triangle, there may be vertices located in the upper and lower lines (3.29) in a cross - section way. There are 6 such variants in (3.29), i.e. more $2C_3^1$. Therefore, the incidence coefficient of a vertex with respect to a two - dimensional element in 4 - cross - polytope is $k_{d_0^i d_2} = 2^2\,\mathrm{C}_{n-1}^2 = 12, i = 1 \div 8$. Multiplying this number by the number of vertices in 4 - cross - polytope (8) you can get the total number of incidents of vertices to two - dimensional elements in 4 - cross - polytope equal to 96. This number coincides with the corresponding number, defined earlier in Figure 3.4. Combination d_0^1 with three the different vertices from (3.29) can get the incidence coefficient of vertex to the three - dimension body (tetrahedron) for condition absent in combination opposite vertices. Can to show that this combination is 8, $k_{d_0^1 d_3} = 2^3 C_3^3 = 8, i = 1 \div 8$. Multiplying this number by the number of vertices in 4 - cross - polytope (8) you can get the total number of incidents of vertices to three - dimensional elements in 4 - cross - polytope equal to 64. This number coincides with the corresponding number, defined earlier in Figure 3.4. Combination two vertices, for example d_0^1 and d_0^2 (the edge $d_0^1\ d_0^2$) with two the different vertices from (3.29) you can get the incidence coefficient of edge to three - dimension body (tetrahedron) for condition absent in combination opposite vertices. Can to show that this combination is 4, $k_{d_1^1 d_3} = 2^2 C_2^2 = 4, i = 1 \div 8$. Multiplying this number by the number of edges in 4 - cross - polytope (8) you can get the total number of incidents of edges to three - dimensional elements in 4 - cross - polytope equal to 96. This number coincides with the corresponding number, defined earlier in Figure 3.4. Combination two vertices (the edge $d_0^1\ d_0^2$) with one the vertex from (3.29) can get the incidence coefficient of edge to triangle for condition absent in combination opposite vertices. Can to show that this combination is 4, $k_{d_1^1 d_2} = 2C_2^1 = 4, i = 1 \div 8$. Multiplying this number by the number of edges in 4 - cross - polytope (8) you can get the total number of incidents of edges to triangle in 4 - cross - polytope equal to 24. This number coincides with the corresponding number, defined earlier in Figure 3.4.

In general case can write of $2n$ vertices of the n - cross - polytope in two lines

$$\begin{matrix} d_0^1 & d_0^2 & d_0^3 & \ldots & d_0^n \\ d_0^5 & d_0^6 & d_0^7 & \ldots & d_0^{2n} \end{matrix},$$

where vertex in the top line is not connected by an edge to the vertex in the bottom line located strictly under this vertex in the top line. The incidence coefficient of a vertex with respect to an edge in n - cross - polytope is $k_{d_0^i d_1} = 2\mathrm{C}_{n-1}^1, i = 1 \div f_0(n)$. The incidence coefficient of a vertex with respect to a two - dimensional element in n - cross - polytope is $k_{d_0^i d_2} = 2^2\ \mathrm{C}_{n-1}^2, i = 1 \div f_0(n)$. The incidence coefficient of a vertex with respect to a three - dimensional edge in n - cross - polytope is $k_{d_0^i d_3} = 2^3\ \mathrm{C}_{n-1}^3, i = 1 \div f_0(n)$.

Go on can to say the incidence coefficient of a vertex with respect to a $(n-1)$ - dimensional element in n - cross - polytope is $k_{d_0^i d_{n-1}} = 2^{n-1}\ \mathrm{C}_{n-1}^{n-1}, i = 1 \div f_0(n)$. Obviously, that the incidence coefficient of a vertex with respect to n - cross - polytope is $k_{d_0^i d_n} = 1, i = 1 \div f_0(n)$.

Multiplying of the incidence coefficients of a vertex with respect to different dimension elements in n - cross - polytope and sum the product you can get the common express for the number of incident vertices to elements of different dimension in n - cross - polytope

$$f_0(2^0 + 2^1 C_{n-1}^1 + 2^2 C_{n-1}^2 + \ldots + 2^{n-1} C_{n-1}^{n-1}) = f_0(n) \sum_{i=0}^{n-1} 2^i C_{n-1}^i = f_0(n) 3^{n-1}.$$

In this way you can get the common express for the number of incident edges to elements of different dimension more one in n - cross - polytope

$$f_1(2^0+2^1C^1_{n-2}+2^2C^2_{n-2}+...+2^{n-2}C^{n-2}_{n-2})=f_1(n)\sum_{i=0}^{n-2}2^iC^i_{n-2}=f_1(n)3^{n-2},$$

and go on.

In result, one can get the common express for the sum of all incidents of elements of a lower dimension with respect to all elements of a higher dimension in n - cross - polytope

$$f_0(n)3^{n-1}+f_1(n)3^{n-2}+...+f_{n-1}3^0=\sum_{i=0}^{n-1}f_i(n)3^{n-1-i},\ f_i(n)=C_n^{n-1-i}2^{1+i}. \tag{3.30}$$

Let $k_{d_h d_j^i}$ be the number of elements of dimension h in a n - cross - polytope belonging to some one element of dimension j ($h<j$), that is d_j^i . Obviously, this number is equal to $f_h(j)_s$ for simplex, and $i=1\div f_j(n)_{cr}$ for n - cross - polytope. So, elements of cross - polytope are simplexes the product $f_h(j)_s f_j(n)_{cr}$ corresponds to the number of elements of dimension h belonging to all elements of dimension j for a simplex. This product is equal to $f_d(n)=C_{n+1}^{d+1}$

$$C_{j+1}^{h+1}2^{1+j}C_n^{n-1-j}=2^{1+j}\frac{(j+1)!}{(h+1)!(j-h)!}\frac{n!}{(n-1-j)!(1+j)!}.$$

Let us compare this number with the number of elements of dimension j, which have elements of dimension h in the n - cross - polytope

$$k_{d_h^i d_j}f_h(n)_{cr}=C_n^{n-1-h}2^{1+h}C_{n-h-1}^{j-h}2^{j-h}=2^{1+j}\frac{n!}{(n-1-h)!(1+h)!}\frac{(n-h-1)!}{(n-j-1)!(j-h)!}.$$

Obviously, these numbers are equal to each other. This proves that the number of elements of dimension h in a n - cross - polytope belonging to all dimensions of j ($j>h$) in a n - cross - polytope is equal to the number of elements of dimension j, which have elements of dimension h as elements.

Thus, the total number of incidents of elements of a smaller dimension with respect to elements of a higher dimension is equal to the total number of incidents of elements of a higher dimension with respect to elements of a smaller dimension. This the total number define expression (3.30).

INCIDENT CONSERVATION LAW FOR N - GOLDEN - HYPER – RHOMBOHEDRON

The smallest dimension of the golden hyper - rhombohedron is 4 (Figure 5) (Zhizhin, 2014 c, 2019 a). From Figure 3.5 it follows that in the case:

$f_0(4)=16, f_1(4)=32, f_2(4)=24, f_3(4)=8$

Figure 5. The 4 - golden - hyper - rhombohedron

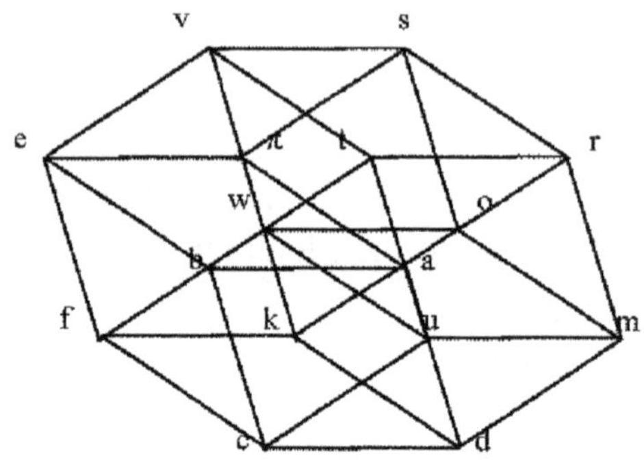

The factors of incidents (from smaller dimension to larger) are:

$$k_{d_0^i d_1}=4, i=1\div16; k_{d_0^i d_2}=6, i=1\div16; k_{d_0^i d_3}=4, i=1\div16; k_{d_0^i d_4}=1, i=1\div16; k_{d_1^i d_2}=3, i=1\div32;$$
$$k_{d_1^i d_3}=3, i=1\div32; k_{d_1^i d_4}=1, i=1\div32; k_{d_2^i d_3}=2, i=1\div24; k_{d_2^i d_4}=1, i=1\div24; k_{d_3^i d_4}=1, i=1\div8.$$

Sum up the incidence coefficients for all vertices, edges, two - dimension faces and three - dimension faces of the 4 - golden - hyper – rhombohedron:

$$\sum_{i=1}^{16} k_{d_0^i d_1}=4\cdot16=64, \sum_{i=1}^{16} k_{d_0^i d_2}=6\cdot16=96, \sum_{i=1}^{16} k_{d_0^i d_3}=4\cdot16=64, \sum_{i=1}^{16} k_{d_0^i d_4}=16, \sum_{i=1}^{32} k_{d_1^i d_2}=3\cdot32=96,$$
$$\sum_{i=1}^{32} k_{d_1^i d_3}=3\cdot32=96, \sum_{i=1}^{32} k_{d_1^i d_4}=32, \sum_{i=1}^{24} k_{d_2^i d_3}=2\cdot24=48, \sum_{i=1}^{24} k_{d_2^i d_4}=24, \sum_{i=1}^{8} k_{d_3^i d_4}=1\cdot8=8.$$
$$\sum_{i=1}^{16} k_{d_0^i d_1}+\sum_{i=1}^{16} k_{d_0^i d_2}+\sum_{i=1}^{16} k_{d_0^i d_3}+\sum_{i=1}^{16} k_{d_0^i d_4}+\sum_{i=1}^{32} k_{d_1^i d_2}+\sum_{i=1}^{32} k_{d_1^i d_3}+\sum_{i=1}^{32} k_{d_1^i d_4}+\sum_{i=1}^{24} k_{d_2^i d_3}+\sum_{i=1}^{24} k_{d_2^i d_4}+\sum_{i=1}^{8} k_{d_3^i d_4}=544. \quad (3.31)$$

The factors of incidents (from larger dimension to smaller, Figure 3.5) are:

$$k_{d_0 d_1^i}=2, i=1\div32; k_{d_0 d_2^i}=4, i=1\div24; k_{d_0 d_3^i}=8, i=1\div8; k_{d_0 d_4}=16; k_{d_1 d_2^i}=4, i=1\div24;$$
$$k_{d_1 d_3^i}=12, i=1\div8; k_{d_1 d_4}=32; k_{d_2 d_3^i}=6, \mathrm{i}=1\div8; k_{d_2 d_4}=24; k_{d_3 d_4}=8.$$

Sum up the incidence coefficients for all elements of the 4 - golden - hyper - rhombohedron with dimension larger of zero:

$$\sum_{i=1}^{32} k_{d_0 d_1^i}{}_0+\sum_{i=1}^{24} k_{d_0 d_2^i}+\sum_{i=1}^{24} k_{d_1 d_2^i}+\sum_{i=1}^{8} k_{d_0 d_3^i}+\sum_{i=1}^{8} k_{d_1 d_3^i}+\sum_{i=1}^{8} k_{d_2 d_3^i}+k_{d_0 d_4}+k_{d_1 d_4}+k_{d_2 d_4}+k_{d_3 d_4}=544. \quad (3.32)$$

Comparing (3.31) and (3.32) you can see that the sum of incidents in a 4 - golden - hyper - rhombohedron from elements with a lower dimension to elements with a higher dimension is equal to the sum of incidents from elements with a higher dimension to elements with a lower dimension. Thus, the sum of incidents retains its value when the direction of the relationship between the elements it is changes. It is easy to see that each incidence coefficient from a smaller dimension to a higher dimension corresponds to its inverse incidence factor from a larger dimension to a smaller dimension. For example, $k_{d_0^i d_2} = 6, k_{d_0 d_2^i} = 4$. Moreover, the number of such incidents from a smaller dimension to a higher dimension is 16, the number of corresponding inverse incidents is 24. So, the product of the incidence coefficient and the corresponding number retains its value, $6 \times 16 = 4 \times 24 = 96$. It can be seen that such situation exists for all incidence rates. Therefore, the sum of incidents in the forward and reverse direction is preserved. In order to write down the sum of incidents for a golden hyper - rhombohedron of any dimension in general, it is necessary to write expressions for the incidence coefficients and numbers of incidences in the forward and reverse directions, as a function of the golden - hyper - rhombohedron dimension. This will also allow us to prove the next statement.

Theorem 3.4

In any golden - hyper - rhombohedron of dimension n, the sum of all incidents of elements of a lower dimension with respect to all elements of a higher dimension is equal to the sum of all incidents of elements of a higher dimension with respect to all elements of a lower dimension and equals the sum of the series:

$$f_0(n)(2^n - 1) + f_1(n)(2^{n-1} - 1) + f_2(n)(2^{n-2} - 1) + ... + f_{n-1}(n), f_d(n) = C_{n+1}^{d+1}, d = 0 \div (n-1).$$

Proof

From the previous it is clear that the incidence coefficients of low - dimensional elements with respect to high - dimensional elements and vice versa in the gold n – hyper -rhombohedron are described by the same formulas as the n - cube. Therefore, the proof of Theorem 3.4 coincides with the proof of Theorem 3.2 relating to the n - cube.

It is significant that equality of incidence (information) flows exists despite the difference in the incidence coefficients of elements of a lower dimension in relation to elements of a higher dimension and the incidence coefficients of elements of a higher dimension in relation to elements of a smaller dimension. Therefore, the incidence conservation law found in the previous section for the n - simplex is also valid for the n - golden - hyper - rhombohedron. The analytical form of this law is preserved, expressed through functions $f_i(n)$ that determine the values of the numbers of the elements of dimension i in the polytope of dimension n.

INCIDENT CONSERVATION LAW FOR CLUSTERS CHEMICAL COMPOUND

Incident Conservation Law Regular and Irregular Polytopes

The polytopes n - simplex, n - cube and n - cross - polytopes considered in early was topological regular. Each vertex in any of it was incident same number of edges. However, the length of edges not put equal. Therefore, the polytopes are not metrical regular. Obviously, incident conservation law is for regular and topological regular polytopes. Now consider incident conservation law for irregular and same - regular polytopes.

Incident Conservation Law for Quadrangular Three - Dimension Pyramid

In this pyramid (Figure 6) the number of vertices are f_0=5, the number of edges are f_1=8, the number of two - dimension faces are f_2=5.

Figure 6. The quadrangular three - dimension pyramid

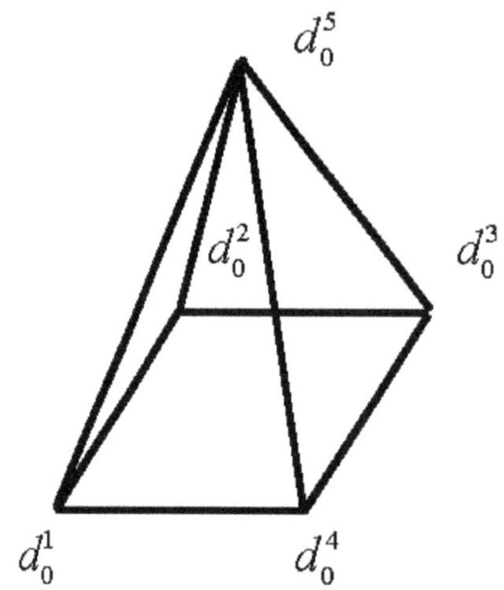

The incident coefficients at vertices of bases to edges are equal 3 $(k_{d_0^i d_1} = 3, i = 1 \div 4)$, the incident coefficient at top of the pyramid to edges is equal 4 $(k_{d_0^5 d_1} = 4)$. The incident coefficients at vertices of bases to two - dimension faces are equal 3 $(k_{d_0^i d_2} = 3, i = 1 \div 4)$, the incident coefficient at top of the pyramid to two - dimension faces is equal 4 $(k_{d_0^5 d_2} = 4)$. The incident coefficients at edges of pyramid to two - dimension faces are equal 2 $(k_{d_1^i d_2} = 2, i = 1 \div 8)$. The incident coefficients of all elements of pyramid to pyramid are 1

$(k_{d_0^i d_3} = 1, i = 1 \div 5; k_{d_1^i d_3} = 1, i = 1 \div 8; k_{d_2^i d_3} = 1, i = 1 \div 5)$.

Multiplying this incident coefficients by the number elements of correspond dimension you can get

$$\sum_{i=1}^{5} k_{d_0^i d_1} = 16, \sum_{i=1}^{5} k_{d_0^i d_2} = 16, \sum_{i=1}^{5} k_{d_0^i d_3} = 5, \sum_{i=1}^{8} k_{d_1^i d_2} = 16, \sum_{i=1}^{5} k_{d_2^i d_3} = 5, \sum_{i=1}^{8} k_{d_1^i d_3} = 8.$$

Sum up the incidence coefficients for all elements of the pyramid (from smaller dimension to larger)

$$\sum_{i=1}^{5} k_{d_0^i d_1} + \sum_{i=1}^{5} k_{d_0^i d_2} + \sum_{i=1}^{5} k_{d_0^i d_3} + \sum_{i=1}^{8} k_{d_1^i d_2} + \sum_{i=1}^{5} k_{d_2^i d_3} + \sum_{i=1}^{8} k_{d_1^i d_3} = 64. \quad (3.33)$$

The factors of incidents (from larger dimension to smaller) are

$k_{d_0 d_1^i} = 2, i = 1 \div 8; k_{d_0 d_2^i} = 3, i = 1 \div 4$ (for triangle faces of the pyramid),

$k_{d_0 d_2} = 4$ (for bases of the pyramid),

$k_{d_0 d_3} = 5, k_{d_1 d_2^i} = 3, i = 1 \div 4$ (for triangle faces of the pyramid),

$k_{d_1 d_2} = 4$ (for bases of the pyramid),

$k_{d_1 d_3} = 8, k_{d_2 d_3} = 5$.

Sum up the incidence coefficients for all elements of the pyramid (from larger dimension to smaller)

$$\sum_{i=1}^{8} k_{d_0 d_1^i} + \sum_{i=1}^{4} k_{d_0 d_2^i} + \sum_{i=1}^{4} k_{d_1 d_2^i} + k_{d_0 d_2} + k_{d_0 d_3} + k_{d_1 d_2} + k_{d_1 d_3} + k_{d_2 d_3} = 64. \quad (3.34)$$

Comparing (3.33) and (3.34) you can see that the sum of incidents in a pyramid from elements with a lower dimension to elements with a higher dimension is equal to the sum of incidents from elements with a higher dimension to elements with a lower dimension. Thus, the sum of incidents retains its value when changes the direction of the relationship between the elements.

Incident Conservation Law for the Mackay Cluster

Consider the Mackay (Mackay, 2001; Lord et al., 2006) cluster (Figure 2.1, Chapter 2). In it, the values of the numbers of elements of different dimensions are determined by the equalitiesf_0=24, f_1=72, f_2=70, f_3=22, f_4=1.

We calculate the values of the incidence coefficients of the elements of lower dimension with respect to the elements of higher dimension. The factors of incidents (from smaller dimension to larger) are

$k_{d_0^i d_1} = 6, i = 1 \div 24; k_{d_0^i d_2} = 10, i = 1 \div 24; k_{d_0^i d_3} = 6, i = 1 \div 24; k_{d_0^i d_4} = 1, i = 1 \div 24; k_{d_1^i d_2} = 3, i = 1 \div 60$ (edges on icosahedron);

$k_{d_1^i d_2} = 5, i = 1 \div 12$ (edges in connected of icosahedrons);

$k_{d_1^i d_3} = 3, i = 1 \div 60$ (edges on icosahedron);

$k_{d_1^i d_2} = 5, i = 1 \div 12$ (edges in connected of icosahedrons);

$k_{d_1^i d_4} = 1, i = 1 \div 72; k_{d_2^i d_3} = 2, i = 1 \div 70; k_{d_2^i d_4} = 1, i = 1 \div 70; k_{d_3^i d_4} = 1, i = 1 \div 22.$

Sum up the incidence coefficients for all vertices, edges, two dimension faces and three dimension faces of the Mackay cluster

$\sum_{i=1}^{24} k_{d_0^i d_1} = 6 \cdot 24 = 144, \sum_{i=1}^{24} k_{d_0^i d_2} = 10 \cdot 24 = 240, \sum_{i=1}^{24} k_{d_0^i d_3} = 6 \cdot 24 = 144, \sum_{i=1}^{24} k_{d_0^i d_4} = 24, \sum_{i=1}^{60} k_{d_1^i d_2} = 3 \cdot 60 = 180,$ (edges on icosahedron);

$\sum_{i=1}^{5} k_{d_1^i d_2} = 5 \cdot 12 = 60$ (edges in connected of icosahedron);

$\sum_{i=1}^{60} k_{d_1^i d_3} = 3 \cdot 60 = 180$ (edges on icosahedron);

$\sum_{i=1}^{12} k_{d_1^i d_3} = 5 \cdot 12 = 60$ (edges in connected of icosahedron);

$$\sum_{i=1}^{72} k_{d_1^i d_4} = 72, \sum_{i=1}^{70} k_{d_2^i d_3} = 2 \cdot 70 = 140, \sum_{i=1}^{70} k_{d_2^i d_4} = 70, \sum_{i=1}^{22} k_{d_3^i d_4} = 22,$$

$$\sum_{i=1}^{24} k_{d_0^i d_1} + \sum_{i=1}^{24} k_{d_0^i d_2} + \sum_{i=1}^{24} k_{d_0^i d_3} + \sum_{i=1}^{24} k_{d_0^i d_4} + \sum_{i=1}^{72} k_{d_1^i d_2} + \sum_{i=1}^{72} k_{d_1^i d_3} + \sum_{i=1}^{72} k_{d_1^i d_4} + \sum_{i=1}^{70} k_{d_2^i d_3} + \sum_{i=1}^{70} k_{d_2^i d_4} + \sum_{i=1}^{22} k_{d_3^i d_4} = 1336. \tag{3.35}$$

The factors of incidents (from larger dimension to smaller) are

$k_{d_0 d_1^i} = 2, i = 1 \div 72; k_{d_0 d_2^i} = 3, i = 1 \div 40$ (at triangle faces);

$k_{d_0 d_2^i} = 4, i = 1 \div 30$ (at quadrangular faces);

$k_{d_0 d_3^i} = 6, i = 1 \div 20$ (at prisms);

$k_{d_0 d_3^i} = 12, i = 1 \div 2$ (at icosahedrons);

$k_{d_0 d_4} = 24; k_{d_1 d_2^i} = 3, i = 1 \div 40$ (at triangle);

$k_{d_1 d_2^i} = 4, i = 1 \div 30$ (at quadrangular faces),(edges in connected of icosahedrons);

$k_{d_1 d_3^i} = 3, i = 1 \div 60$ (edges on icosahedron);

$k_{d_1 d_2^i} = 5, i = 1 \div 12$ (edges in connected of icosahedrons);

$k_{d_1 d_4} = 1, i = 1 \div 72; k_{d_2 d_3^i} = 2, i = 1 \div 70; k_{d_2 d_4} = 1, i = 1 \div 70; k_{d_3 d_4} = 1, i = 1 \div 22.$

Sum up the incidence coefficients for all vertices, edges, two - dimension faces and three - dimension faces of the Mackay cluster

$$\begin{aligned}
&\sum_{i=1}^{72} k_{d_0 d_1^i} = 2 \cdot 72 = 144, \sum_{i=1}^{70} k_{d_0 d_2^i} = 3 \cdot 40 + 4 \cdot 30 = 240, \sum_{i=1}^{22} k_{d_0 d_3^i} = 6 \cdot 20 + 2 \cdot 12 = 144, \\
&\sum_{i=1}^{24} k_{d_0 d_4} = 24, \sum_{i=1}^{70} k_{d_1 d_2^i} = 3 \cdot 40 + 4 \cdot 30 = 240, \sum_{i=1}^{22} k_{d_1 d_3^i} = 9 \cdot 20 + 2 \cdot 20 = 240, \sum_{i=1}^{72} k_{d_1 d_4} = 72, \\
&\sum_{i=1}^{22} k_{d_2 d_3^i} = 5 \cdot 20 + 2 \cdot 20 = 140, \sum_{i=1}^{70} k_{d_2 d_4} = 70, \sum_{i=1}^{22} k_{d_3 d_4} = 22, \sum_{i=1}^{72} k_{d_0 d_1^i} + \sum_{i=1}^{70} k_{d_0 d_2^i} + \sum_{i=1}^{22} k_{d_0 d_3^i} \\
&+ \sum_{i=1}^{24} k_{d_0 d_4} + \sum_{i=1}^{70} k_{d_1 d_2^i} + \sum_{i=1}^{22} k_{d_1 d_3^i} + \sum_{i=1}^{72} k_{d_1 d_4} + \sum_{i=1}^{22} k_{d_2 d_3^i} + \sum_{i=1}^{70} k_{d_2 d_4} + \sum_{i=1}^{22} k_{d_3 d_4} = 1336.
\end{aligned} \tag{3.36}$$

Comparing (3.35) and (3.36) you can see that the sum of incidents in the Mackay cluster from elements with a lower dimension to elements with a higher dimension is equal to the sum of incidents from elements with a higher dimension to elements with a lower dimension. Thus, the sum of incidents retains its value when changes the direction of the relationship between the elements.

The Law of Conservation of Incidence in Clusters with a Central Metal Atom

Let us verify the implementation of the law of conservation of incidence in clusters with a central metal atom. As an example, take an octahedron with a center. A metal atom is located in the center; ligands are located at the vertices of the octahedron. In this polytope (Figure 7) the values of the numbers of elements of different dimensions are determined by the equalities f_0=7, f_1=18, f_2=16, f_3=9, f_4=1.

Figure 7. The octahedron with a centrum

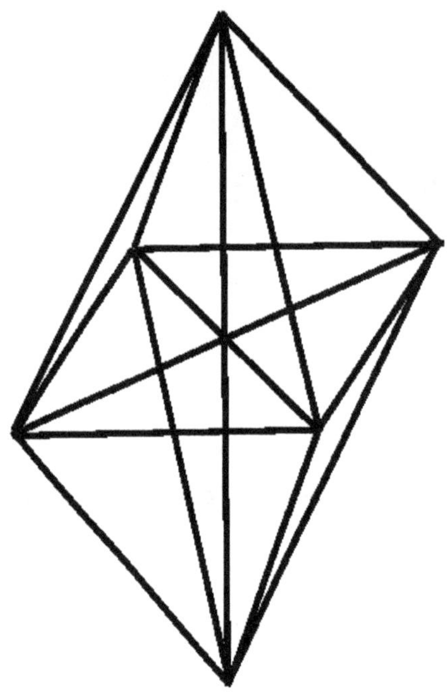

We calculate the values of the incidence coefficients of the elements of lower dimension with respect to the elements of higher dimension. The factors of incidents (from smaller dimension to larger) are

$k_{d_0^i d_1} = 5, i = 1 \div 6$ (for vertices of octohedron);

$k_{d_0 d_1} = 6$ (for atom at the center);

$k_{d_0^i d_2} = 8, i = 1 \div 6$ (for vertices of octohedron);

$k_{d_0 d_2} = 12$ (for atom at the center);

$k_{d_0^i d_3} = 5, i = 1 \div 6$ (for vertices of octohedron);

$k_{d_0 d_3} = 8$ (for atom at the center);

$k_{d_0^i d_4} = 1, i = 1 \div 7; k_{d_1^i d_2} = 4, i = 1 \div 6$ (edges from the center);

$k_{d_1^i d_2} = 3$ (for edges of octahedron);

$k_{d_1^i d_3} = 4, i = 1 \div 6$ (edges from the center);

$k_{d_1^i d_3} = 3$ (for edges of octahedron);

$k_{d_1^i d_4} = 1, i = 1 \div 18; k_{d_2^i d_3} = 2, i = 1 \div 16; k_{d_2^i d_4} = 1, i = 1 \div 16; k_{d_3^i d_4} = 1, i = 1 \div 9.$

Sum up the incidence coefficients for all vertices, edges, two - dimension faces and three - dimension faces

$$\sum_{i=1}^{7} k_{d_0^i d_1} + \sum_{i=1}^{7} k_{d_0^i d_2} + \sum_{i=1}^{7} k_{d_0^i d_3} + \sum_{i=1}^{7} k_{d_0^i d_4} + \sum_{i=1}^{18} k_{d_1^i d_2} + \sum_{i=1}^{18} k_{d_1^i d_3} + \sum_{i=1}^{18} k_{d_1^i d_4} + \sum_{i=1}^{16} k_{d_2^i d_3} + \sum_{i=1}^{16} k_{d_2^i d_4} + \sum_{i=1}^{9} k_{d_3^i d_4} = 324. \quad (3.37)$$

The factors of incidents (from larger dimension to smaller) are

$k_{d_0 d_1^i} = 2, i = 1 \div 18; k_{d_0 d_2^i} = 3, i = 1 \div 16; k_{d_0^i d_3} = 4, i = 1 \div 9$ (at tetrahedrons);

$k_{d_0 d_3^i} = 6, i = 1$ (at octahedron);

$k_{d_0 d_4} = 7; k_{d_1 d_2^i} = 3, i = 1 \div 16; k_{d_1 d_3^i} = 6, i = 1 \div 8; k_{d_1 d_3^i} = 12, i = 1$ (for octahedron);

$k_{d_1 d_4} = 18; k_{d_2 d_3^i} = 4, i = 1 \div 8; k_{d_2 d_3} = 8$ (for octahedron);

$k_{d_2 d_4} = 16; k_{d_3 d_4} = 9$

Sum up the incidence coefficients for all vertices, edges, two - dimension faces and three -dimension faces

$$\sum_{i=1}^{18} k_{d_0 d_1^i} + \sum_{i=1}^{16} k_{d_0 d_2^i} + \sum_{i=1}^{10} k_{d_0 d_3^i} + k_{d_0 d_4} + \sum_{i=1}^{16} k_{d_1 d_2^i} + \sum_{i=1}^{9} k_{d_1 d_3^i} + k_{d_1 d_4} + \sum_{i=1}^{9} k_{d_2 d_3^i} + k_{d_2 d_4} + k_{d_0 d_4} = 324. \quad (3.38)$$

Comparing (3.37) and (3.38) you can see that the sum of incidents in the octahedron with center of elements with a lower dimension to elements with a higher dimension is equal to the sum of incidents from elements with a higher dimension to elements with a lower dimension. Thus, the sum of incidents retains its value when changes the direction of the relationship between the elements.

THE LAW OF CONSERVATION OF INCIDENTS FOR CONVEX POLYTOPES

Theorem 3.5

The sum of the incidence coefficients from the elements of lower dimension to the elements of high dimension in any convex polytope is equal to the sum of the incidence coefficients from the elements of high dimension to the elements of low dimension. It depends on the type of polytope and its dimension.

Proof

From the studies of incidents carried out in this chapter between elements of different dimensions in the n - simplex, n - cross polytope, n - cube, n – hyper - rhombohedron, clusters of chemical compounds, it follows that the sum of the incidence coefficients between elements of arbitrary dimension j and elements of another arbitrary the dimensions of h in the forward and reverse directions are always equal to each other. This means that the sum of the incidence coefficients of all elements of dimension j in a given polytope with respect to all elements of dimension h is equal to the sum of the incidence coefficients of all elements of dimension h with respect to all elements of dimension j in this polytope. In this case, the incidence coefficient of a given element of dimension j with respect to an element of dimension h is generally not equal to the incidence coefficient of an element of dimension h with respect to an element of dimension j. Here are some examples.

In 4 - cross - polytopes, the sum of incidence coefficients from elements of zero dimension to elements of dimension 1 is equal

$$\sum_{i=1}^{8} k_{d_0^i d_1} = k_{d_0^i d_1} \cdot i = 6 \cdot 8 = 48.$$

At the same time, the sum of incidence coefficients from elements of dimension 1 to elements of dimension 0 is

$$\sum_{i=1}^{24} k_{d_0 d_1^i} = k_{d_0 d_1^i} \cdot i = 2 \cdot 24 = 48.$$

The sum of incidence coefficients from elements of dimension 3 to elements of dimension 4 is equal

$$\sum_{i=1}^{16} k_{d_3^i d_4} = k_{d_3^i d_4} \cdot i = 1 \cdot 16 = 16.$$

At the same time, of incidence coefficient from element of dimension 4 to elements of dimension 3 is $k_{d_3 d_4} = 16$.

In 4 – cube (or 4 – hyper – rhombohedron), the sum of incidence coefficients from elements of zero dimension to elements of dimension 1 is equal

$$\sum_{i=1}^{16} k_{d_0^i d_1} = k_{d_0^i d_1} \cdot i = 4 \cdot 16 = 64.$$

At the same time, the sum of incidence coefficients from elements of dimension 1 to elements of dimension 0 is equal

$$\sum_{i=1}^{32} k_{d_0 d_1^i} = k_{d_0 d_1^i} \cdot i = 2 \cdot 32 = 64.$$

The sum of incidence coefficients from elements of dimension 3 to elements of dimension 4 is equal

$$\sum_{i=1}^{8} k_{d_3^i d_4} = k_{d_3^i d_4} \cdot i = 1 \cdot 8 = 8.$$

At the same time, of incidence coefficient from element of dimension 4 to elements of dimension 3 is $k_{d_3 d_4} = 8$.

In the cluster of Mackay, the sum of incidence coefficients from elements of zero dimension to elements of dimension 1 is equal

$$\sum_{i=1}^{24} k_{d_0^i d_1} = k_{d_0^i d_1} \cdot i = 6 \cdot 24 = 144.$$

At the same time, the sum of incidence coefficients from elements of dimension 1 to elements of dimension 0 is equal

$$\sum_{i=1}^{72} k_{d_0 d_1^i} = k_{d_0 d_1^i} \cdot i = 2 \cdot 72 = 144.$$

The sum of the coefficients from elements of dimension 1 to elements of dimension 3 consists of two parts: for edges of icosahedrons and edges connecting icosahedrons. For the edges of icosahedrons, this part is equal to

$$\sum_{i=1}^{60} k_{d_1^i d_3} = k_{d_1^i d_3} \cdot i = 3 \cdot 60 = 180.$$

For the edges connecting icosahedrons, this part is equal

$$\sum_{i=1}^{12} k_{d_1^i d_3} = k_{d_1^i d_3} \cdot i = 5 \cdot 12 = 60.$$

Thus, the total sum of incidence coefficients from elements of dimension 1 to elements of dimension 3 in Mackay clusters is

$$\sum_{i=1}^{60} k_{d_1^i d_3} + \sum_{i=1}^{12} k_{d_1^i d_3} = 180 + 60 = 240.$$

At the same time the sum of incidence coefficients from elements of dimension 3 to elements of dimension 1 in the cluster of Mackay consist of two parts too: for edges of icosahedrons and edges connecting icosahedrons. For the edges of icosahedrons, this part is equal to

$$\sum_{i=1}^{60} k_{d_1 d_3^i} = k_{d_1 d_3^i} \cdot i = 3 \cdot 60 = 180.$$

For the edges connecting icosahedrons, this part is equal

$$\sum_{i=1}^{12} k_{d_1 d_3^i} = k_{d_1 d_3^i} \cdot i = 5 \cdot 12 = 60.$$

Thus, the total sum of incidence coefficients from elements of dimension 3 to elements of dimension 1 in Mackay clusters is

$$\sum_{i=1}^{60} k_{d_1 d_3^i} + \sum_{i=1}^{12} k_{d_1 d_3^i} = 180 + 60 = 240.$$

As we can see, the sums of incidence coefficients from low - dimensional elements to high -dimensional elements and from high - dimensional elements to low - dimensional elements consist of terms linking elements of different dimensions, and these terms are equal for the same dimension pair. Therefore, the total amounts in the direct (from the smaller to the larger) and the inverse (from the larger to the smaller) are equal to each other.

Obviously, such equalities hold for any convex polytope. This reflects the general property of belonging as a contact: if an element with a certain dimension contacts in some extent with an element of another dimension, then the latter contacts in this extent with the first.

It is also obvious from the studies that the total sum of the incidence coefficients calculated over the entire set of elements of different dimensions of a polytope depends on the type of polytope and its dimension.

CONCLUSION

In the modern scientific world, information exchange is of great importance. This is characteristic for various fields of knowledge. In mathematics, there is the concept of incidence, i.e. belonging of some mathematical element to another mathematical element. If mathematical elements are models of some material bodies, then incidence can be interpreted as the transfer of information from one material body to another. In the framework of this monograph, the polytopes of higher dimension are the mathematical elements that model material bodies. With their help, the structure of clusters of chemical compounds is described, i.e. bonded to each other by chemical bonds of atoms, molecules, functional groups. The geometric elements in this case are the faces of different dimensions of polytopes. Of interest is the question of transferring information from elements of lower dimension to elements of higher dimension and vice versa from elements of higher dimension to elements of lower dimension. This chapter first establishes the existence of integral equality in relation to the issue of the transmission of information by elements of lower and higher dimensions that describe natural objects (Zhizhin, 2019 e). This integral equality is called the law of conservation of incidents. At the beginning, using simple geometric examples, a direct calculation of incidents from images of geometric shapes shows that the sum of incidents of elements of lower dimension with respect to elements of higher dimension is equal to the sum of incidents of elements of higher dimension with respect to elements of lower dimension. Then, the fulfillment of the law of conservation of incidents for the n – simplex, of the n – cube, the n - cross – polytope and n – hyper – rhombohedron is proved in general terms. It is shown that the law of conservation of incidents is valid for both regular bodies and irregular bodies, which can be clusters of chemical compounds. In addition, summarizing the results of considering polytopes of various kinds, a general theorem is proved on the fulfillment of the law of conservation of incidents for any convex polytopes. It is significant that the flow of incidents from elements of lower dimension to elements of higher dimension and vice versa increases sharply with increasing dimension of the polytope. It is significant that the flow of incidents from elements of lower dimension to elements of higher dimension and vice versa increases sharply with increasing dimension of the polytope. Interestingly, the flow of incidents (information) is of greatest importance to the n - cross - polytope.

REFERENCES

Lord, E. A., Mackay, A. L., & Ranganathan, S. (2006). *New Geometry for New Materials*. Cambridge University Press.

Mackay, A. (2001). On Complexity. *Crystallography Reports*, *46*(4), 524–526. doi:10.1134/1.1387117

Mancuso, S., & Viola, A. (2013). *Sensibilità e intelligenza del mondo vegetale*. GUIUTI EDITORE.

Mancuso, S. (2017). *Plant Revolution. Le piante hanno già inventato il nostro futuro*. GUIUTI EDITORE.

Sanbonmatsu, K. Y. (2016). COOLAIR antisense RNAs from evolutionarily conserved elaborate secondary structures. *Cell Reports*, *16*(12), 3087–3096. doi:10.1016/j.celrep.2016.08.045 PMID:27653675

Zhizhin, G. V. (2012, October). *Hierarchical filling of spaces with polytopes.* Paper presented at "St. Petersburg Scientific Forum: Science and Human Progress," St. Petersburg, Russia.

Zhizhin, G. V. (2013). Relations for the numbers of faces of various dimensions in the tower of n - dimensional convex regular polytopes. In *Proceedings of the 9th All - Russian School "Mathematical Studies in the Natural Sciences"*. Apatity: Geological Institute, Russian Academy of Sciences.

Zhizhin, G. V. (2014a). The fractal nature of disproportionate phases. *Materials (Basel)*, 10.

Zhizhin, G. V. (2014b). Incommensurable and fluctuating structures in the terrestrial space. *Biosphere*, *3*, 211–221.

Zhizhin, G. V. (2014c). *World - 4D*. Polytechnic Service.

Zhizhin, G. V. (2015, November). *Polytopic prismahedrons – fundamental regions of the n-dimension nanostructures.* Paper presented at the International conference "Nanoscience in Chemistry, Physics, Biology and Mathematics," Napoca, Romania.

Zhizhin, G. V. (2016). The structure, topological and functional dimension of biomolecules. *J. Chemoinformatics and Chemical Engineering*, *5*(2), 44–58. doi:10.4018/IJCCE.2016070104

Zhizhin, G. V. (2018). *Chemical Compound Structures and the Higher Dimensional of Molecules: Emerging Research and Opportunities*. IGI Global. doi:10.4018/978-1-5225-4108-0

Zhizhin, G. V. (2019a). *The Geometry of Higher - Dimensional Polytopes*. IGI Global. doi:10.4018/978-1-5225-6968-8

Zhizhin, G. V. (2019b). The metric structure of nucleic acids and the higher dimension of their constituents. *J. Chemoinformatics and Chemical Engineering*, *7*(2), 1–15.

Zhizhin, G. V. (2019c). *Higher Dimension of Clusters of Intermetallic Compounds*. J. Applied Nanotechnology Research. doi:10.4018/IJANR.2019010102

Zhizhin, G. V. (2019d). *Attractors and Higher Dimensions in Population and Molecular Biology: Emerging Research and Opportunities*. IGI Global. doi:10.4018/978-1-5225-9651-6

Zhizhin, G. V. (2019e). The Law of Conservation of Incidents in the Space of Nanoworld. *International Journal of Chemoinformatics and Chemical Engineering*, *8*(1), 25–46. doi:10.4018/IJCCE.2019010103

Zhizhin, G. V. (2019f). The Polytope of Hereditary Information the Structure, Location, Signification. *Biochemistry and Modern Applications*, *2*(1), 56–62. doi:10.33805/2638-7735.124

Zhizhin, G. V., & Diudea, M. V. (2016). Space of Nanoworld. In M. V. Putz & M. C. Mirica (Eds.), *Sustainable Nanosystems, Development, Properties, and Applications* (pp. 214–236). IGI Global.

Zhizhin, G. V., Khalaj, Z., & Diudea, M. V. M. V. (2016). Geometrical and topological dimensions of the diamond. In M. V. Diudea & C. L. Nagy (Eds.), *Distance, symmetry and topology in carbon nanomaterials* (pp. 167–188). Springer. doi:10.1007/978-3-319-31584-3_12

KEY TERMS AND DEFINITIONS

Dimension of the Space: The member of independent parameters needed to describe the change in position of an object in space.

Incidence Coefficients of Elements of Lower Dimension With Respect to Elements of Higher Dimension: The number of elements of a certain higher dimension to which the given element of a lower dimension belongs.

Incidence Coefficients of Elements of Higher Dimension With Respect to Elements of Lower Dimension: The number of elements of a given lower dimension that are included in a particular element of a higher dimension.

N-Cross-Polytope: The convex polytope of dimension n in which opposite related of centrum edges do not have connection of edge.

N-Cube: The convex polytope of dimension n in which each vertex incident to n edges.

N-Simplex: The convex polytope of dimension n in which each vertex is joined by edges with all remain vertices of polytope.

Nanocluster: A nanometric set of connected atoms, stable either in isolation state or in building unit of condensed matter.

Polytope: Polyhedron in the space of higher dimension.

Chapter 4
Higher Dimensions in the Theory of Heredity

ABSTRACT

On the basis Mendel's experiments, a mathematical model is constructed that describes the results of these experiments in a wide range of parameters. There is shown that in the mathematical model of Mendel's experiments, based on real patterns of plant development, there are equilibrium positions between the dominant and recessive forms. This equilibrium position is stable and located in the multidimensional space of system phenotypes. This newly discovered behavior of the dominant and recessive forms in the vicinity of the equilibrium position (true) differs significantly from the logistic equilibrium position in the Hardy-Weinberg principle, built without taking into account the real patterns in the plant population. The geometry of the neighborhood of the compound of two nucleic acid helices with nitrogen bases was investigated. It is proved that this neighborhood is a cross-polytope of dimension 13 (polytope of hereditary information), in the coordinate planes of which there are complementary hydrogen bonds of nitrogenous bases.

INTRODUCTION

In 1865, Mendel made a presentation at the Society of Naturalists in Brynn about experiments on plant hybrids (Mendel, 1965). This presentation gave birth to the development of genetics as a science, although not immediately the content of the speech became of known to the scientific community and was appreciated (Gaisinovich, 1988). Understanding of its significant results took place in the struggle of opposing scientific trends and was accompanied by dramatic events in human relations. At different times, scientists have seen in these experiments different, sometimes opposite, results. More than 30 years after this speech, when the results work of Mendel were reopened and confirmed experimentally in the works of Correns (Correns, 1900), and De Vries (De Vries, 1904).

Over the past 150 years after the speech of Mendel, a chromosome theory of heredity was created, which gave a molecular explanation to the results of the experiments of Mendel (Weismann, 1885; Johannsen, 1933; Morgan, 1937; Koltsov, 1935; Chetverikov, 1926; Watson & Crick, 1953 a, b; Zhizhin,

DOI: 10.4018/978-1-7998-6768-5.ch004

2018, 2019 a). The same time after of the second discovery of Mendel's experiments, there appeared works in which it was noted that in Mendel's experiments there was a steady increase in the number of dominant alleles in populations, which indicated the absence of equilibrium positions in plant populations that obeyed Mendel's law.

In this regard, Yule (Yule, 1902) purely mathematically proved that in the case of absolutely random crossing in the population of heterozygous forms, there is an equilibrium between the number of dominant and recessive forms. Continued these studies Hardy (Hardy, 1908), who derived the formula for the distribution of genotypes in freely crossbreeding populations. Regardless of him and even earlier, Weinberg (Weinberg, 1908) established the same formula. This formula was called as principle the Hardy - Weinberg and became widespread. However, for the mathematical derivation of this formula, very strong assumptions are used: lack of choice in organisms, infinity of the population, accidental crossing of the population's organisms with each other, uniform distribution of male and female individuals, absence of mutation and genetic drift. The totality of these assumptions precludes the possibility of realizing such populations in nature. Therefore, this principle cannot be confirmed experimentally. Its only advantage is that it has equilibrium positions (a finite ratio of the numbers of dominant and recessive forms). However, these equilibrium positions are formed for any initial contents of these forms. Consequently, the set of equilibrium positions is a continuous manifold, and therefore they are asymptotically unstable, since any small perturbation can translate the system from one equilibrium position to another (Zhizhin, 1972, 2004 a). The combination of necessary mathematical conditions in the derivation of the Hardy -Weinberg formula does not alleviate this advantage.

In this Chapter, a mathematical analysis of the sequences of obtaining the values of the number of phenotypes of organisms in Mendel's experiments was carried out. It is found that the patterns of inheritance of constant - differing sings, experimentally established by Mendel, are described by special algebraic relations, the basis of which are geometric progressions. Mendel and subsequent researchers did not pay attention to this. Naturally, Mendel could not trace the patterns of inheritance on experiments with a large number of generations. However, the mathematical model obtained in this Chapter (Zhizhin, 2019), accurately confirmed by Mendel's opaque in the field of conditions of their conduct, allows one to cram into the inheritance processes of sings with a large number of generations. A numerical and qualitative analysis of the equations of the mathematical model with an increase in the number of generations with an arbitrary number of pairs of constant - differing features was carried out.

There is shown that in the mathematical model of Mendel's experiments, based on real patterns of plant development, there are equilibrium positions between the dominant and recessive forms. It is shown that with an increase in the number of generations of the number all of dominant and recessive phenotypes of organisms with any number of sings quickly equalize and then synchronously (in the absence of death of organisms) increase together. In this case, the system asymptotically strive to achieve an isolated stable equilibrium position of the node type in a multidimensional space with coordinates in the form of quantities inversely proportional to the number of different phenotypes of organisms.

Thus, for finding equilibrium positions there is no need to use such stringent conditions which definition of the Hardy – Weinberg principle. Moreover, the regularities obtained by Mendel are based on clear biological processes in cells, in chromosomes. Moreover, the behavior of the system in the vicinity of the equilibrium position of the mathematical model of Mendel's experiments has a clear physical meaning. While one of the authors of the Hardy -Weinberg principle wrote that, a purely mathematical probabilistic description of random crossing is caused by the impossibility of having any physical justification for it (Yule, 1902).

In the experiments of Mendel, the plants were pollinated of the self – pollination, since many plants are pollinated by this method (Takhtadzhyan, 1980). In addition, recently in the neurobiology of plants convincingly shown (Mancuso, & Viola, 2013) that even in cross - pollination, for example, bees, insects pollinate plants not in a random way. The so - called "law of place" appears. The bees pollinate the plants of the species from which they started to pollinate in the early morning. This manifests the intellectual ability of plants to manage the surrounding nature for their needs (Pollan, 2001; Pereira et al., 2012).

In 1953, James Watson and Francis Crick based on the analysis of diffractograms proposed a three - dimensional model of a DNA molecule consisting of two chains twisted into a spiral (Watson, & Crick, 1953 a, b). At the same time, sugar molecules and phosphoric acid residues included in the DNA molecule were also considered three - dimensional. However, it was not indicated which three - dimensional figures correspond to sugar molecules and phosphoric acid residues. According to this model, flat nitrogenous bases, being in different chains, using two complementary hydrogen bonds connected two spirals into a single whole. But how exactly the flat nitrogenous bases are located in space remained unknown. And this is despite the fact that it is these relations that determine the most important issue of the transmission of hereditary information. Recently, it was proved that sugar molecules and phosphoric acid residues have the highest dimension (Zhizhin, 2017, 2018). The phosphoric acid residue has a dimension of 4, and the sugar molecule has a dimension of 12. In this regard, the geometry of nucleic acids was considered taking into account the highest dimension of the components (Zhizhin, 2018, 2019 b, d). In addition, for the convenience of images, a simplified three - dimensional model of a sugar molecule was built on the basis of a full twelve - dimensional model of a sugar molecule (Zhizhin, 2019 d). This model can be used in further constructions, not forgetting the dimension of its complete model. This model was used to construct images of single nucleic acid helices. When passing to the analysis of nucleic acids consisting of two linked helices, it is necessary to take into account the emerging anti - parallelism of geometric elements in linked spirals (Zhizhin, 2019 f). In this article, continuing the research begun in the works (Zhizhin, 2018, 2019 d), it will be shown that sugar molecules located in linked nucleic acid helices form a polytope of dimension 13 of the type of cross - polytope in which there are exactly 12 coordinate planes. In these coordinate planes outlined by rectangles with antiparallel edges, exactly 12 compounds of nitrogen bases currently known can be located (Spirin, 2019). The image of the polytope and its coordinate planes is obtained, its structure is determined. This polytope can be called the polytope of hereditary information, since it transmits hereditary information from one spiral to another using a sequential alternation of nitrogenous bases. The highest dimension of the interaction region of nucleic acids allows us to establish a previously unknown order of their interaction and to the unequivocal correspondence of the number of nucleotides and amino acids. This in turn leads to the cancellation of the genetic code. The structure of the polytope of hereditary information is defined, its image is given. This destroys the idea of the correspondence of triplets to certain amino acids, introduced logistically, in order to avoid the contradiction between a small number of nucleotides and a large number of amino acids. It turned out that the polytope of hereditary information has an internal (hidden) degree of freedom in the form of a different arrangement of pairs of flat nitrogen base in the space of higher dimension. The latent order of interaction of nucleic acids is also important when using analog nucleic acids, in particular, in xenobiology.

It has been shown that the polytopic of hereditary information is characterized by a powerful stream of incidents between geometric elements of different dimensions (Zhizhin, 2019 f), providing an extremely intense exchange of information between the components of nucleic acids. The evidence of the existence of such an exchange of information can be found in the recently discovered inheritance of changes not related to the modification of sequences in DNA, i.e. with methylation - the binding of a methyl group CH_3 to the nitrogenous base of DNA (Mancuso, 2017; Lindquist, et al., 2016; Sanbonmatsu, et al., 2016).

MONO - HYBRID CROSSING

Mendel carefully selected plants so that they consistently had differing signs (constant -distinguishing signs) to determine the inheritance patterns of plant sings. Now can to say that, in accordance with the chromosome theory of heredity, two homozygous organisms were crossed, the homologous chromosomes of each of which contain two identical genes. One pair of genes (*AA*) in one organism corresponds to one sing (dominant), and the other pair of genes (*aa*) in another organism corresponds to another sing (recessive). This crossing is called monohybrid. In accordance with Mendel's first law, when crossing homozygous parent forms that differ in one pair of sings, all hybrids of the first generation will be uniform both in genotype (gene composition, *Aa*) and in phenotype (appearance). When the first generation of hybrids are crossed in accordance with the second Mendelian law, the predominance of the dominant sing over the recessive one in the ratio 3: 1 for the phenotype and 1: 2: 1 for the genotype will be observed in the offspring. This means that organisms of the first generation F_1 are characterized by a pair of *Aa* genes, and the second generation F_2 is characterized by pairs *AA, Aa, aa* in the ratio 1: 2: 1. Moreover, organisms with *AA* and *Aa* genes have the same phenotype, since the influence of recessive genes affects only when there are two, and not one, in homologous chromosomes.

Suppose that in the generation of F_1 there are m organisms and their mean fecundity is ν. Then in the generation of F_2 there are $m\nu$ organisms. Taking into account Mendel's second law is necessary to conclude that $m\nu$ is a multiple of four. Then in the generation of F_2 there are $\frac{m\nu}{4}$ homozygous organisms with *AA* genes, as many homozygous organisms with genes *aa* and $\frac{m\nu}{2}$ of heterozygous organisms with genes *Aa*

$$F_2 = \frac{m\nu}{4}AA + \frac{m\nu}{2}Aa + \frac{m\nu}{4}aa.$$

Each of the constituent parts of the second generation gives offspring to the next generation F_3. At the same time, homozygous organisms have homozygous organisms in the progeny, but the total number of corresponding organisms increases by a factor of ν. The heterozygous organisms, which have both dominant and recessive genes, also breed homozygous and heterozygous organisms. Taking into account the number of heterozygous organisms in the generation F_2 and their fertility, one has

$$\frac{m\nu}{2}Aa \rightarrow \frac{m\nu^2}{8}AA + \frac{m\nu^2}{4}Aa + \frac{m\nu^2}{8}aa.$$

In sum, homozygous and heterozygous organisms of generation F_2 give the following distribution of organisms in generation F_3

$$F_3 = mv^2(\frac{1}{4}+\frac{1}{8})AA + \frac{mv^2}{4}Aa + mv^2(\frac{1}{4}+\frac{1}{8})aa.$$

It is easy to determine that the ratio of the number of organisms of the dominant phenotype to the number of organisms of the recessive phenotype in this generation is 5: 3. In the generation F_4, heterozygous organisms also breed homozygous and heterozygous organisms

$$\frac{mv^2}{4}Aa \to \frac{mv^3}{16}AA + \frac{mv^3}{8}Aa + \frac{mv^3}{16}aa.$$

Homozygous organisms increase their numbers according to their fertility. In the sum in generation four, the following distribution of progeny is obtained

$$F_4 = mv^3(\frac{1}{4}+\frac{1}{8}+\frac{1}{16})AA + \frac{mv^3}{8}Aa + mv^3(\frac{1}{4}+\frac{1}{8}+\frac{1}{16})aa.$$

The ratio of the numbers of the dominant and recessive phenotypes in this generation is 9: 7.

In the n generation, arguing in a similar way, one get the following expression for the distribution of organisms

$$F_n = mv^{n-1}(\frac{1}{2^2}+\frac{1}{2^3}+...+\frac{1}{2^n})AA + \frac{mv^{n-1}}{2^{n-1}}Aa + mv^{n-1}(\frac{1}{2^2}+\frac{1}{2^3}+...+\frac{1}{2^n})aa.$$

The numbers of the dominant N_{df} and recessive N_{rf} phenotypes in the generation n are determined by the following equalities

$$N_{df}(n) = mv^{n-1}\frac{1+2+2^2+...+2^{n-2}+2}{2^n},$$

$$N_{rf}(n) = mv^{n-1}\frac{1+2+2^2+...+2^{n-2}}{2^n}.$$

The numerator on the right - hand side of the last equality is a geometric progression

$$\sum_{k=1}^{n-1} 2^{k-1} = 2^{n-1} - 1.$$

It should be noted that in the book A.E. Gaisinovich (Gaisinovich, 1988) mistakenly asserted that the numbers of phenotypes of organisms in Mendel's law are expressed in terms of Bin Newton. In accordance with the last equality, the numbers of the dominant and recessive phenotype are equal, respectively

$$N_{df}(n) = mv^{n-1}\frac{2^{n-1}+1}{2^n}, N_{rf}(n) = mv^{n-1}\frac{2^{n-1}-1}{2^n}. \quad (4.1)$$

The ratio of these numbers is

$$\frac{N_{df}}{N_{rf}} = \delta_1(n) = \frac{2^{n-1}+1}{2^{n-1}-1}. \quad (4.2)$$

Equation (4.2) represents a general expression for the ratio of the numbers of the dominant and recessive phenotypes in a mono - hybrid crossing for any number of generation $n > 1$. It depends only on n. When $n = 2$, the ratio (4.2) is 3: 1, which corresponds to Mendel's second law. It follows from (4.1), (4.2) when the number of the generation increases, the numbers of the dominant and recessive phenotypes quite quickly occur (their ratio tends to unity). Absolute values of these numbers grow steadily with increasing number of generations. This is natural, since conditions that are not too remote from the initial state are considered, when the death of organisms is not taken into account due to external influences (aggression of surrounding, climatic and seasonal causes (Zhizhin, 2004 b, 2005). The number of hybrids $m\left(\frac{v}{2}\right)^{n-1}$, which is a component of the number of dominant phenotypes, also grows steadily as the number of the generation increases (ν is usually greater than 2). Thus, the assertion that with the passage of time there is a restoration of parental populations and the disappearance of hybrids is erroneous.

Since the numbers of phenotypes grow with increasing n, it is convenient to go over to variables, inverses of the numbers of phenotypes, in order to analyze the nature of the desire of the system for the equilibrium position. The set of solutions of the system for various initial values of the number of organisms m and the constant value of fecundity ν in these variables are represented by trajectories in the phase plane (N_{df}^{-1}, N_{rf}^{-1}) (Figure 1).

Figure 1. Phase diagram of changes in the numbers of dominant and recessive phenotypes in coordinates of inverses numbers of phenotypes N_{df}^{-1} and N_{rf}^{-1}

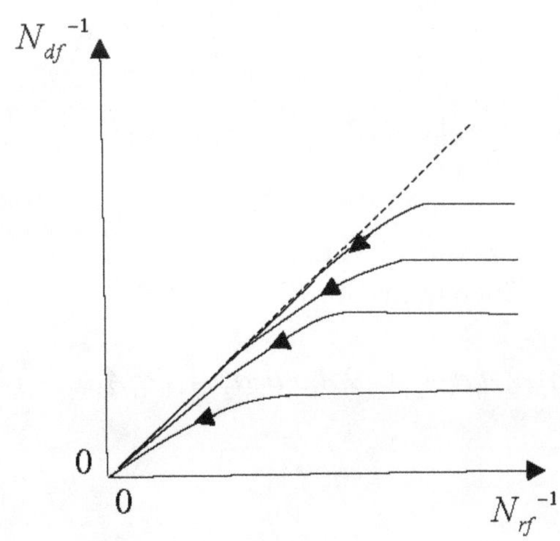

From which we see that the ratio (4.2) defines in this plane an intrinsic direction, along which the solutions tend to the equilibrium position at the origin of coordinates of the phase plane. The equilibrium position is isolated, stable and has a node type.

BE – HYBRID CROSSING

Let the organisms have two pairs of constant - differing signs. In one pair, the A_1A_1 genes *correspond* to the dominant sign, and the genes a_1a_1 correspond to the recessive sign. In another pair, the A_2A_2 genes correspond to the dominant sign, and the genes a_2a_2 correspond to the recessive sign. In each pair of signs, Mendel's second law is independent and at the same time acts

$$A_1a_1 \to \left(A_1A_1 + 2A_1a_1 + a_1a_1\right)$$
$$A_2a_2 \to \left(A_2A_2 + 2A_2a_2 + a_2a_2\right).$$

This corresponds to the product of expansions

$$\left(A_1A_1 + 2A_1a_1 + a_1a_1\right)\left(A_2A_2 + 2A_2a_2 + a_2a_2\right), \tag{4.3}$$

since the probability of the product of independent events is equal to the product of the probability of these events. The result of this work are 16 terms

$$\begin{aligned} &A_1A_1A_2A_2 + A_1A_1a_2a_2 + a_1a_1A_2A_2 + a_1a_1a_2a_2 \\ &+2A_1A_1A_2a_2 + 2a_1a_1A_2a_2 + 2A_1a_1A_2A_2 + 2A_1a_1a_2a_2 + 4A_1a_1A_2a_2. \end{aligned} \tag{4.4}$$

Therefore, if in the first generation F_1 there are m hybrid organisms $A_1a_1A_2a_2$, then in the second generation F_2, taking into account their fertility ν, there are $m\nu$ organisms and this number must be a gradual number 16. Consequently, the second generation is the sum of the following organisms

$$\begin{aligned} F_2 &= \frac{m\nu}{16}A_1A_1A_2A_2 + \frac{m\nu}{16}A_1A_1a_2a_2 + \frac{m\nu}{16}a_1a_1A_2A_2 + \frac{m\nu}{16}a_1a_1a_2a_2 \\ &+ \frac{m\nu}{8}A_1A_1a_2a_2 + \frac{m\nu}{8}a_1a_1A_2a_2 + \frac{m\nu}{8}A_1a_1A_2A_2 + \frac{m\nu}{8}A_1a_1a_2a_2 + \frac{m\nu}{4}A_1a_1A_2a_2. \end{aligned} \tag{4.5}$$

One distinguish among the 16 terms in (4.4) those terms that correspond to organisms with the phenotype on the sing of A_1A_1 genes. Such terms are 12

$$A_1A_1A_2A_2,\ A_1A_1a_2a_2,\ 2A_1A_1A_2a_2,\ 2A_1a_1A_2A_2,\ 2A_1a_1a_2a_2,\ 4A_1a_1A_2a_2.$$

In (4.4) are terms that correspond to organisms with the phenotype on the sing of a_1a_1 genes. This are 4 terms $a_1a_1A_2A_2, a_1a_1a_2a_2, 2a_1a_1A_2a_2$. Thus, the ratio of the number of organisms in the second generation (4.5) with the A_1A_1 phenotype and with the phenotype a_1a_1 is the same as in the mono - hybrid crossing 3:1. Obviously, the same 3: 1 ratio of the number of organisms with phenotypes of A_2A_2 and of a_2a_2. This manifests the third law of Mendel, discovered by him in experiments with bi - hybrid crossing: the law of independent splitting of sings in the ratio of 3: 1 in the second generation. In this Chapter, this law follows from a mathematical treatment of bi - hybrid crossing. From the same consideration follows the second part of the third law of Mendel.

Indeed, if one select the terms corresponding to the organisms with the phenotype for the A_1A_2 genes from the sum (4.4), then there are nine such terms

$$A_1A_1A_2A_2, 2A_1A_1A_2a_2, 2A_1a_1A_2A_2, 4A_1a_1A_2a_2.$$

The terms in (4.4) corresponding to the organisms with the phenotype in the product a_1A_2 are three $a_1a_1A_2A_2, 2a_1a_1A_2a_2$. The same number of terms in (4.4) for phenotype a_2A_1. These are the terms $A_1A_1a_2a_2, 2A_1a_1a_2a_2$. One term in (4.4), corresponding to the organisms with phenotype a_1a_2, this is $a_1a_2a_1a_2$. Thus, in the second generation (4.5), the ratio of the number of organisms to the simultaneous presence of some two of the four sings is 9: 3: 3: 1. This is the second part of the third law of Mendel, observed by him in the experiments on bi - hybrid crossing in the second generation.

Mendel did not succeed in elucidating the regularity of bi - hybrid crossing in the following after the second generation. This can be done using a mathematical description, the correctness of which has already been proved. In the transition to the third generation, heterozygous organisms $A_1a_1A_2a_2$ in (4.3) are split according to Mendel's second law. If we focus on the ratio of the phenotype counts by the dominant A_1A_1 gene and the recessive a_1a_1 gene, then after splitting the heterozygous organism A_1a_1 in the first bracket of the product (4.3), this product takes the form $(3A_1A_1 + 4A_1a_1 + 3a_1a_1)f$ where, for brevity, f denotes the second bracket in (4.3) $f = (A_2A_2 + 2A_2a_2 + a_2a_2)$. It follows that the ratio of the number of organisms in the phenotypes A_1A_1 and a_1a_1 in the third generation is 7: 3.

Similarly, in the fourth generation, the product (4.3) takes the form $(7A_1A_1 + 8A_1a_1 + 7a_1a_1)f$, that is, in the fourth generation, the ratio of the numbers of organisms according to the phenotypes A_1A_1 and a_1a_1 is 15: 7 and so on. Expanding the sequence of the obtained digits in the relationship, we conclude that in the generation n the number of organisms according to the phenotype A_1A_1 in the product (4.3) is a geometric progression

$$1 + 2 + 2^2 + \ldots + 2^{n-1} = \sum_{k=1}^{n} 2^{k-1} = 2^n - 1,$$

In addition, the number of organisms according to the phenotype a_1a_1 is a geometric progression

$$1 + 2 + 2^2 + \ldots + 2^{n-2} = \sum_{k=1}^{n-1} 2^{k-1} = 2^{n-1} - 1.$$

Thus, one obtain that the ratio of the number of organisms in the generation n according to the phenotypes A_1A_1 and a_1a_1 in the bi - hybrid crossing is

$$\delta_2(n) = \frac{2^n - 1}{2^{n-1} - 1}. \tag{4.6}$$

It is clear that this formula is also suitable for the ratio of the numbers of organisms according to the phenotypes A_2A_2 and a_2a_2 in the bi - hybrid crossing in the generation n, in view of the equality of rights of brackets in the product (4.3).

If it is necessary to consider the ratio of the number of organisms to phenotypes with the simultaneous presence of two pairs of sings in the generation after the second, then it is necessary to split both brackets in the product (4.3) sequentially. Denote the organisms by the phenotype A_1A_1 by the symbol φ_{A_1}. These organisms include organisms with A_1A_1 and A_1a_1 genes. Similarly, φ_{A_2} denotes the organisms according to the phenotype of A_2A_2, φ_{a_1} - organisms according to the phenotype of a_1a_1, φ_{a_2} - organisms according to the phenotype of a_2a_2. The number of organisms with a phenotype A_iA_i is equal to the sum of the number of organisms with genes A_iA_i and the number of organisms with genes A_ia_i, (i = 1, 2). Then, expression (4.3) in the third generation takes the form

$$\left(7\varphi_{A_1} + 3\varphi_{a_1}\right)\left(7\varphi_{A_2} + 3\varphi_{a_2}\right) = 7^2\varphi_{A_1}\varphi_{A_2} + 7\cdot 3\varphi_{a_1}\varphi_{A_2} + 7\cdot 3\varphi_{A_1}\varphi_{a_2} + 3^2\varphi_{a_1}\varphi_{a_2}.$$

From this it follows that the ratio of the number of organisms according to the phenotypes A_1A_1 and A_2A_2, a_1a_1 and A_2A_2, $a_2a_2$and A_1A_1, a_1a_1 and a_2a_2 in the third generation is

7^2: 7•3: 7•3: 3^2.

In the fourth generation, the ratio of the number of organisms by the same phenotypes is obtained similarly when the brackets are split in expression (4.3)

$$\left(15\varphi_{A_1} + 7\varphi_{a_1}\right)\left(15\varphi_{A_2} + 7\varphi_{a_2}\right) = 15^2\varphi_{A_1}\varphi_{A_2} + 15\cdot 7\varphi_{a_1}\varphi_{A_2} + 15\cdot 7\varphi_{A_1}\varphi_{a_2} + 7^2\varphi_{a_1}\varphi_{a_2}.$$

From this it follows that the ratio of the number of organisms according to the phenotypes A_1A_1 and A_2A_2, a_1a_1 and A_2A_2, a_2a_2 and A_1A_1, a_1a_1 and a_2a_2 in the fourth generation is

15^2: 15•7: 15•7: 7^2.

The construction of these relations continues in the next generation. The numbers participating in the relationships (1, 3, 7, 15,...) are numerators and denominators (4.6) for the corresponding values of n. The indicators and the number of repetitive products in the relationship indicate the number of independent feature pairs (in this case 2). This leads to the following ratios of the number of organisms according to the phenotypes, with the simultaneous presence of two pairs of characters for an arbitrary generation number n

$$(2^n-1)^2:(2^n-1)(2^{n-1}-1):(2^n-1)(2^{n-1}-1):(2^{n-1}-1)^2. \quad (4.7)$$

We introduce the notation of the absolute numbers of organisms in the generation n and phenotypes with two dominant genes $N_{d_1d_2}$, with one dominant and one recessive gene $N_{d_1r_2}$ and $N_{d_2r_1}$, and with two recessive genes $N_{d_1r_2}$. These numbers, according to (4.7), are equal

$$N_{d_1d_2}=mv^{n-1}\frac{(2^n-1)^2}{2^{2n}}, N_{d_1r_2}=N_{d_2r_1}=mv^{n-1}\frac{(2^n-1)(2^{n-1}-1)}{2^{2n}}, N_{r_1r_2}=mv^{n-1}\frac{(2^{n-1}-1)^2}{2^{2n}}.$$

For large values of n, these numbers tend to the same value mv^n, that is, there is an alignment of the number of organisms with purely dominant genes, purely recessive genes and hybrid genes. Just as with mono - hybrid crossing in the case of bi - hybrid crossings, solutions with an increase in the number of generations can be represented in space with coordinates in the form of inverse numbers $N^{-1}_{d_1d_2}, N^{-1}_{d_1r_2}, N^{-1}_{d_2r_1}, N^{-1}_{r_1r_2}$. In this case, there are four such coordinates. The equilibrium position of the given system is the origin of the coordinates of the four - dimensional space. This is a special point of the type of a considerate four - dimensional node. The solutions of the system enter into a position of equilibrium in its own direction, defined by the equality of a unit of relation pairwise coordinates of space.

POLY - HYBRID CROSSING

If organisms with more than two pairs of characteristics are crossing, for example, t pairs, then the distribution of organisms in the progeny can be obtained from the product of t factors

$$\left(A_1A_1+2A_1a_1+a_1a_1\right)\left(A_2A_2+2A_2a_2+a_2a_2\right)\ldots\left(A_tA_t+2A_ta_t+a_ta_t\right). \quad (4.8)$$

If the ratio of the numbers of phenotype organisms according to the dominant and recessive members of one of the i pairs of sings is of interest ($1 \le i \le t$), then in the second generation the product (4.8) can be written in the form $\left(A_iA_i+2A_ia_i+a_ia_i\right)f$, where f includes all the other factors of the product (4.8). From this expression, it follows that in the second generation the ratio of the numbers of organisms according to the dominant and recessive members of the pair i is a ratio of 3: 1. When splitting heterozygous organisms A_ia_i with formed of a third generation the ratio of the number of organisms to phenotypes A_iA_i and a_ia_i is 7: 3 and so on. Therefore, the ratio of the numbers of organisms in the generation n to the phenotypes A_iA_i and a_ia_i in the case of poly - hybrid crossing has the same form as in the mono - hybrid crossing

$$\delta_t(n)=\frac{2^n-1}{2^{n-1}-1}. \quad (4.9)$$

This ratio is valid for any pair of sings i from t pairs due to their independence.

If it is necessary to consider the ratio of the number of organisms to a phenotype with the simultaneous presence of several pairs of sings, then it is necessary to split all the brackets in the product (4.8) sequentially. For example, in the case of three pairs of sings ($t = 3$), using the notations introduced in the previous section, product (4.8) takes in this case the form

$$\left(3\varphi_{A_1}+\varphi_{a_1}\right)\left(3\varphi_{A_2}+\varphi_{a_2}\right)\left(3\varphi_{A_3}+\varphi_{a_3}\right)=27\varphi_{A_1}\varphi_{A_2}\varphi_{A_3}+9\varphi_{a_1}\varphi_{A_2}\varphi_{A_3}$$
$$+9\varphi_{A_1}\varphi_{a_2}\varphi_{A_3}+9\varphi_{A_1}\varphi_{A_2}\varphi_{a_3}+3\varphi_{A_1}\varphi_{a_2}\varphi_{a_3}+3\varphi_{a_1}\varphi_{A_2}\varphi_{a_3}+3\varphi_{a_1}\varphi_{a_2}\varphi_{A_3}+\varphi_{a_1}\varphi_{a_2}\varphi_{a_3}. \tag{4.10}$$

It follows from equality (4.10) that in the case of crossing with three pairs of characteristics in the second generation, the ratio between the numbers of organisms according to the phenotypes of different possible pairs of genes has the form

$3^3 : 3^2 : 3^2 : 3^2 : 3 : 3 : 3 : 1$.

In the third generation, the product (4.8) is transformed so

$$\left(7\varphi_{A_1}+3\varphi_{a_1}\right)\left(7\varphi_{A_2}+3\varphi_{a_2}\right)\left(7\varphi_{A_3}+3\varphi_{a_3}\right)=7^3\varphi_{A_1}\varphi_{A_2}\varphi_{A_3}+7^2\cdot 3\varphi_{a_1}\varphi_{A_2}\varphi_{A_3}$$
$$+7^2\cdot 3\varphi_{A_1}\varphi_{a_2}\varphi_{A_3}+7^2\cdot 3\varphi_{A_1}\varphi_{A_2}\varphi_{a_3}+7\cdot 3^2\varphi_{A_1}\varphi_{a_2}\varphi_{a_3}+7\cdot 3^2\varphi_{a_1}\varphi_{A_2}\varphi_{a_3}+7\cdot 3^2\varphi_{a_1}\varphi_{a_2}\varphi_{A_3}+3^3\varphi_{a_1}\varphi_{a_2}\varphi_{a_3}.$$

From the last expression, it follows that the relationship between the numbers of organisms with different genes when crossing organisms with three pairs of characters in generation 3 has the form

$7^3 : (7^2 \bullet 3) : (7^2 \bullet 3) : (7^2 \bullet 3) : (7 \bullet 3^2) : (7 \bullet 3^2) : (7 \bullet 3^2) : 3^3$.

In the fourth generation, the product (4.8) is transformed so

$$\left(15\varphi_{A_1}+7\varphi_{a_1}\right)\left(15\varphi_{A_2}+7\varphi_{a_2}\right)\left(15\varphi_{A_3}+7\varphi_{a_3}\right)=15^3\varphi_{A_1}\varphi_{A_2}\varphi_{A_3}+15^2\cdot 7\varphi_{a_1}\varphi_{A_2}\varphi_{A_3}$$
$$+15^2\cdot 7\varphi_{A_1}\varphi_{a_2}\varphi_{A_3}+15^2\cdot 7\varphi_{A_1}\varphi_{A_2}\varphi_{a_3}+15\cdot 7^2\varphi_{A_1}\varphi_{a_2}\varphi_{a_3}+15\cdot 7^2\varphi_{a_1}\varphi_{A_2}\varphi_{a_3}+15\cdot 7^2\varphi_{a_1}\varphi_{a_2}\varphi_{A_3}+7^3\varphi_{a_1}\varphi_{a_2}\varphi_{a_3}.$$

From the last expression, it follows that the relationship between the numbers of organisms with different genes when crossing organisms with three pairs of characters in generation 3 has the form

$15^3 : (15^2 \bullet 7) : (15^2 \bullet 7) : (15^2 \bullet 7) : (15 \bullet 7^2) : (15 \bullet 7^2) : (15 \bullet 7^2) : 7^3$

In the same way, it is possible to obtain the corresponding ratios of the numbers of organisms in the crossing of organisms with 3 pairs of characters in the generation of n

$$(2^n-1)^3:\left[(2^n-1)^2(2^{n-1}-1)\right]_1:\left[(2^n-1)^2(2^{n-1}-1)\right]_2:\left[(2^n-1)^2(2^{n-1}-1)\right]_3:$$
$$\left[(2^n-1)\ (2^{n-1}-1)^2\right]_1:\left[(2^n-1)\ (2^{n-1}-1)^2\right]_2:\left[(2^n-1)\ (2^{n-1}-1)^2\right]_3:(2^{n-1}-1)^3.$$

Here, the subscript in square brackets shows that the number of repetitions of identical expressions is equal to the number of pairs of constant - distinct characters.

By indexing repetitive products by the number of pairs of attributes, these relations take the form

$$\begin{aligned}&(2^n-1)^t:\left[(2^n-1)^{t-1}(2^{n-1}-1)\right]_1:\left[(2^n-1)^{t-1}(2^{n-1}-1)\right]_2:\ldots:\left[(2^n-1)^{t-1}(2^{n-1}-1)\right]_t:\\&\left[(2^n-1)^{t-2}(2^{n-1}-1)^2\right]_1:\left[(2^n-1)^{t-2}(2^{n-1}-1)^2\right]_2:\ldots:\left[(2^n-1)^{t-2}(2^{n-1}-1)^2\right]_t:\ldots:\\&\left[(2^n-1)(2^{n-1}-1)^{t-1}\right]_1:\left[(2^n-1)(2^{n-1}-1)^{t-1}\right]_2:\ldots:\left[(2^n-1)(2^{n-1}-1)^{t-1}\right]_t:(2^{n-1}-1)^t.\end{aligned}\tag{4.11}$$

Just as in the case of bi – hybrid crossing, here it is possible to introduce inverse numbers of organisms according to phenotypes with different sets of genes. As can be seen from the expression (4.11), the number of these variables in the general case is $(t - 1)\, t + 2$. Each of the variables is proportional to $\frac{1}{mv^{n-1}}$ and tends to zero as the number of generations increases. The ratio of the pairs of any two variables with increasing n tends to unity. Therefore, in poly - hybrid crossing, with increasing n, the number of organisms is equalized by phenotypes with different sets of genes. Solutions for different values of m tend in the space of the reciprocal numbers of organisms to the proper direction of the equilibrium position at the origin, which has the type of an isolated multidimensional stable node.

Thus, it is shown that the patterns of inheritance of features of organisms in their crossing are based on a special kind of the mathematical relationships that allow us to derive algebraically the experimental patterns studied 150 years ago by Mendel. The core of these relations are geometric progressions. An expression is obtained for the general relationship between the numbers of organisms with different phenotypes for any number of pairs of constant - distinct characters and any number of generations. Using these relationships is possible to predict and investigate the inheritance processes outside the intervals of the initial data of Mendel's experiments (for any number of pairs of constant - distinguishing signs and any number of generations). The general formulas obtained are transformed continuously into the relations of Mendel in the conditions of the experiments carried out by him.

It has been established that with an increase in the number of generations in a plant population, the phenotypes of organisms in different gene sets are leveled quickly. These numbers, with an increase in the number of generations and the absence of death of organisms, synchronously increase while remaining practically equal to each other. Mathematically, this can be represented by trajectories in a multidimensional phase space with coordinates in the form of inverse numbers of phenotypes of organisms. These trajectories tend along their own direction to an isolated stable equilibrium position of the type of a multidimensional node at the origin. With the results obtained, based on G. Mendel's experiments, the Hardy - Weinberg principle is fundamentally inconsistent, constructed purely logistically without taking into account the real laws in living nature.

SPATIAL MODES OF SUGAR MOLECULE AND NUCLEIC ACID HELICES

From a sugar molecule with five carbon atoms in the Fischer form (Lehninger, 1982), we can move on to representing the molecule as a closed chain. Having chosen, for example, Form B from the two enantimorphic forms A and B, the closed chain of the sugar molecule has the form in Figure 2 (Zhizhin, 2017; 2019 b, d).

Figure 2. Closed chain sugar molecule (form B)

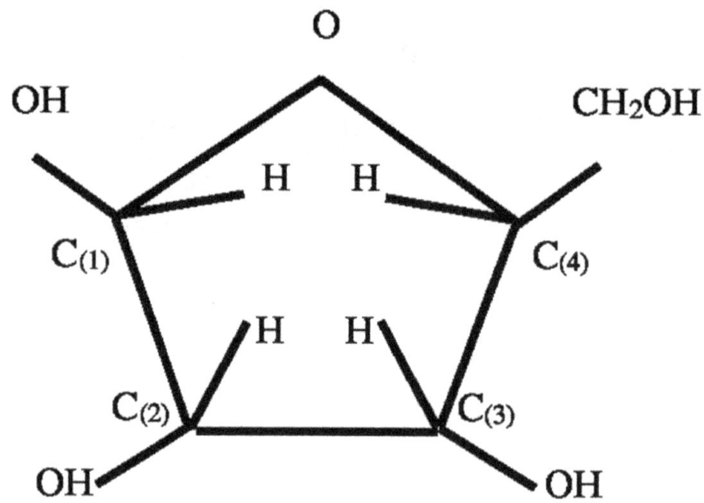

To obtain a spatial figure of a sugar molecule (B - ribose), considering the atoms and functional groups shown in the projection onto the plane in Figure 4.2 as the vertices of the corresponding spatial figure, it is necessary to introduce additional edges in addition to the edges shown in Figure 4.2 (Zhizhin, 2019 d). These edges will have only spatial significance, unlike the edges in Figure 4.2, which also depict chemical covalent bonds. To distinguish between edges, we will depict edges corresponding to chemical covalent bonds with thick solid black lines. Edges that have only spatial significance in the images of the figures will be indicated by thin dashed black lines.

Since the number of atoms and functional groups is 13, the closest convex closed figure that can be obtained from Figure 4.2 is a simplex of dimension 12. The dimension of the simplex is n - 1, where n is the number of vertices (Zhizhin, 2019 c). Moreover, each vertex is connected to all other vertices by edges, i.e. 12 edges emanate from each vertex. A spatial image of the B - ribose molecule is given in the article (Zhizhin, 2017). If, as a simplification, we leave, in addition to the edges corresponding to chemical bonds, only the edges of the external contour in the projection of the spatial figure onto the plane, we get an image of the B - ribose molecule, as shown in Figure 3.

Since the lengths of the carbon – carbon and carbon – oxygen chemical bonds are close, we can approximately consider them equal to a = 0.15 nm. In addition, it can be assumed that the edges of chemical bonds emanating from four carbon atoms in the carbon cycle are located symmetrically with respect to the plane of the carbon cycle. Then, as mathematically proved (Zhizhin, 2019 d), chemical bonds emanating from four carbon atoms of the carbon cycle are perpendicular to the plane of the cycle. This leads to a simplified spatial image of the B - ribose molecule in Figure 4.

Figure 3. Schematic representation of a sugar molecule on a plane

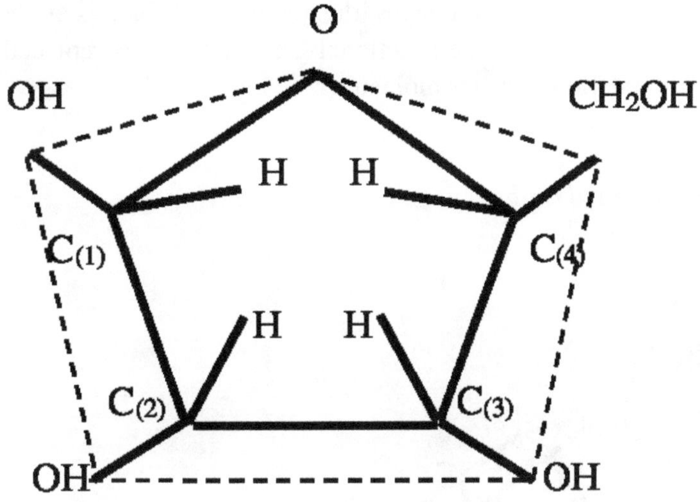

In Figure 4, of the additional edges needed to construct a convex closed spatial figure and to fulfill the Euler - Poincare equation (Poincare, 1895), only the edges are left to give the figure a closed form. In this image, the B - ribose molecule is three - dimensional. But we must remember that the number of edges for fulfilling the Euler - Poincare equation should be much larger - 12 edges from each vertex.

Figure 4. A simplified spatial image of a B - ribose molecule

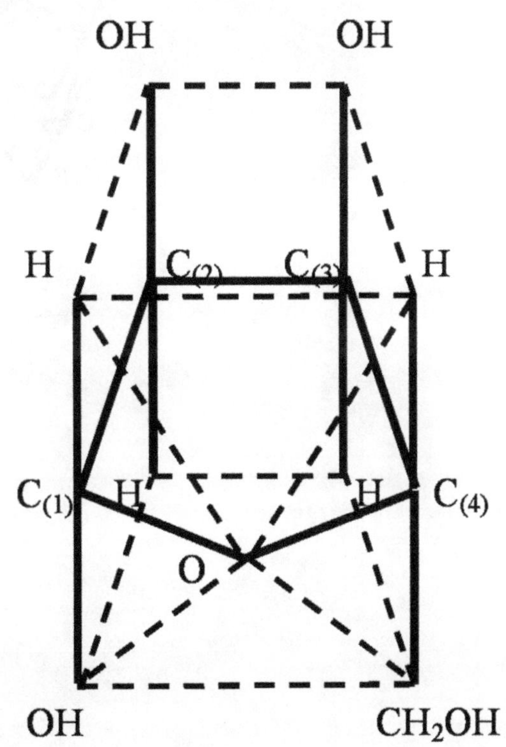

The remainder of phosphoric acid H_3PO_4 is geometrically a tetrahedron with a center. Its dimension is 4 (Zhizhin, 2019 b). Having taken such images, the nucleic acid chain is as shown in Figure 5.

In Figure 5, in the sugar molecules, the functional group CH_2OH is replaced by some nitrogenous bases F_i as a result of the isolation of water molecules.

Figure 5. Simplified nucleic acid chain

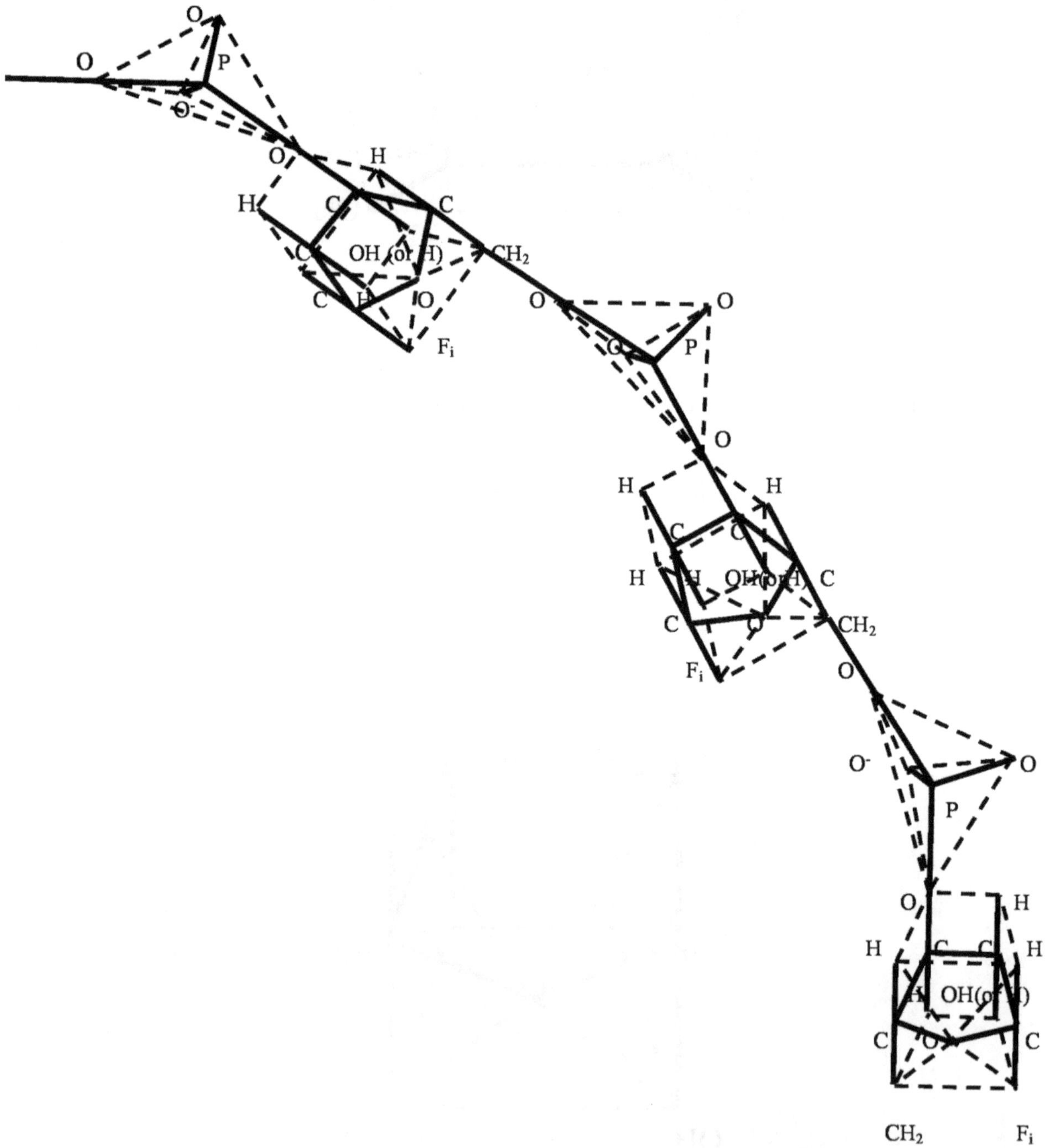

POLYTOPES WITH ANTIPARALLEL EDGES

In single - stranded and double - stranded nucleic acids (RNA, DNA), the constituents of acids (residues of phosphoric acid and sugar molecules) interact with each other (Watson & Crick, 1953a, b; Spirin & Gavrilova, 1971; Frank – Kamenetskiy, 1986, 1988). Phosphoric acid residues are connected by divalent metal ions, mainly magnesium ions, due to the interaction of negative charges of phosphoric acid residues with positive charges ions (Spirin & Gavrilova, 1971). This interaction is essential for the stability of nucleic acid structures, especially in the ribosomes. Sugar molecules interact with each other due to the hydrogen bond between the nitrogenous bases attached to the sugar molecules. Being geometric forms, the constituents of nucleic acids interact with each other to form new geometric forms - new polytopes. However, it is not known how flat nitrogenous bases are oriented in space, whether their orientation depends on the type of nitrogenous base. Currently there is no information on this. There is also no information on how exactly the metal ions are located, connecting the phosphoric acid residues. It should be remembered here that the adopted three - dimensional model of the components and the nucleic acid molecule itself is only a model for visual perception. As it was shown earlier, the phosphoric acid residue is a polytope of dimension 4, and the sugar molecule has a dimension of 12.

The movement of triangles along a helix leads to the formation of polytopes with antiparallel edges. The patterns of formation of polytopes with anti - parallel edges and their properties were investigated in detail in the first chapter. In this chapter, we will use this analysis to examine the polytope of hereditary information.

THE POLYTOPE OF HEREDITARY INFORMATION

Let us consider in detail the formation of a polytope of two sugar molecules with anti - parallel edges. Here, as in the case of the tetrahedron, to form a polytope with antiparallel edges from two sugar molecules, you must have one sugar molecule on one helix to turn this helix together with the sugar molecule and move the sugar molecule along this reversed helix in opposite direction to the original helix direction. When the sugar molecule rotates 180 degrees to the right while moving, then the original sugar molecule and the sugar molecule on the reversed helix are two polytopes with anti - parallel edges. Both of these sugar molecules in a simplified form are represented in Figure 6.

When the five - carbon sugar molecule is rotated to the left by 180 degrees, the nitrogenous bases are on opposite sides of both molecules, so that for their connection it is necessary to cross the entire set of atoms of two molecules. This is un realistic, therefore this option is not considered. When the sugar molecule is rotated 180 degrees to the left, it is possible to connect the sugar molecules through nitrogenous bases only between two different chains of nucleic acids or in the case of turning the chain itself in the opposite direction.

In full, the sugar molecules have a dimension of 12, in the corresponding polytope each vertex must have an edge connection with all the other vertices. Knowledge of this now one need. For the formation of a polytope of dimension 13, it is necessary to connect each vertex of one polytope with the vertices of another polytope so that there are no vertex connections with the same letters. All connecting edges break into pairs of antiparallel edges. At the same time, a set of coordinate a two - dimensional planes emanates from the center of the formed polytope as from the origin of coordinates. Their number is equal to the number of combinations from 13 to 2, i.e. 48 coordinate planes. In order to clarify the pos-

Figure 6. The two sugar molecules in a simplified form with anti – parallel edges

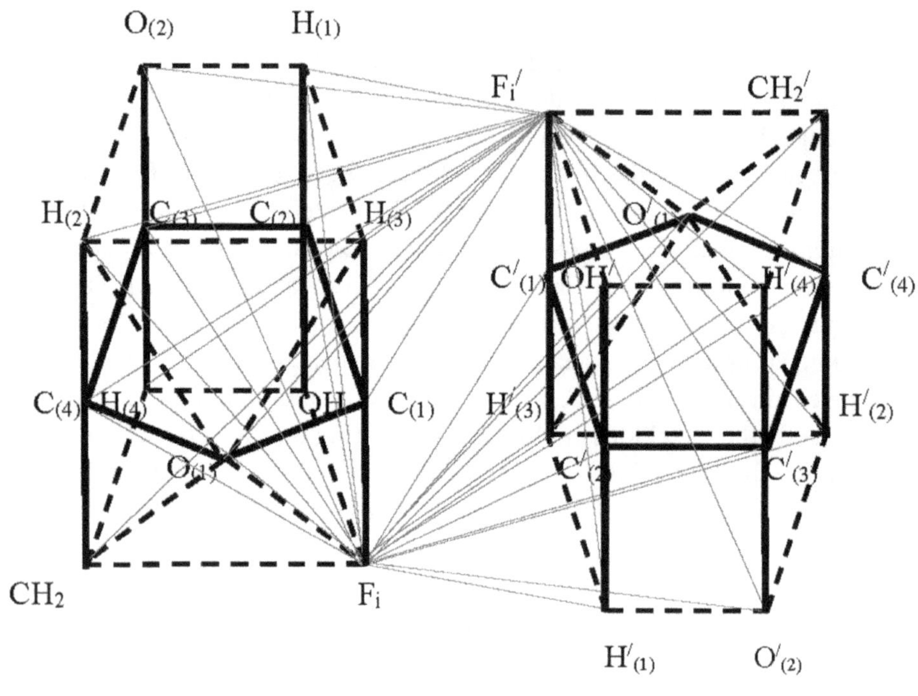

sible geometrical circumstances of the connection of helices in double - helix nucleic acid molecules with nitrogenous bases, we are primarily interested in the coordinate planes containing these nitrogenous bases $F_i, F_i^{/}$. There are exactly 12 such coordinate planes in the obtained polytope of dimension 13. They are depicted in Figure 6 by blue solid lines and are indicated below by the vertices of the polytopes contained in them

$$F_i^{/}H_{(1)} \quad F_i^{/}O_{(2)} \quad F_i^{/}H_{(2)} \quad F_i^{/}C_{(3)} \quad F_i^{/}C_{(2)} \quad F_i^{/}H_{(3)} \quad F_i^{/}C_{(4)} \quad F_i^{/}H_{(4)} \quad F_i^{/}OH \quad F_i^{/}C_{(1)} \quad F_i^{/}O_{(1)} \quad F_i^{/}CH_2$$
$$H_{(1)}^{/}F_i\text{'}\ O_{(2)}^{/}F_i\text{'}\ H_{(2)}^{/}F_i\text{'}\ C_{(3)}^{/}F_i\text{'}\ C_{(2)}^{/}F_i\text{'}\ H_{(3)}^{/}F_i\text{'}\ C_{(4)}^{/}F_i\text{'}\ H_{(4)}^{/}F_i\text{'}\ (OH)^{/}F_i\text{'}\ C_{(1)}^{/}F_i\text{'}\ O_{(1)}^{/}F_i\text{'}\ (CH_2)^{/}F_i$$

Other edges of the polytope of dimension 13 are not shown in Figure 4.6, so as not to ignite the figure. In the center of each parallelogram, indicated by its four vertices, is the origin of coordinates and the corresponding pair of coordinate axes (they are not shown). To identify the different hydrogen and oxygen atoms at vertices of polytope, they indicated by numbers in brackets at the lower indices.

It is surprising that the number of coordinate planes containing vertices is exactly equal to the number of possible compounds of nitrogenous bases 12 (Spirin, 2019)

A: *U*,*U*: *A*,*G*: *C*,*C*: *G*,*G*: *U*,*G*: *A*,*U*: *U*,*U*: *C*,*A*•*A*,*A*•*C*,*U*: *G*,*A*: *G*

Since each coordinate plane designated by the vertices of the parallelograms has a specific atomic environment, it can be assumed that each of the 12 possible compounds of nitrogenous bases is located on one particular coordinate plane out of 12 possible. This solves the question of the possible orientation of the bond of nitrogenous flat bases in nucleic acids using ideas about the high dimensionality of the constituent nucleic acids. It is also surprising that in order to create 13 – cross - polytopes, providing the connection with nitrogenous bases $F_i, F_i^/$, nature specially created double - stranded nucleic acids with oppositely directed spirals. This is realized in DNA and RNA when creating regions with inverted helices.

In a variety of nucleic acid molecules, the issue of chain interaction is important. In ribosomes, RNA interacts with each other due to bivalent metal ions (mainly magnesium ions). Positively charged magnesium ions attract negative charges of phosphoric acid residues, ensuring the stability of the ribosomes. In double - stranded nucleic acids, the helices are connected to each other by means of nitrogenous bases complementarily interacting with each other by a hydrogen bond. However, the magnesium ions and nitrogen bases in nucleic acids could not be specifically located. It has been established that magnesium ions and flat nitrogenous bases can be located inside special polytopes of higher dimension. Here knowledge is needed of the higher dimension of phosphoric acid residues and sugar molecules. Such polytopes are polytopes with anti - parallel edges, i.e. cross - polytopes of higher dimension. Binding agents are located on the free coordinate planes of these polytopes in the vicinity the center of the polytope.

In this case, the two - dimensional coordinate plane on the boundary of the polytope should contain the objects to be joined. In the case of magnesium ions, there are four specific coordinate planes inside the 5 – cross - polytope, in which an ion can accommodate, combining negative charges. In the case of nitrogenous bases, the existence of 12 coordinate planes inside a cross - polytope of dimension 13, in which flat nitrogenous bases can be located, connecting the helix of nucleic acids, was discovered. Exactly as much as there are options for combining nitrogenous bases. It is given, that each coordinate plane of these 12 planes has a specific environment of atoms. It should be assumed that only one of the 12 possible compounds of nitrogenous bases are placed in each of these planes. It is surprising that the existence of higher - dimensional polytopes with anti - parallel edges is possible only in the case of the opposite direction of interacting helices, and this is exactly what nature provides in the double - helix DNA and in the RNA segments with self - inversion of the helix in the opposite direction.

To build the polytopic of hereditary information, we need to supplement Figure 6 with edges that create a closed and convex figure. In this case, it is not necessary to connect the vertices symmetrically relative to the center of Figure 4.6 with the edges. Each vertex will be connected by an edge to the other vertices. In this case, the polytope will be a cross - polytope and its dimension is equal to half the number of vertices. Indeed, according to (Zhizhin, 2019 c), there is a relation between f_i the number of dimension elements i in the cross - polytope and the dimension d of the cross - polytope itself (see Chapter 1)

$$f_i(d) = 2^{1+i} C_d^{d-1-i} . \qquad (4.12)$$

At the vertex $i = 0$, therefore, according to (4.12) $f_0 = 2C_d^{d-1} = 2d, d = f_0 / 2 = 13$.

Thus, the polytope of the hereditary information has dimension 13. When portraying this polytope, we will use the technique that is used in portraying cross - polytopes. We distribute all 26 vertices on the circle so that the vertices opposite in Figure 4.5 remain opposite and there is no edge between them (Figure 7).

The edges corresponding to the chemical bonds are indicated in this figure by thick black solid lines. Anti - parallel edges highlighting coordinate planes with vertices corresponding to nitrogenous bases are indicated by bluey lines. The remaining edges are indicated by thin dash - dotted black lines. We

Figure 7. The polytope hereditary information

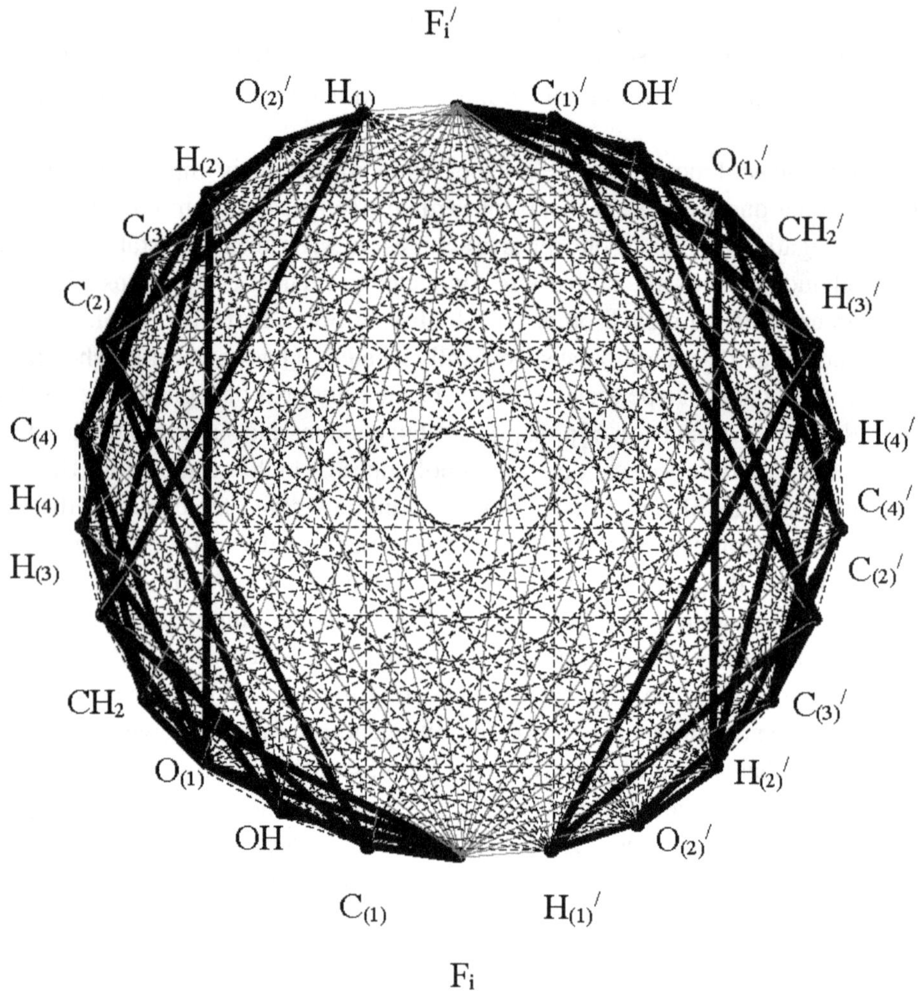

emphasize that there is a chemical bond between the nitrogenous bases, but it is not covalent. Therefore, this connection is not indicated by an edge.

When the polytope of hereditary information moves along the common axis of the double helix of nucleic acids in accordance with the sequence of bases in DNA, the connection between the vertices $F_i^{/}, F_i^{//}$ is carried out by one of the twelve base pairs, which occupies one of the 12 coordinate planes of the polytope.

By relation (4.12), one can determine the number of elements of different dimensions in the polytope of hereditary information.

the number of edges is $f_1(13) = 2^2 C_{13}^{11} = 312$,
the number of triangles is $f_2(13) = 2^3 C_{13}^{10} = 2288$,
the number of tetrahedrons is $f_3(13) = 2^4 C_{13}^{9} = 10560$,
the number of fourth - dimensional faces (simplexes) is $f_4(13) = 2^5 C_{13}^{8} = 41184$,

the number of fives - dimensional faces (simplexes) is $f_5(13) = 2^6 C_{13}^7 = 109824$,
the number of six - dimensional faces (simplexes) is $f_6(13) = 2^7 C_{13}^6 = 219648$,
the number of seven - dimensional faces (simplexes) is $f_7(13) = 2^8 C_{13}^5 = 164736$,
the number of eight – dimensional faces (simplexes) is $f_8(13) = 2^9 C_{13}^4 = 183040$,
the number of nine - dimensional faces (simplexes) is $f_9(13) = 2^{10} C_{13}^3 = 292864$,
the number of ten - dimensional faces (simplexes) is $f_{10}(13) = 2^{11} C_{13}^2 = 159744$,
the number of eleven - dimensional faces (simplexes) is $f_{11}(13) = 2^{12} C_{13}^1 = 26624$,
the number of twelve - dimensional faces (simplexes) is $f_{12}(13) = 2^{13} = 4096$.

The obtained numbers determine the structure of the polytope of hereditary information.

HIDDEN NUCLIEC ACID BOND ORDER

From the two sugar molecules in a simplified form Figure 6 it follows, however, that not all 12 coordinate planes can have flat nitrogenous bases. They could be located only between the vertices of the polytope belonging to different sugar molecules closest to each other, i.e. between the planes $OH\text{-}H_{(3)}\text{-}H_{(1)}\text{-}F_i$ and $OH'\text{-}H_{(3)}{}' - H_{(1)}{}' - F_i'$. Between the distant peaks located on the opposite planes of sugar molecules, the location of flat nitrogenous bases is difficult, due to the obstacles encountered in the form of sugar molecule atoms and covalent chemical bonds. Therefore,out of the 12 possible coordinate planes, only five coordinate planes designated by the vertices of the polytope

$$\frac{F_i^/ H_{(1)}}{H_{(1)}^/ F_i}, \frac{F_i^/ C_{(2)}}{C_{(2)}^/ F_i}, \frac{F_i^/ H_{(3)}}{H_{(3)}^/ F_i}, \frac{F_i^/ C_{(1)}}{C_{(1)}^/ F_i}, \frac{F_i^/ OH}{(OH)^/ F_i}. \tag{4.13}$$

If among 12 pairs (Spirin, 2019) canonical (Watson & Crick, 1953a, b) base pairs for linked DNA molecules are distinguished, then there are only 4 of these pairs

A: *U*,*G*: *C*,*C*: *G*,*U*: *A* (4.14)

If RNA molecules are linked, then in (4.14), base *U* must be replaced with base *T*.

Considering that there are 4 canonical base pairs, and there are 5 planes in which each pair can be located, the total number of base pairs location options is 20, i.e. exactly as many different amino acids exist. Thus, each variant of the arrangement of base pairs corresponds to a certain amino acid connected in the ribosome to transport RNA. As is known (Spirin, 2019), the attachment of an amino acid to transport RNA carries out a specific enzyme that recognizes the amino acid and its corresponding transport RNA when they are combined. In this case, transport RNA is associated with messenger RNA, and between these RNAs at the junction point there again exists a polytopic of hereditary information. Thus, information on the state of messenger RNA is transmitted by transport RNA, on which the given amino acid joins. As we see in this process, there is no need to use the concept of a genetic code to describe it. The latter was introduced as an attempt to explain the difference in the number of canonical base pairs from the number of amino acids. However, this concept together with the concept of the codon does not explain anything. Taking into account the dimension of the polytope of genetic information allows us to determine the internal (hidden) order in the location of the bases. The situation is similar to the situation with the discovery of quasicrystals. X-ray diffraction patterns of some intermetallic compounds revealed a lack of translational symmetry in them (Shehtman, 1984). This caused a stormy reaction of the scientific community. Such intermetallic compounds were called quasicrystals. However, in works (Zhizhin, 2014, 2018) it was shown that translational symmetry immediately appears if we consider diffractograms as projections onto a two-dimensional plane of a structure of higher dimension (a hidden order of diffractograms is revealed).

Currently, there are no results of special experiments to determine a unique correspondence between the location of this pair of bases in the polytopic of hereditary information and a specific amino acid. However, one can qualitatively confirm the existence of five different arrangements of base pairs using experimental data on the correspondence of codons (nucleotide triplets) to amino acids (Spirin, 2019). If we assume that the average nucleotide in the codon plays the main influence in this correspondence, then five amino acids correspond to triplets with an average nucleotide *U: Phe, Leu, Ile, Met, Val.* Since each amino acid corresponds to one of the positions of nucleotide *U*, it turns out that these directions exist 5. A similar situation is repeated with triplets with an average nucleotide *C*. They correspond to 5 amino acids: *Ser, Pro, Thr, Ala, Cus*. For triplets with an average nucleotide *G* there correspond the amino acids: *Ser, Cys, Trp, Arg, Gly*. Triplets with an average nucleotide A correspond to 7 amino acids: *Tyr, His, Glu, Asn, Lys, Gln, Asp*. Some violations of the strict unambiguous equality of 5 amino acid numbers to each nucleotide can be explained by the influence of lateral nucleotides in triplets and noncanonical parasitic bases.

It should be noted that taking into account the highest dimension of the polytopic of hereditary information is also necessary in the analysis of pandemics in which viral RNA penetrates the cell and forces the ribosomes to produce proteins necessary for viruses to reproduce. The antiviral drugs currently under development are acting on one or another phase of the development of a viral infection (Kozlov, & Stragunsky, 2002; De Clerey, 2004; Razonable, 2011; Crotty, et al., 2002; Caly, et al.,2020). Currently, clinical trials of antiviral drugs based on nucleic acid analogs that are structurally similar to natural RNA and DNA, but have altered any of the main components of nucleic acids: phosphat backbone, pentose sugar, or one of the canonical nucleic bases, are currently undergoing clinical trials. In particular, analog nucleic bases give, among other things, various properties of base pairing and base folding. This is especially significant in light of the established high dimension of the interaction region of nucleic acids. For example, there are universal bases that can mate with all four canonical bases and thereby stop the replication of viral RNA. An example of analog RNA is a peptide nucleic acid, which has a high binding strength and does not contain a faceted backbone. Analog nucleic acids are also called xenonucleic acids. They represent one of the main pillars of xenobiology, the construction of new life forms based on alternative biochemistry. In all these cases of noncanonical nucleic acids, it is of interest to study their geometry taking into account the higher dimensionality of polytopes and to analyze the effect of higher dimensionality on the features of their interaction.

THE LAW OF CONSERVATION OF INCIDENTS IN POLYTOPE OF HEREDITARY INFORMATION

The monograph (Zhizhin, 2019 c, see too Chapter 3) introduced the concept of the incidence coefficients of elements of lower dimension with respect to elements of the higher dimension and elements of higher dimension with respect to elements of the lower dimension. The first characterizes the number of elements of a certain higher dimension to which the given element of a lower dimension belongs. The second characterizes the number of elements of a given lower dimension that are included in a particular element of a higher dimension. Here we must remember that the vertices of geometric elements of various dimensions are atoms, molecules or functional groups. Therefore, the incidence of geometric elements to a friend means contact between particles of the matter, including living matter. The contact between particles of matter can be interpreted as the transfer of information on material structures, including biological structures. We introduce the notation: $k_{d_j^i d_u}$ is the number of elements of dimension *u*, which include an element of dimension *j* ($u > j$) with number *i*. Thus, $k_{d_j^i d_u}$ is the incidence factor of element *i* with dimension *j* relative to elements with dimension *u*. We introduce the notation also: $k_{d_j d_u^i}$ is the number of elements of dimension *j*, which included in element *i* with a dimension *u* ($u > j$). Thus, $k_{d_j d_u^i}$ is the incidence factor of element *i* with dimension *u* relatively to elements with dimension *j*.

In Chapter 3, the law of conservation of incidents was formulated and proved for all convex political scientists. For *n* - cross - polytope this law is based on a proven statement: In any cross - polytope of dimension *n*, the sum of all incidents of elements of a lower dimension with respect to all elements of a higher dimension is equal to the sum of all incidents of elements of a higher dimension with respect to all elements of a lower dimension and equals the sum of the series

$$f_0(n)3^{n-1} + f_1(n)3^{n-2} + f_2(n)3^{n-3} + ... + f_{n-1}(n)3^0, f_d(n) = C_n^{n-1-d}2^{1+d}, d = 0 \div (n-1).$$

In result, one can get the common express for the sum of all incidents of elements of a lower dimension with respect to all elements of a higher dimension in *n* - cross - polytope

$$f_0(n)3^{n-1} + f_1(n)3^{n-2} + ... + f_{n-1}3^0 = \sum_{i=0}^{n-1} f_i(n)3^{n-1-i}, f_i(n) = C_n^{n-1-i}2^{1+i}.$$

The polytope of hereditary information is a cross - polytope of dimension 13. Substituting the value $n = 13$ and the values of $f_i(n)$ calculated in the previous section in this expression, we find the total value of the incident flow in the polytope of hereditary information

$$\sum_{i=0}^{12} f_i(13)3^{12-i} = 1.78 \cdot 10^9 .$$

Interestingly, the flow of incidents (information) is of greatest importance to the *n* - cross - polytope. If we compare the values of these flows calculated from the obtained expressions, then with the same dimension value (13) for 13 - simplex this value is $4.75 \bullet 10^6$, for 13 – cube (and13 - hyper – rhombohedron) this value is a thousand times greater ($1.2 \bullet 10^9$), and in the 13 - cross - polytope is 1.5 times more than the 13 - cube and is $1.8 \bullet 10^9$. This corresponds to a significant flow of information between the elements of the DNA molecule. The latter can serve as a mathematical basis for the recently discovered epigenetic principle (Lindquist, 2016;). of the transmission of hereditary information without changing the sequence of genes in DNA and RNA molecules (Zhizhin, 2019 f).

CONCLUSION

It is shown that the patterns of inheritance of features of organisms in their crossing are based on a special kind of mathematical relationships that allow us to derive algebraically the experimental patterns studied 150 years ago by Mendel. Using these relationships is possible to predict and investigate the inheritance processes outside the intervals of the initial date of Mendel`s experiments. The general formulas obtained are transformed continuously into the relations of Mendel. It has been established that with an increase in the number of generations in a plant population, the phenotypes of organisms in different gene sets are leveled quickly. These numbers, with an increase in the number of generations and the absence of death of organisms, synchronously increase while remaining practically equal to each other. Mathematically, this can be represented by trajectories in a multidimensional space with coordinates in form of inverse numbers of phenotypes of organisms. These trajectories tend along their own direction to an isolated stable equilibrium position of the type of a multidimensional node at the origin. With the results obtained, based on G. Mendel`s experiments, the Hardy - Weinberg principle is fundamentally inconsistent, constructed purely logistically without taking into account the real laws in living nature.

The representations of the sugar molecule and the residue of phosphoric acid in the form of polytopes of higher dimension are used. Based on these ideas and their simplified three-dimensional images, a three - dimensional image of nucleic acids is constructed. The geometry of the neighborhood of the compound of two nucleic acid helices with nitrogen bases has been investigated in detail. It is proved that this neighborhood is a polytope with anti - parallel edges of dimension 13 (13 – cross - polytope). This polytope is called of the polytope of hereditary information. The geometry of the polytope of hereditary information is investigated. It is shown that in the flat coordinate planes of the polytope of hereditary information there are located flat complementary hydrogen compounds of the nitrogenous bases of two nucleic acid helices. It turned out that the number of these coordinate planes (12) is exactly as many as there are various options for hydrogen compounds of nitrogenous bases. Thus, in each of these coordinate planes one of the possible types of bonding of nitrogenous bases is located. Thus, the possible orientation of flat nitrogenous bases in the space of higher dimension is determined. An image of the polytope of hereditary information with a specific indication of its coordinate planes is constructed. A detailed examination of the geometry of the nucleic acid compound region led to the construction of a hereditary information polytope and then to the determination of an almost unambiguous correspondence between nucleic acid nucleotides and amino acids. This destroys the idea of the correspondence of triplets to certain amino acids, introduced logistically, in order to avoid the contradiction between a small number of nucleotides and a large number of amino acids (Zhizhin, 2020 a, b). It turned out that the polytope of hereditary information has an internal (hidden) degree of freedom in the form of a different arrangement of pairs of flat nitrogen base in the space of higher dimension. Consideration of this internal degree of freedom is necessary when analyzing the transmission of hereditary information and analyzing the spread of viral infectious diseases.

Currently, clinical trials of antiviral drugs based on nucleic acid analogs that are structurally similar to natural RNA and DNA, but have altered any of the main components of nucleic acids.

Analog nucleic acids represent one of the main pillars of xenobiology, the construction of new life forms based on alternative biochemistry. In all these cases of noncanonical nucleic acids, it is of interest to study their geometry taking into account the higher dimensionality of polytopes and to analyze the effect of higher dimensionality on the features of their interaction.

The incidence of low - dimensional elements to the highest - dimensional elements of this polytope is studied, as well as the incidence of higher - dimensional elements to low - dimensional elements of this polytope. The values of the total incidence flow from low - dimensional elements to higher - dimensional elements and vice versa are determined. These values turned out to be equal (the law of conservation of incidents) and exceeding one billion. This indicates an intensive flow of information between the elements of the polytopic of hereditary information, ensuring the transmission of hereditary information. Which can serve, in particular, to explain the existence of the transmission of hereditary changes without changing the sequence of genes (epigenetics).

REFERENCES

Caly, L. (2020). The FDA – approved Drug Invermectin inhibits the replication of SARS – CoV – 2 in vitro. *Elsevier Enhanced Reader, 13*(4).

Chetverikov, S. S. (1926). About some moments of the evolutionary process from the point of view of modern genetics. *Journal of Experimental Biology A, 2*, 3–54.

Correns, C. (1900). Mendel's Regel über das Verhalten der Nachkommenschaft der Rassenbastarde. *Berichte der Deutschen Botanischen Gesellschaft, 18*, 158–168.

Crotty, S., Cameron, C., & Audino, R. (2002). Rubavirin's antiviral mechanism of action: Lethal mutagenesis? *Journal of Molecular Medicine, 80*(2), 86–95. PMID:11907645

De Clerey, E. (2004). Antiviral drugs in current clinical use. *Journal of Clinical Virology, 30*(2).

De Vries, G. (1904). Mutation theory: Mutations and mutational periods in the origin of species. In *Development Theory*. Brockhaus and Efron.

Frank-Kamenetskiy, M. (1986). DNA structure: A simple solution to the stability of the double helix? *Nature, 324*, 325.

Frank-Kamenetskiy, M. (1988). H – form DNA and the hairpin – triplex model. *Nature, 333*, 214.

Gaisinovich, A. E. (1988). *Origin and development of genetics*. Science.

Grunbaum, B. (1967). *Convex Polytopes*. Springer.

Hardy, G. H. (1908). Mendelian proportions in a mixed population. *Science, 28*, 49–50. PMID:17779291

Johannsen, V. L. (1933). *Elements of the exact doctrine of variability and heredity with the basics of biological variation statistics*. Sel'khozgiz.

Koltsov, N. K. (1935). Hereditary molecules. *Science and Life*, *5*, 4–13.

Kozlov, S. N., & Stragunsky, L. S. (2002). *Modern antimicrobial chemotherapy. A guide for doctors*. Borges.

Lehninger, A. L. (1982). *Principles of Biochemistry*. Worth Publishers, Inc.

Lindquist, S. (2016). Luminidependens (LD) is an arabidopsis protein with prion behavior. *Proceedings of the National Academy of Sciences of the United States of America*, *113*(21), 6065–6070. PMID:27114519

Mancuso, S. (2017). *Plant Revolution. Le piante hanno già inventato il nostro futuro*. GUIUTI EDITORE.

Mancuso, S., & Viola, A. (2013). *Verde Brillante. Sensibilita e intellidenza del mondo vegetale*. GIUNTI EDITORE.

Mendel, G. (1965). *Experiments on plant hybrids*. Science.

Morgan, T. G. (1937). *Selected works on genetics*. Sel'khozgiz.

Pereira, C. G. (2012). Underground Leavers of Philcoxia Trap and Digest Nematodes. *Proceedings of the National Academy of Science of the United States of America*.

Poincare, A. (1895). Analysis situs. *J de e'Ecole Polytechnique.*, *1*, 1–121.

Pollan, M. (2001). *The Botany of Desire: A Plant's – Eye View of the World*. Random House.

Razonable, R. R. (2011). Antiviral Drugs for Viruses Other Than Humanimmunodeficiency virus. *Mayo Clinic Proceedings*, *86*(10), 1009–1026. PMID:21964179

Sanbonmatsu, K. Y. (2016). COOLAIR antisense RNAs from evolutionarily conserved elaborate secondary structures. *Cell Reports*, *16*(12), 3087–3096. PMID:27653675

Shechtman, D. (1984). Metallic phase with longerange orientational order and no translational symmetry. *Physical Review Letters*, *53*(20), 1951–1953. doi:10.1103/physRevLett.53.1951

Spirin, A. S. (2019). *Molecular biology. The ribosomes and protein biosynthesis*. Knowledge Lab.

Spirin, A. S., & Gavrilova, L. P. (1971). *The Ribosome*. Science.

Takhtadzhyan, A. L. (1980). Life of plants. T. 5. Flowering plants. Moscow: Education.

Watson, J. D., & Crick, F. H. C. (1953a). Molecular structure of nucleic acids. *Nature*, *171*, 738–740. PMID:13054693

Watson, J. D., & Crick, F. H. C. (1953b). Genetical implications of the structure of deoxyribose nucleic acid. *Nature*, *171*, 964–967. PMID:13063483

Weinberg, W. (1908). Ueber den Nachweis der Vererbung beim Menschen. *Jahreshefte Verreins vaterländ, 64*, 368 – 382.

Weismann, A. (1885). *Die Continuität des Keimplasmas als Grundlage einer Theore der Vererbung*. Fisheri.

Yule, G. U. (1902). Mendel's laws and their probable relations to intra-racial heredity. *The New Phytologist, 1*, 193–207, 222–238.

Zhizhin, G. V. (1972). *Qualitative investigation of one - dimensional stationary flows. Thesis of the candidate of physical and mathematical sciences*. Leningrad: Polytechnic Institute.

Zhizhin, G.V. (2004a). Synergetics of flows. *Life and Safety, 2-3*, 366-385.

Zhizhin, G. V. (2004b). *Self - regulating waves of chemical reactions and biological populations*. Science.

Zhizhin, G. V. (2005). The Dissipative Structures in Chemical, Geological and Ecological Systems. St. - Petersburg. *Science*.

Zhizhin, G. V. (2014). *World - 4D*. Polytechnic Service.

Zhizhin, G. V. (2017). The Structure, Topological, and Functional Dimension of Biomolecules. *International Journal of Chemoinformatics and Chemical Engineering, 5*(2), 44–58. doi:10.4018/IJCCE.2016070104

Zhizhin, G. V. (2018). Chemical Compound Structures and the Higher Dimension of Molecules: Emerging Research and Opportunities. Hershey, PA: IGI Global.

Zhizhin, G. V. (2019a). About the True Equilibrium in the Population of the Plants. *Journal of Biochemistry and Molecular Biology Research, 4*(1), 214–224.

Zhizhin, G. V. (2019b). The Metric Structure of Nucleic Acids and Higher Dimension of Their Constituents. *J. Chemoinformatics and Chemical Engineering, 7*(2), 1–15.

Zhizhin, G. V. (2019c). *The Geometry of Higher – Dimensional Polytopes*. IGI Global.

Zhizhin, G. V. (2019d). *Attractors and Higher Dimensions in Population and Molecular Biology: Emerging Research and Opportunities*. IGI Global.

Zhizhin, G. V. (2019e). The Law of Conservation of Incidents in the Space of Nanoworld. *J. Chemoinformatics and Chemical Engineering., 8*(1), 1–15.

Zhizhin, G. V. (2019f). The Polytope of Hereditary Information: Structure, Location, Signification. *Biochemistry and Modern Applications., 2*, 56–62.

Zhizhin, G. V. (2020a). Hidden Nucleic Acid Bond Order? *Acta Scientific Biotechnology, 1*(4), 34–36.

Zhizhin, G. V. (2020b). Interaction of Nucleic Acids: Hidden Order of Interaction. *Int. Journal of Applied Research in Biotechnology, 11*(2), 1–15.

KEY TERMS AND DEFINITIONS

Dimension of the Space: The member of independent parameters needed to describe the change in position of an object in the space.

Dominant Traits: The traits that predominate in the first generation.

Gene: A hereditary factor; functionally indivisible unit of genetic material; section of the DNA molecule encoding the primary structure of the polypeptide.

Genotype: A set of genes of a given cell or organism.

Heterozygous Individuals: The individuals that produce cleavage in the offspring.

Homozygous Individuals: The individuals that do not produce cleavage in the offspring.

Incidence Coefficients of Elements of Lower Dimension With Respect to Elements of Higher Dimension: The number of elements of a certain higher dimension to which the given element of a lower dimension belongs.

Incidence Coefficients of Elements of Higher Dimension With Respect to Elements of Lower Dimension: The number of elements of a given lower dimension that are included in a particular element of a higher dimension.

Monohybrid Crossing: A crossing in which the manifestation of only one trait is examined.

N-Cross-Polytope: The convex polytope of dimension *n* in which opposite related of centrum edges do not have connection of edge.

N-Simplex: The convex polytope of dimension *n* in which each vertex is joined by edges with all remain vertices of the polytope.

Phenotype: The totality of all traits and properties of an individual, which are formed in the process of interaction between its genetic structure and the external environment.

Polyhybrid Crossing: A crossing in which explores the manifestation of several signs.

Recessive Traits: The traits that do not appear in the first generation.

Polytope: The polyhedron in the space of the higher dimension.

Chapter 5
The Geometric Structure of the Product of Polytopes

ABSTRACT

The structure of polytopes of higher dimension (polytopic prismahedrons), which are products of polytopes of lower dimensionality, is investigated. The products of polytopes do not belong to the well-studied class of simplicial polytopes, and therefore, their investigations are of independent interest. Analytical dependencies characterizing the structure of the product of polytopes are obtained as a function of the structures of polytope factors. Images of a number of specific polytopic prismahedrons are obtained, and tables of structures of polytopic prismahedrons are compiled depending on the types of polytopes of the factors. The geometric properties of the boundary complexes of polytopic prismahedrons are investigated.

INTRODUCTION

In the previous chapters, the general conditions for the existence of polytopes of the highest dimension were formulated and the types of polytopes of the highest dimensionality, which are found in nature in the form of molecules of chemical compounds, clusters, biomolecules, were considered. The consideration of polytopes was short enough only to note the features necessary for the further solution of the main task of the monograph, i.e. investigation of the normal partition of an n - dimensional space (18 Hilbert's problem) and the hierarchical filling of an n - dimensional space. Details of the geometry of polytopes of higher dimension can be found in the author's previously published monographs (Zhizhin, 2018; 2019 a, b). None of the higher -dimensional polytopes found in nature and listed in Chapters 1 – 4, with the exception of the hyper - rhombohedron, can, as one can see, form a normal partition of n - dimensional space. The hyper - rhombohedron appeared as a result of the analysis of diffraction patterns of quasi -crystals (Zhizhin, 2018) as projections from multidimensional space onto a two - dimensional plane. Therefore, by construction, a hyper - rhombohedron provides a normal partition of an n - dimensional space. A distinctive feature of the hyper - rhombohedron is the formation of parallel lines by its families. Therefore, to solve the problem of the normal partitioning of n - dimensional space, it is necessary to find ways of transition from various types of polytopes of the highest dimension, found

DOI: 10.4018/978-1-7998-6768-5.ch005

in nature, to polytopes of the highest dimension based on families of parallel lines. Such a transition is possible if we multiply polytopes by each other. Then naturally polytopes of higher dimension appear with systems of parallel edges - polytopic prismahedrons (Zhizhin, 2019 a). In this regard, this chapter is devoted to the analysis of the geometry of high - dimensional polytopes resulting from the product of lower - dimensional polytopes.

A polytopic prismahedron is the product of a polytope to another polytope, or in a particular case to a one - dimensional segment. Pontryagin mentioned the structures resulting from the product of a polyhedron by a one - dimensional segment the cylinder ones (Pontryagin, 1976). We can say that the product of a polytope by a segment is a prism (Robertson, 1984) with a base in form of a polytope. To distinguish it from a usual three - dimensional prism, can call it polytopic prism. Ziegler noted that the product of polytopes is not a simplex even if the factors are simplexes, so the polytopes are of considerable interest (Ziegler, 1995). Especially taking into account that a multi - dimensional world has its own peculiarities having no analogues in the three - dimensional world (Ziegler, 1995), contrary to some opposite erroneous statements (Panina, 2006). In this regard, the theory of simplicial polytopes for the analysis of polytopes product becomes inapplicable, especially in the case of high - dimensional factors.

The product of two polytopes is result of the product of one of them by one - dimensional edges of another polytope. Thus, the product of polytopes is a complex of polytopic prism. Let`s call this complex a polytopic prismahedron. The polytopic prism is particular case of polytopic prismahedron. The interest to the study of polytopic prismahedrons is connected, in first place, with the novelty of this field and, secondly, with the fact that polytopic prismahedrons, due to their construction, can be "bricks" to fill the spaces of higher dimension face in face.

The definition of polytopes product (Ziegler, 1995) does not give the possibility to specify the structure of product of polytopes as a function of the structures of factors.

THE STRUCTURE OF PRODUCT OF POLYTOPES AS FUNCTION OF FACTORS STRUCTURES

There is the structure of product of polytopes having different structures of their factors is determined.

Theorem 5.1. (Zhizhin, 2015). If there are convex polytopes of dimensions n and m, respectively denoted P^n and Q^m (or simply P and Q), then their product $P^n \times Q^m$, (or simply $\times$, when it is clear what polytopes are multiplied) has a face $F_\times^k$ with numbers

$$f_\times^k = \sum_{i=0}^{j} f_P^{k-i} f_Q^i , \tag{5.1}$$

$j=k$, if $0 \le k < m$; $j=m$, if $m \le k \le n+m$; $n \ge m$

Here the symbol f indicates the number of faces, the superscript of f and F indicates the dimension of faces, the lower index indicates belonging of a face to the respective polytope.

Proof. To find the product of polytopes P and Q one has find the product of geometric elements of different dimensions of one of the polytopes by geometric elements of different dimensions of another polytope. According to the definition of polytopes product (Ziegler, 1995), the product of vertices (elements of zero - dimension) of one of the polytopes by the vertices of another polytope is a set of vertices, the number of which is equal to the product of the number of vertices of one of the polytopes by the number of vertices of another polytope

$$f_\times^0 = f_P^0 f_Q^0 .$$

The number of elements dimension 1 of polytopes two product is sum from two parts: a product of the number of elements dimension 1 of one polytope by the number of elements dimension 0 of another polytope and a product of the number of elements dimension 0 of one polytope by the number of elements dimension 1 of another polytope

$$f_\times^1 = f_P^1 f_Q^0 + f_P^0 f_Q^1 .$$

The number of elements dimension 2 of polytopes two product is sum from three parts:

a product of the number of elements dimension 2 of one polytope by the number of elements dimension 0 of another polytope, a product of the number of elements of dimension 0 of one polytope by the number of elements dimension 1 of another polytope, and a product of the number of elements dimension 0 of one polytope by the number of elements of dimension 2 of another polytope

$$f_\times^2 = f_P^2 f_Q^0 + f_P^1 f_Q^1 + f_P^0 f_Q^2 .$$

Continuing this process, can obtain the expression for the number of elements of dimension k in the product of polytopes P and Q

$$f_\times^k = f_P^k f_Q^0 + f_P^{k-1} f_Q^1 + \dots + f_P^0 f_Q^k . \tag{5.2}$$

The specific form of equation (5.2) depends on the ratio of k with the dimensions of polytopes P and Q. If k is greater than m, i.e. the dimension of the polytope Q, then in the series (5.2) several last items are lost because of the absence of faces of the polytope Q with dimension greater than m. If k is greater than n, i.e. the dimension of the polytope P, in the series (5.2) several first items are lost because of the absence of faces of the polytope P with the dimension greater than n.

If $k < m$ the series (5.2) can be written in the form

$$f_\times^k = \sum_{i=0}^{k} f_P^{k-i} f_Q^i . \tag{5.3}$$

When $k \geq m$ summation by i in the series (5.3) is carried out only up to the value $i = m$

$$f_\times^k = \sum_{i=0}^{m} f_P^{k-i} f_Q^i . \tag{5.4}$$

At that, in the row (5.4) the items corresponding to the values of $k - i > n$ equal to zero. Equations (5.3) and (5.4) can be written as a single equation

$$f_{\times}^{k} = \sum_{i=0}^{j} f_{P}^{k-i} f_{Q}^{i}, \tag{5.5}$$

where $j=k$, if $0 \leq k < m$; $j=m$, if $m \leq k \leq n+m$, $n \geq m$.

Theorem 5.1 determines the number of figures of different dimensions in the product of polytopes, but it cannot identify in general terms what is the form of these figures. It follows from theorem 5.1 that the form of the figures of polytopes product depends on the form of the polytopes, which are the factors of this product.

THE PRODUCTS OF POLYTOPES BY ONE – DIMENSIONAL SEGMENT

It is obvious that the product of a point a by a one - dimensional segment is the same segment with point on end coinciding with point a. The product of segment by segment is a quadrangle. The product triangle on the segment is a triangular prism with the bases equal to the triangle and generating lines equal to the segment.

The Product of a Tetrahedron by a Segment

For the designation of a polytope let's agree to indicate not only its dimension but the number of its vertices by lower index of polytope symbol. Besides, one will also indicate the number of its faces in brackets after the polytope sign, face dimension and the number of its vertices (for the faces with dimension greater or equal to 2). For example, a tetrahedron in this record has the form $P_4^3(4F_3^2)$. If a polytope it is considered in this case as a face one will denote it by the letter F. Let's call such a record a structural polytope formula. One need to get from structural formulas of polytopes as factors the structural formula of the product of polytopes. Thus, in this case, we consider the product $P_4^3(4F_3^2) \times Q_2^1$. According to theorem 1 and the ratios (5.5) there are

$$f_{\times}^{0} = f_{P}^{0} f_{Q}^{0} = 4 \cdot 2 = 8,$$

$$f_{\times}^{1} = f_{P}^{1} f_{Q}^{0} + f_{P}^{0} f_{Q}^{1} = 6 \cdot 2 + 4 \cdot 1 = 16..$$

When calculating the number of faces of dimension greater than 1 one will indicate in brackets after the sign of the faces number the designation of the corresponding face. This will give us an opportunity to get the composition of faces in the product of polytopes. Thus,

$$f_{\times}^{2} = f_{P}^{2}(F_{P3}^{2}) f_{Q}^{0}(F_{Q1}^{0}) + f_{P}^{1}(F_{P2}^{1}) f_{Q}^{1}(F_{Q2}^{1}) = 4 \cdot 2 + 6 \cdot 1 = 14. \tag{5.6}$$

For receiving of the composition of the faces in the dimension 2 polytopes product we multiply the faces in brackets in the product of the numbers of faces, observing the same rules as at the definition of the product of polytopes, i.e. dimensions are added and the numbers of vertices are multiplied. At that, indexes P and Q are replaced by index ×. Then 14 faces obtained in (5.6), are deciphered so

$$14 = 8F_{\times 3}^{2} + 6F_{\times 4}^{2}. \tag{5.7}$$

This means that 14 faces of dimension 2 include eight faces $F_{\times 3}^{2}$ (i.e., 8 triangles) and six faces $F_{\times 4}$ (i.e., 6 quadrangles). In future, equations similar (5.6), (5.7) will be recorded in a single line.

For the faces of dimension 3, there is

$$f_{\times}^{3} = f_{P}^{3}(F_{P4}^{3})f_{Q}^{0}(F_{Q1}^{0}) + f_{P}^{2}(F_{P3}^{2})f_{Q}^{1}(F_{Q2}^{1}) = 1\cdot 2 + 4\cdot 1 = 6 = 2F_{\times 4}^{3} + 4F_{\times 6}^{3}.$$

Thus, 6 faces of dimension 3 consist of two tetrahedron $P_{4}^{3}(4F_{3}^{2})$ and four triangular prisms $P_{6}^{3}(3F_{4}^{2}, 2F_{3}^{2})$.

For the faces of dimension 4, there is

$$f_{\times}^{4} = f_{P}^{3}(F_{P4}^{3})f_{Q}^{1}(F_{Q2}^{1}) = 1\cdot 1 = 1 = P_{8}^{4}.$$

According to the preceding equation, can obtain the structural formula of the product

$$P_{4}^{3}(4F_{3}^{2}) \times Q_{2}^{1} = P_{8}^{4}[4P_{6}^{3}(3F_{4}^{2}, 2F_{3}^{2}), 2F_{4}^{3}(4F_{3}^{2})]. \tag{5.8}$$

Therefore, the product of the tetrahedron by the segment is polytope 4*D*, faces which includes four triangular prisms and 2 tetrahedrons. You can call a polytope (5.8) tetrahedral prism. The picture of the polytope (5.8) is shown in Figure 1.

Figure 1. The tetrahedral prism

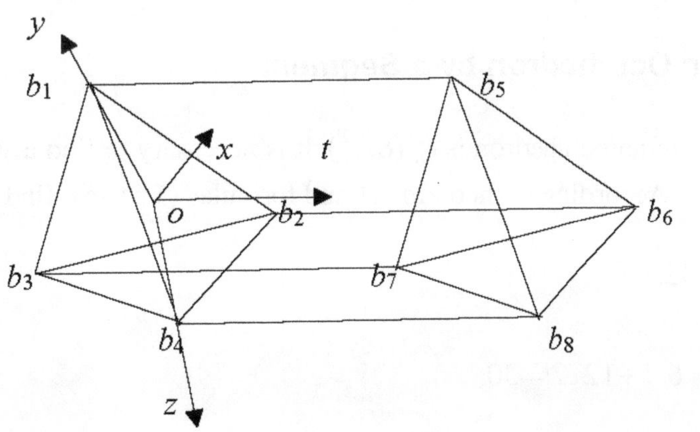

The equation of the Euler - Poincaré (Poincaré, 1895) in this case has view

$$\sum_{i=0}^{d-1}(-1)^i f^i(P) = 1+(-1)^{d-1}.$$

In this equation $f^i(P)$ is the number of faces with dimension i in polytope P with dimension d.

Substitute obtained numbers $f^i(P)$ in the equation of the Euler - Poincaré it is easy to verify that this equation for the polytope (5.8) is satisfied 8 – 16 + 14 – 6 = 0.

In future, the verification of the Euler – Pioncaré (Pioncaré, 1895) equation satisfaction is omitted, as a reader can easily do it on his own.

The Product of a Cube by a Segment

The structural formula of a cube is $P_8^3(6F_4^2)$ It is necessary to find a structural formula of the product $P_8^3(6F_4^2)\times Q_2^1$. According to theorem 5.1 and formulas (5.5), can find

$$f_\times^0 = f_P^0 f_Q^0 = 8 \cdot 2 = 16,$$

$$f_\times^1 = f_P^1 f_Q^0 + f_P^0 f_Q^1 = 12\cdot 2 + 8\cdot 1 = 32,.$$

$$f_\times^2 = f_P^2(F_4^2) f_Q^0(F_{Q1}^0) + f_P^1(F_{P2}^1) f_Q^1(F_{Q2}^1) = 6\cdot 2 + 12\cdot 1 = 24 = 24F_{\times 4}^2,$$

$$f_\times^3 = f_P^3(F_{P8}^2) f_Q^0(F_{Q1}^0) + f_P^2(F_{P4}^2) f_Q^1(F_{Q2}^1) = 1\cdot 2 + 6\cdot 1 = 8 = 8F_{\times 8}^3,$$

$$f_\times^4 = f_P^3(F_{P8}^3) f_Q^1(F_{Q2}^1) = 1\cdot 1 = 1 = P_{16}^4(8F_8^3). \tag{5.9}$$

Thus, easy to see that the product of a cube by a segment is 4 – cube (hyper - cube) containing 8 facets as cubes and 24 quadrangles.

The Product of an Octahedron by a Segment

The structural formula of an octahedron is $P_6^3(8F_3^2)$ It is necessary to find a structural formula of the product $P_6^3(8F_3^2)\times Q_2^1$. According to theorem 5.1 and formulas (5.5), can find

$$f_\times^0 = f_P^0 f_Q^0 = 6 \cdot 2 = 12,$$

$$f_\times^1 = f_P^1 f_Q^0 + f_P^0 f_Q^1 = 6\cdot 1 + 12\cdot 2 = 30,$$

$$f_{\times}^{2}=f_{P}^{2}(F_{P3}^{2})f_{Q}^{0}(F_{Q1}^{0})+f_{P}^{1}(F_{P2}^{1})f_{Q}^{1}(F_{Q2}^{1})=8\cdot 2+12\cdot 1=28=16F_{\times 3}^{2}+12F_{\times 4}^{2},$$

$$f_{\times}^{3}=f_{P}^{3}(F_{P6}^{3})f_{Q}^{0}(F_{Q1}^{0})+f_{P}^{2}(F_{P3}^{2})f_{Q}^{1}(F_{Q2}^{1})=1\cdot 2+8\cdot 1=8F_{\times 6}^{3}(3F_{4}^{2},2F_{3}^{2})+2F_{\times 6}^{3}(8F_{3}^{2}),$$

$$f_{\times}^{4}=f_{P}^{3}(F_{P6}^{3})f_{Q}^{1}(F_{Q2}^{1})=1\cdot 1=1=P_{12}^{4}.$$

According to the last equality can obtain the structural formula of the product

$$P_{6}^{3}(8F_{3}^{2})\times Q_{2}^{1}=P_{12}^{4}[8F_{6}^{3}(3F_{4}^{2},2F_{3}^{2}),2F_{6}^{3}(8F_{3}^{2})]. \tag{5.10}$$

Thus, the product of an octahedron by a segment is a polytope 4*D*. It has 12 vertices, 30 edges, 16 triangles, 12 quadrangles and facets which includes 8 triangular prisms and two octahedrons. One can call the polytope (5.10) an octahedral prism.

The Product of an Icosahedron by a Segment

The structural formula of an icosahedron is $P_{12}^{3}(20F_{3}^{2})$ It is necessary to find the structural formula of the product $P_{12}^{3}(20F_{3}^{2})\times Q_{2}^{1}$ According to theorem 5.1 and formulas (5.5), one can to find $f_{\times}^{0}=f_{P}^{0}f_{Q}^{0}$ = 12 · 2 = 24,

$$f_{\times}^{1}=f_{P}^{1}f_{Q}^{0}+f_{P}^{0}f_{Q}^{1}=30\cdot 2+12\cdot 1=72,.$$

$$f_{\times}^{2}=f_{P}^{2}(F_{P3}^{2})f_{Q}^{0}(F_{Q1}^{0})+f_{P}^{1}(F_{P2}^{1})f_{Q}^{1}(F_{Q2}^{1})=20\cdot 2+30\cdot 1=70=40F_{\times 3}^{2}+30F_{\times 4}^{2},$$

$$f_{\times}^{3}=f_{P}^{3}(F_{P12}^{3})f_{Q}^{0}(F_{Q1}^{0})+f_{P}^{2}(F_{P3}^{2})f_{Q}^{1}(F_{Q2}^{1})=1\cdot 2+20\cdot 1=22=20F_{\times 6}^{3}(3F_{4}^{2},2F_{3}^{2})+2F_{\times 12}^{3},$$

$$f_{\times}^{4}=f_{P}^{3}(F_{P12}^{3})f_{Q}^{1}(F_{Q2}^{1})=1\cdot 1=1=P_{24}^{4}.$$

According to the last equality, one can obtain the structural formula of the product

$$P_{12}^{3}(20F_{3}^{2})\times Q_{2}^{1}=P_{24}^{4}[20F_{6}^{3}(3F_{4}^{2},2F_{3}^{2}),2F_{12}^{3}(20F_{3}^{2})]. \tag{5.11}$$

Thus, the product of an icosahedron by a segment is a polytope 4*D*. It has 24 vertices, 72 edges, 30 quadrangles, 40 triangles, facets which includes 20 triangular prisms and 2 icosahedrons. One can call a polytope (5.11) icosahedral prism.

The Product of a Dodecahedron on a Segment

The structural formula of a dodecahedron is $P_{20}^{3}(12F_{5}^{2})$. It is necessary to find the structural formula of product $P_{20}^{3}(12F_{5}^{2})\times Q_{2}^{1}$. According to theorem 5.1 and formulas (5.5), one can find

$$f_{\times}^{0} = f_{P}^{0} f_{Q}^{0} = 20 \cdot 2 = 40,$$

$$f_{\times}^{1} = f_{P}^{1} f_{Q}^{0} + f_{P}^{0} f_{Q}^{1} = 30\cdot 2+20\cdot 1=80,.$$

$$f_{\times}^{2} = f_{P}^{2}(F_{P5}^{2}) f_{Q}^{0}(F_{Q1}^{0}) + f_{P}^{1}(F_{P2}^{1}) f_{Q}^{1}(F_{Q2}^{1}) = 12\cdot 2+30\cdot 1=54=24F_{\times 5}^{2}+30F_{\times 4}^{2},$$

$$f_{\times}^{3} = f_{P}^{3}(F_{P20}^{3}) f_{Q}^{0}(F_{Q1}^{0}) + f_{P}^{2}(F_{P5}^{2}) f_{Q}^{1}(F_{Q2}^{1}) = 1\cdot 2+12\cdot 1=14=12F_{\times 10}^{3}(5F_{4}^{2},2F_{5}^{2})+2F_{\times 20}^{3}(12F_{5}^{2}),$$

$$f_{\times}^{4} = f_{P}^{3}(F_{P20}^{3}) f_{Q}^{1}(F_{Q2}^{1}) = 1\cdot 1=1=P_{40}^{4}.$$

According to the last equality, one can to obtain the structural formula of the product

$$P_{20}^{3}(12F_{5}^{2})\times Q_{2}^{1} = P_{40}^{4}[12F_{10}^{3}(5F_{4}^{2},2F_{5}^{2}),2F_{20}^{3}(12F_{5}^{2})]. \tag{5.12}$$

Thus, the product of a dodecahedron by a segment is a polytope 4*D*. It has 40 vertices, 80 edges, 24 pentagons, 30 quadrangles, facets of him including 12 pentagonal prisms and 2 dodecahedrons. One can call the polytope (5.12) a dodecahedral prism.

The Product of an Egyptian Pyramid by a Segment

The structural formula of an Egyptian pyramid is $P_{5}^{3}(4F_{3}^{2},\mathrm{F}_{4}^{2})$. It is necessary to find the structural formula of product $P_{5}^{3}(4F_{3}^{2},\mathrm{F}_{4}^{2})\times Q_{2}^{1}$. According to theorem 5.1 and formulas (5.5), one can find

$$f_{\times}^{0} = f_{P}^{0} f_{Q}^{0} = 5 \cdot 2 = 10,$$

$$f_{\times}^{1} = f_{P}^{1} f_{Q}^{0} + f_{P}^{0} f_{Q}^{1} = 5\cdot 1+8\cdot 2=21,.$$

$$f_{\times}^{2} = f_{P}^{2}(4F_{P3}^{2},\mathrm{F}_{P4}^{2}) f_{Q}^{0}(F_{Q1}^{0}) + f_{P}^{1}(F_{P2}^{1}) f_{Q}^{1}(F_{Q2}^{1}) = 5\cdot 2+8\cdot 1=18=8F_{\times 3}^{2}+10F_{\times 4}^{2},$$

$$f_{\times}^{3} = f_{P}^{3}(F_{P5}^{3}) f_{Q}^{0}(F_{Q1}^{0}) + f_{P}^{2}(F_{P3}^{2},\mathrm{F}_{P4}^{2}) f_{Q}^{1}(F_{Q2}^{1}) = 1\cdot 2+5\cdot 1=7=2F_{\times 5}^{3}+4F_{\times 6}^{3}+F_{\times 8}^{3},$$

$$f_{\times}^{4} = f_{P}^{3}(F_{P5}^{3})f_{Q}^{1}(F_{Q2}^{1}) = 1 \cdot 1 = 1 = P_{10}^{4}.$$

According to the last equality, one can obtain the structural formula of the product

$$P_{5}^{3}(4F_{3}^{2}, F_{4}^{2}) \times Q_{2}^{1} = P_{10}^{4}[4F_{6}^{3}(3F_{4}^{2}, 2F_{3}^{2}), F_{8}^{3}(6F_{4}^{2}), 2F_{5}^{3}(4F_{3}^{2}, F_{4}^{2})].$$

Thus, the product of an Egyptian pyramid by segment is a polytope *4D*. It has 10 vertices, 21 edges, 8 triangles, 10 quadrangles, facets of it including four triangular prisms, 1 quadrangular prism and 2 Egyptian pyramids (Figure 2). One can call the polytope a pyramidal prism.

Figure 2. The pyramidal prism

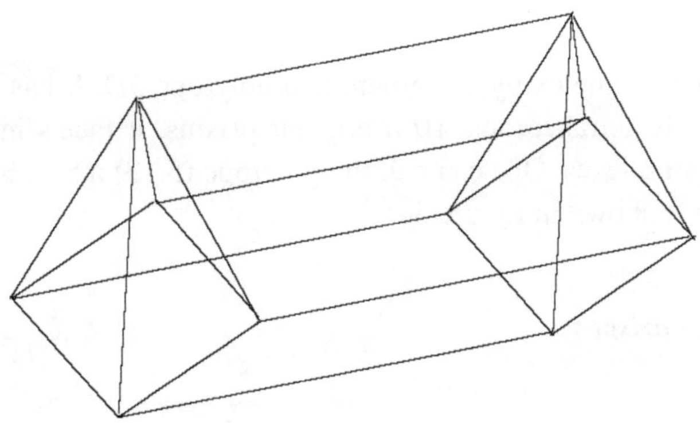

It is interesting that the pyramids of ancient Egypt had a strict geometric form (the pyramids of Cheops, Khafre, Mikerin) and were built in groups not far from each other (Snegirev & Franzev, 1938). In aggregate, this led, according to the analysis carried out, to an increase in the dimensionality of the space in the vicinity of the pyramids.

The Product of a 4 - Simplex by a Segment

The structural formula of a 4 - simplex is $P_{5}^{3}[5F_{4}^{3}(4F_{3}^{2})]$. It is necessary to find the structural formula of the product $P_{5}^{3}[5F_{4}^{2}(4F_{3}^{2})] \times Q_{2}^{1}$. According to theorem 5.1 and formulas (5.5), one can find

$$f_{\times}^{0} = f_{P}^{0} f_{Q}^{0} = 5 \cdot 2 = 10,$$

$$f_{\times}^{1} = f_{P}^{1} f_{Q}^{0} + f_{P}^{0} f_{Q}^{1} = 10 \cdot 2 + 5 \cdot 1 = 25,.$$

$$f_{\times}^{2} = f_{P}^{2}(F_{P3}^{2}) f_{Q}^{0}(F_{Q1}^{0}) + f_{P}^{1}(F_{P2}^{1}) f_{Q}^{1}(F_{Q2}^{1}) = 10 \cdot 2 + 10 \cdot 1 = 30 = 20F_{\times 3}^{2} + 10F_{\times 4}^{2},$$

$$f_\times^3 = f_P^3(F_{P4}^3) f_Q^0(F_{Q1}^0) + f_P^2(F_{P3}^2) f_Q^1(F_{Q2}^1) = 5 \cdot 2 + 10 \cdot 1 = 20 = 10F_{\times 4}^3(4F_3^2) + 10F_{\times 6}^3(3F_4^2, 2F_3^2),$$

$$f_\times^4 = f_P^4[5F_{P4}^3(4F_3^3)] f_Q^0(F_{Q1}^0) + f_P^3(F_{P4}^3) f_Q^1(F_{Q2}^1) = 1 \cdot 2 + 5 \cdot 1 = 7$$
$$= 5F_{\times 8}^4[4F_6^3(3F_4^2, 2F_3^2), 2F_4^3(4F_3^2)] + 2F_{\times 5}^4[5F_4^3(4F_3^2)],$$

$$f_\times^5 = f_P^4(F_{P5}^3) f_Q^1(F_{Q2}^1) = 1 \cdot 1 = 1 = P_{10}^5.$$

According to the last equality, one can obtain the structural formula of the product

$$P_5^3[5F_4^3(4F_3^2)] \times Q_2^1 = P_{10}^4\{5F_8^4[4F_6^3(3F_4^2, 2F_3^2), 2F_4^3(4F_3^2)], 2F_5^4[5F_4^3(4F_3^2)]\}. \tag{5.13}$$

Thus, the product of a 4 - simplex by a segment is a polytope *5D*. It has 10 vertices, 25 edges, 20 triangles, 10 quadrangles, 10 tetrahedrons, 10 triangular prisms, it facets including 5 the tetrahedral prisms (5.8) and 2 the 4 - simplexes. One can call the polytope (5.13) a 5 – simplex - prism. The image of the 5 – simplex - prism is shown in Figure.3.

Figure 3. The 5 – simplex - prism

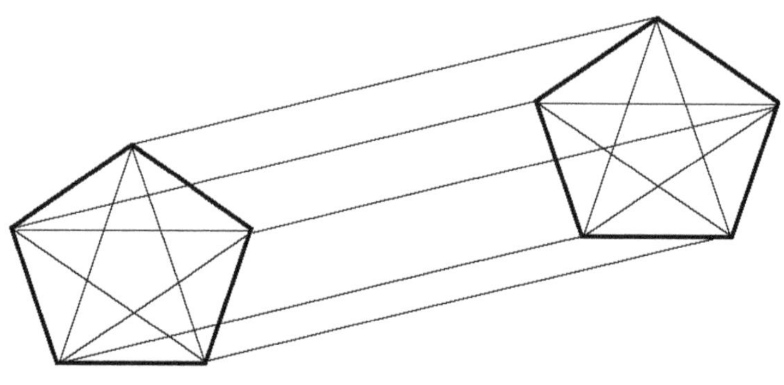

The Product of a 4 – Cross - Polytope by a Segment

The structural formula of a 4 – cross - polytope is $P_8^3[16F_4^3(4F_3^2)]$. It is necessary to find the structural formula of the product $P_8^3[16F_4^2(4F_3^2)] \times Q_2^1$. According to theorem 5.1 and formulas (5.5), there are

$$f_\times^0 = f_P^0 f_Q^0 = 8 \cdot 2 = 16,$$

$$f_\times^1 = f_P^1 f_Q^0 + f_P^0 f_Q^1 = 24 \cdot 2 + 8 \cdot 1 = 56, .$$

$$f_{\times}^{2} = f_{P}^{2}(F_{P3}^{2})f_{Q}^{0}(F_{Q1}^{0}) + f_{P}^{1}(F_{P2}^{1})f_{Q}^{1}(F_{Q2}^{1}) = 32 \cdot 2 + 24 \cdot 1 = 88 = 64F_{\times 3}^{2} + 24F_{\times 4}^{2},$$

$$f_{\times}^{3} = f_{P}^{3}(F_{P4}^{2})f_{Q}^{0}(F_{Q1}^{0}) + f_{P}^{2}(F_{P3}^{2})f_{Q}^{1}(F_{Q2}^{1}) = 16 \cdot 2 + 32 \cdot 1 = 64 = 32F_{\times 4}^{3}(4F_{3}^{2}) + 32F_{\times 6}^{3}(3F_{4}^{2}, 2F_{3}^{2}),$$

$$f_{\times}^{4} = f_{P}^{4}\{\mathrm{F}_{8}^{4}[16F_{P4}^{3}(4F_{3}^{3})]\}f_{Q}^{0}(F_{Q1}^{0}) + f_{P}^{3}(F_{P4}^{2})f_{Q}^{1}(F_{Q2}^{1}) = 1 \cdot 2 + 16 \cdot 1 = 18$$
$$= 16F_{\times 8}^{4}[4F_{6}^{3}(3F_{4}^{2}, 2F_{3}^{2}), 2F_{4}^{3}(4F_{3}^{2})] + 2F_{\times 8}^{4}[16F_{4}^{3}(4F_{3}^{2})],$$

$$f_{\times}^{5} = f_{P}^{4}\{\mathrm{F}_{4}^{3}[16F_{4}^{3}(4F_{3}^{3})f_{Q}^{1}(F_{Q2}^{1}) = 1 \cdot 1 = 1 = P_{16}^{5}.$$

According to the last equality, we obtain the structural formula of the product

$$P_{5}^{3}[5F_{8}^{3}[16F_{4}^{3}(4F_{3}^{2})] \times Q_{2}^{1} = P_{16}^{4}\{16F_{8}^{4}[4F_{6}^{3}(3F_{4}^{2}, 2F_{3}^{2}), 2F_{4}^{3}(4F_{3}^{2})], 2F_{8}^{4}[16F_{4}^{3}(4F_{3}^{2})]\}. \quad (5.14)$$

Thus, the product of a 4 – cross - polytope by a segment is a polytope *5D*. It has 16 vertices, 56 edges, 64 triangles, 24 quadrangles, 32 tetrahedrons, 32 triangular prisms. Facets of it including 16 the tetrahedral prisms (5.8) and 2 the 4 – cross - polytopes. One can to call the polytope (5.14) a 5 – cross - prism (Figure 4).

Figure 4. The 5 - cross - prism

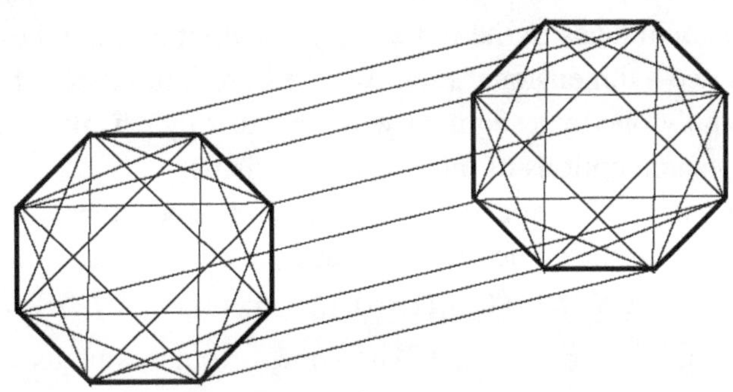

The General Structural Formula of Polytopes Product by a Segment

One will call prisms with bases in the form of polytopes the polytopic prisms.

Theorem 5.2. The product of a polytope by a segment, if all its faces of dimension 2 are equal, is a polytopic prism with the structure defined by equation

$$P_{K_n}^{n}\{f_{n-1}F_{K_{n-1}}^{n-1} \ldots [\mathrm{f}_3 F_{K_3}^{3}(f_2 F_{K_2}^{2})] \ldots\} \times Q_{2}^{1} = P_{2K_n}^{n+1}\{f_{n-1}F_{2K_{n-1}}^{n} \ldots [f_3 F_{2K_3}^{4}(f_2 F_{2K_2}^{3}, 2F_{K_2}^{2})2F_{K_3}^{3}] \ldots 2F_{K_n}^{n}\}. \quad (5.15)$$

There $P^n_{K_n}$ is the polytope of dimension n with the number of vertices K_n; $F^{n-1}_{K_{n-1}}$ are hyper - faces with the number f_{n-1}, each having K_{n-1} vertices; accordingly $F^2_{K_2}$ are hyper - faces of dimension 2 at faces $F^3_{K_3}$; f_2 is the number of faces $F^2_{K_2}$ with the number of vertices K_2 and so on; $P^{n+1}_{2K_n}$ is a polytopic prism of dimension $n+1$ with the number of vertices $2K_n$, the bases of which are the polytopes $P^n_{K_n}$. The polytopic prism $P^{n+1}_{2K_n}$ includes f_{n-1} polytopic prisms of dimension n and the number of the vertices $2K_{n-1}$, and also two polytopes $P^n_{K_n}$ which are the bases of polytopic prisms and so on, up to prisms $F^3_{2K_2}$ with a number f_2 and with bases $F^2_{K_2}$

Proof. Theorem 5.1 follows from the carried examinations of the product of polytopes of different dimensions by a segment. General laws which allowed formulate the theorem can be traced in all these products. Whatever the dimension n of the polytope was, at multiplying it by a segment from each vertex of the polytope this segment as if grows. This forms polytopic prism, the bases of which are two copies of the initial polytope. The dimension of this polytopic prism is by 1 more than the dimension of the initial polytope, i.e. $n + 1$. This polytopic prism includes polytopic prisms built on the hyper - faces of the initial polytope and their number is equal to the number of hyper-faces of the initial polytope f_{n-1}. The dimension of these politopic prisms is n, since the dimension of the initial polytope hyper - faces is equal to n - 1, but when multiplied by a segment the dimension is increased by 1. In addition, a polytopic prism of dimension $n + 1$ consists of two copies of the initial polytope of dimension n, as the bases of the prism. In its turn, the polytopic prisms of dimension n include the polytopic prisms of dimension n - 1, built on the hyper - faces of the polytope of dimension n - 1, and two copies of the polytope of dimension n - 1. This process of the polytopic prisms inclusion in each other happens until the formation of three - dimensional prisms built on two - dimensional facets of the initial polytope.

The equality (5.15) is an analytical expression of these constructions. This process of formation occurs when all two - dimensional faces of the initial polytope are the same, i.e. they are equal polygons. If the initial polytope has two - dimensional faces with different numbers of vertices so the analytical record of the product this polytope by a segment will be added by prisms built on different bases. The ratios (5.15) do not include such options of the polytopic prisms.

Supplement. Values of f_i in (5.15) depend on the form and dimension of the polytope. According to Zhizhin (2019 a) the formulas for their calculation in the case of a simplex, a cube and a cross - polytope will be the following: for simplex $f_i(n) = C^{i+1}_{n+1}; i = 0,1,...,n;$ for cube $f_i(n) = 2^{n-i} C^i_n; i = 0,1,...,n;$ for cross - polytope $f_i(n) = 2^{1+i} C^{n-1-i}_n; i = 0,1,...,n.$ The same formulas were used to calculate the numbers of elements of different dimensions in the product of polytopes with the participation of the n - simplex, n - cube and n - cross - polytope.

Naturally, the results of research about products of polytopes by a segment are to be taken into account when analyzing the products of polytopes by more complex figures, since they all have edges are factors in these products.

THE PRODUCTS OF POLYTOPES BY A TRIANGLE

The Product of a Triangle by a Triangle

There we consider the product $P_3^2 \times Q_3^2$ According to theorem 5.1 and ratios (5.5) there are

$$f_\times^0 = f_P^0 f_Q^0 = 3 \cdot 3 = 9,$$

$$f_\times^1 = f_P^1 f_Q^0 + f_P^0 f_Q^1 = 3 \cdot 3 + 3 \cdot 3 = 18, .$$

$$f_\times^2 = f_P^2 (4F_{P3}^2) f_Q^0 (F_{Q1}^0) + f_P^1 (F_{P2}^1) f_Q^1 (F_{Q2}^1) + f_P^0 (F_{P1}^0) f_Q^2 (F_{Q3}^2) = 1 \cdot 3 + 3 \cdot 3 + 3 \cdot 1 = 15 = 6F_{\times 3}^2 + 9F_{\times 4}^2,$$

$$f_\times^3 = f_P^3 (F_{P3}^2) f_Q^1 (F_{Q2}^1) + f_P^1 (F_{P2}^1) f_Q^2 (F_{Q3}^2) = 1 \cdot 3 + 3 \cdot 1 = 6 = 6F_{\times 6}^3,$$

$$f_\times^4 = f_P^2 (F_{P3}^2) f_Q^2 (F_{Q3}^2) = 1 \cdot 1 = 1 = P_{\times 9}^4.$$

Thus, the product of a triangle by a triangle is 4 – polytope. It has 9 vertices, 18 edges, 6 triangles, 9 quadrangles and 6 triangle prisms. The existence of such polytope was proved of Ziegler (1995) independently.

In particular, the fundamental difference between this polytope and simplicial polytopes and between this polytope and three - dimensional polyhedrons because of its multidimensionality is indicated of Ziegler (1995). The structural formula of this polytope (the triangular prismahedron) has the form

$$P_3^2 (F_{P3}^2) \times Q_3^2 (F_{Q3}^2) = P_9^4 [6P_6^3 (3F_4^2, F_3^2)]. \quad (5.16)$$

The image of a triangular prismahedron (Zhizhin, 2017) is shown in Figure 5.

Figure 5. The triangular prismahedron

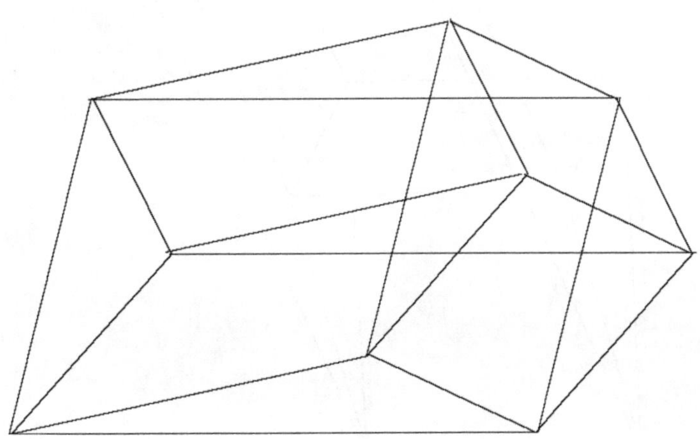

The Product of a Polygon by a Triangle

According to theorem 5.1 and ratios (5.5) for the product $P_i^2(F_{Pi}^2) \times Q_3^2(F_{Q3}^2)$ (i is the number of the polygon sides) there are

$$f_\times^0 = f_P^0 f_Q^0 = i \cdot 3 = 3i,$$

$$f_\times^1 = f_P^1 f_Q^0 + f_P^0 f_Q^1 = i \cdot 3 + 3 \cdot i = 6i, .$$

$$f_\times^2 = f_P^2(F_{Pi}^2) f_Q^0(F_{Q1}^0) + f_P^1(F_{P2}^1) f_Q^1(F_{Q2}^1) + f_P^0(F_{P1}^0) f_Q^2(F_{Q3}^2) = 1 \cdot 3 + i \cdot 3 + i \cdot 1 = 3 + 4i = 3F_{\times i}^2 + 3iF_{\times 4}^2 + iF_{\times 3}^2,$$

$$f_\times^3 = f_P^3(F_{Pi}^2) f_Q^1(F_{Q2}^1) + f_P^1(F_{P2}^1) f_Q^2(F_{Q3}^2) = 1 \cdot 3 + i \cdot 1 = 3 + i = iF_{\times 6}^3 + 3F_{\times 2i}^3,$$

$$f_\times^4 = f_P^2(F_{Pi}^2) f_Q^2(F_{Q3}^2) = 1 \cdot 1 = 1 = P_{\times 3i}^4.$$

*Figure 6. The 4*3 – angular prismahedron*

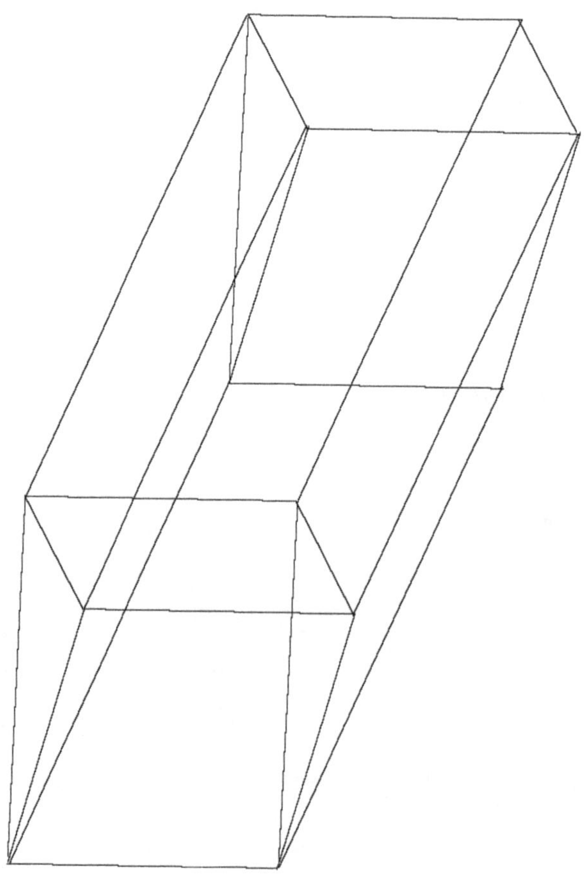

Thus, the product of the polygon F_i^2 by the triangle F_3^2 is a 4 – polytope. It has 3i vertices, 6i edges, i triangles, 3i quadrangles, 3 i-angles, 3 prisms with bases F_i^2 and i prisms with triangle bases. The structural formula of this product has the form

$$P_i^3(F_{Pi}^2)\times Q_3^2(F_{Q3}^2)=P_{3i}^4[3F_{2i}^3(\mathrm{i}\,F_4^2,2F_3^2),\mathrm{i}\,F_6^3(3F_4^2,2F_3^2)]. \qquad (5.17)$$

Examples of products of a square, a pentagon and a hexagon by a triangle are shown in Figures 6-8.

*Figure 7. The 5*3 – angular prismahedron*

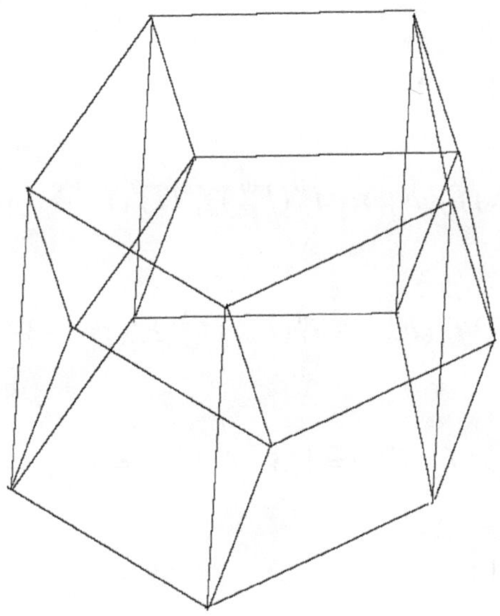

*Figure 8. The 6*3 – angular prismahedron*

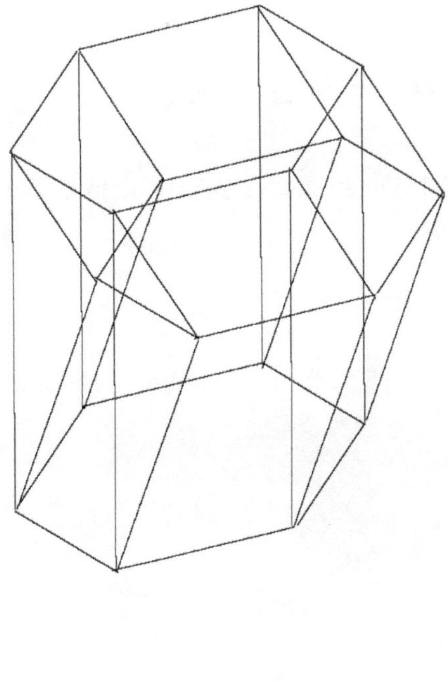

The polytope (5.17) is a generalization of the polytope (5.16) in the case when one of the multiplied polygons has the number of sides greater than 3. This polytope in the general case can be called 3 *i - angular prismahedron. Therefore, the polytope (5.16) can be described more accurately as 3*3 - angular prismahedron.

The Product of a Tetrahedron by a Triangle

According to theorem 5.1 and ratios (5.5) for the numbers of elements of different dimension in the product $P_4^3(4F_3^2)\times Q_3^2$ there are

$$f_\times^0 = f_P^0 f_Q^0 = 4 \cdot 3 = 12,$$

$$f_\times^1 = f_P^1 f_Q^0 + f_P^0 f_Q^1 = 6\cdot 3 + 3\cdot 4 = 30,.$$

$$f_\times^2 = f_P^2(F_{P3}^2)f_Q^0(F_{Q1}^0) + f_P^1(F_{P2}^1)f_Q^1(F_{Q2}^1) + f_P^0(F_{P1}^0)f_Q^2(F_{Q3}^2) = 4\cdot 1 + 6\cdot 3 + 4\cdot 1 = 26 = 8F_{\times 3}^2 + 18F_{\times 4}^2,$$

$$f_\times^3 = f_P^3(F_{P4}^3)f_Q^0(F_{Q1}^0) + f_P^2(F_{P3}^2)f_Q^1(F_{Q2}^1) + f_P^1(F_{P2}^1)f_Q^2(F_{Q3}^2) = 1\cdot 3 + 4\cdot 3 + 6\cdot 1 = 21 = 18F_{\times 6}^3 + 3F_{\times 4}^3,$$

$$f_\times^4 = f_P^3(F_{P4}^3)f_Q^1(F_{Q2}^1) + f_P^2(F_{P3}^2)f_Q^2(F_{Q3}^2) = 1\cdot 3 + 4\cdot 1 = 7 = 3F_{\times 8}^4 + 4F_{\times 9}^4,$$

$$f_\times^5 = f_P^3(F_{P4}^3)f_Q^2(F_{Q3}^2) = 1\cdot 1 = 1 = P_{\times 12}^5.$$

Thus, the product of a tetrahedron by a triangle is a 5-polytope. It has 12 vertices, 30 edges, 8 triangles, 18 quadrangles, 18 triangular prisms, 3 tetrahedrons. Facets of it including 3 tetrahedron prisms (5.8) and 4 triangular prismahedrons (5.16). The structural formula of this polytope has the form

$$P_4^3(4F_3^2)]\times Q_3^2 = P_{12}^5\{4F_9^4[6F_6^3(3F_4^2, 2F_3^2), 2F_8^4[4F_6^3(3F_4^2, 2F_3^2), 2F_4^3(4F_3^2)]\}. \qquad (5.18)$$

One can to call the polytope (5.18) a tetrahedral prismahedron. The image of this polytope on a two-dimension plane is shown in Figure 9.

The Product of an Octahedron by a Triangle

According to theorem 5.1 and ratios (5.5) for the numbers of elements of different dimension in the product $P_6^3(8F_3^2)\times Q_3^2$ there are

$$f_\times^0 = f_P^0 f_Q^0 = 6 \cdot 3 = 18,$$

Figure 9. The tetrahedral prismahedron

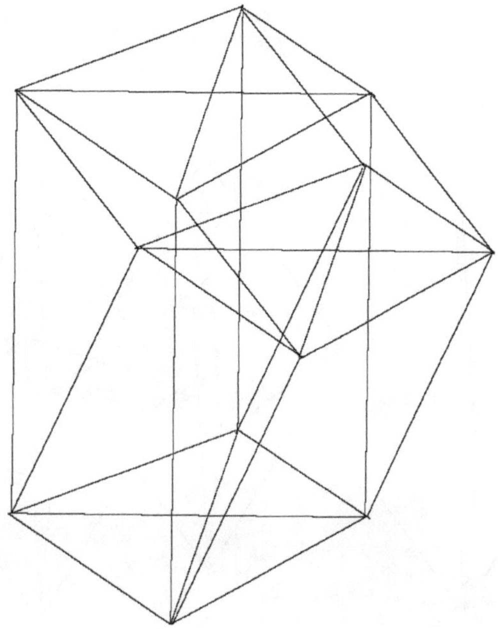

$$f_{\times}^{1}=f_{P}^{1}f_{Q}^{0}+f_{P}^{0}f_{Q}^{1}=12\cdot 3+6\cdot 3=54,.$$

$$f_{\times}^{2}=f_{P}^{2}(F_{P3}^{2})f_{Q}^{0}(F_{Q1}^{0})+f_{P}^{1}(F_{P2}^{1})f_{Q}^{1}(F_{Q2}^{1})+f_{P}^{0}(F_{P1}^{0})f_{Q}^{2}(F_{Q3}^{2})=8\cdot 1+12\cdot 3+6\cdot 1=66=30F_{\times 3}^{2}+36F_{\times 4}^{2},$$

$$f_{\times}^{3}=f_{P}^{3}(F_{P6}^{3})f_{Q}^{0}(F_{Q1}^{0})+f_{P}^{2}(F_{P3}^{2})f_{Q}^{1}(F_{Q2}^{1})+f_{P}^{1}(F_{P2}^{1})f_{Q}^{2}(F_{Q3}^{2})$$
$$=1\cdot 3+8\cdot 3+12\cdot 1=39=36F_{\times 6}^{3}(3F_{4}^{2},2F_{3}^{2})+3F_{\times 6}^{3}(8F_{3}^{2}),$$

$$f_{\times}^{4}=f_{P}^{3}[F_{P6}^{3}(8F_{3}^{2})]f_{Q}^{1}(F_{Q2}^{1})+f_{P}^{2}(F_{P3}^{2})f_{Q}^{2}(F_{Q3}^{2})=1\cdot 3+8\cdot 1=11$$
$$=3F_{\times 12}^{4}[8F_{6}^{3}(3F_{4}^{2},2F_{3}^{2}),2F_{6}^{3}(8F_{3}^{2})]+8F_{\times 9}^{4}[6F_{6}^{3}(3F_{4}^{2},2F_{3}^{2})],$$

$$f_{\times}^{5}=f_{P}^{3}[F_{P6}^{3}(8F_{3}^{2})]f_{Q}^{2}(F_{Q3}^{2})=1\cdot 1=1=P_{\times 18}^{5}.$$

Thus, the product of an octahedron by a triangle is a 5 – polytope. It has 18 vertices, 54 edges, 30 triangles, 36 quadrangles, 3 octahedrons, 36 triangular prisms. Facets of it including 8 the 3*3 - angular prismahedrons (5.16) and 3 the octahedral prisms (5.10).

The structural formula of the polytope P_{18}^{5} has the form

$$P_{6}^{3}(8F_{3}^{2})]\times Q_{3}^{2}=P_{18}^{5}\{8F_{9}^{4}[6F_{6}^{3}(3F_{4}^{2},2F_{3}^{2}),3F_{12}^{3}[8F_{6}^{3}(3F_{4}^{2},2F_{3}^{2}),2F_{6}^{3}(8F_{3}^{2})]\}. \quad (5.19)$$

One can to call the polytope (5.19) an octahedral prismahedron. The image of the polytope P_{18}^5 is shown in Figure 10.

Figure 10. The octahedral prismahedron

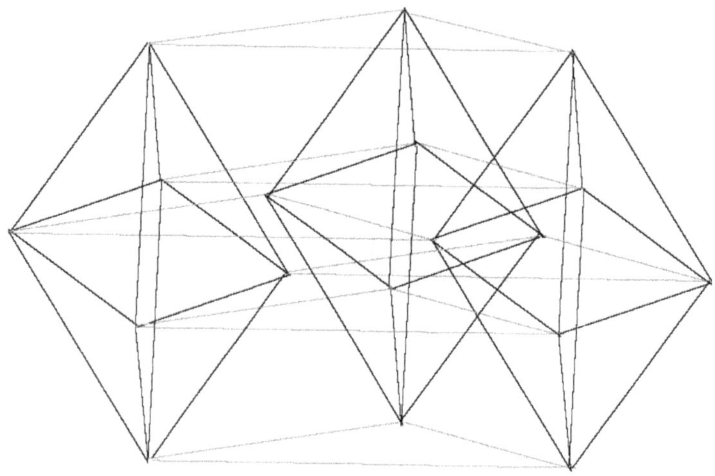

The Product of a Cube by a Triangle

According to theorem 5.1 and ratios (5.5) for the numbers of elements of different dimension in the product $P_8^3(6F_4^2)\times Q_3^2$ there are

$$f_\times^0 = f_P^0 f_Q^0 = 8 \cdot 3 = 24,$$

$$f_\times^1 = f_P^1 f_Q^0 + f_P^0 f_Q^1 = 12\cdot 3 + 8\cdot 3 = 60,.$$

$$f_\times^2 = f_P^2(F_{P4}^2) f_Q^0(F_{Q1}^0) + f_P^1(F_{P2}^1) f_Q^1(F_{Q2}^1) + f_P^0(F_{P1}^0) f_Q^2(F_{Q3}^2) = 6\cdot 3 + 12\cdot 3 + 8\cdot 1 = 62 = 8F_{\times 3}^2 + 54F_{\times 4}^2,$$

$$f_\times^3 = f_P^3(F_{P8}^3) f_Q^0(F_{Q1}^0) + f_P^2(F_{P4}^2) f_Q^1(F_{Q2}^1) + f_P^1(F_{P2}^1) f_Q^2(F_{Q3}^2) = 1\cdot 3 + 6\cdot 3 + 12\cdot 1 = 33 = 21F_{\times 8}^3 + 12F_{\times 6}^3,$$

$$f_\times^4 = f_P^3(F_{P8}^3) f_Q^1(F_{Q2}^1) + f_P^2(F_{P3}^2) f_Q^2(F_{Q3}^2) = 1\cdot 3 + 6\cdot 1 = 9 = 3F_{16}^4 + 6F_{12}^4,$$

$$f_\times^5 = f_P^3(F_{P8}^3) f_Q^2(F_{Q3}^2) = 1\cdot 1 = 1 = P_{\times 24}^5.$$

Thus, the product of a cube by a triangle is a 5 – polytope. It has 24 vertices, 60 edges, 8 triangles, 54 quadrangles, 21 cubes, 12 triangular prisms. Facets of it including 3 the hyper - cubes (5.9) and 6 the 3*4 - angular prismahedrons (5.17).

The structural formula of the polytope P_{24}^5 has the form

$$P_{24}^5\{3F_{16}^4(80F_4^2),6F_{12}^4[3F_8^3(4F_4^2,2F_3^2),4F_6^3(3F_4^2,2F_3^2)]\}. \tag{5.20}$$

One can call the polytope (5.20) a cube - polytopic prismahedron. The image of the polytope P_{24}^5 is shown in Figure 11.

Figure 11. The cube - polytopic prismahedron

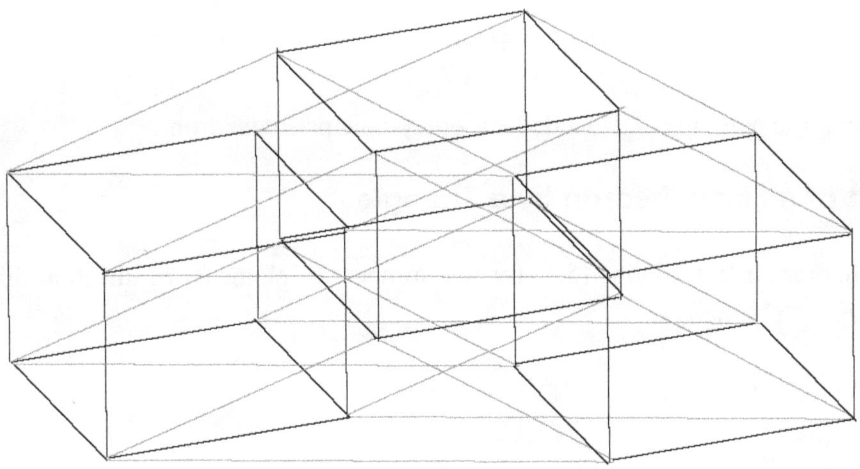

The Product of a Dodecahedral by a Triangle

According to theorem 5.1 and ratios (5.5) for the numbers of elements of different dimension in the product $P_{20}^3(12F_5^2)\times Q_3^2$ there are

$$f_\times^0 = f_P^0 f_Q^0 = 20 \cdot 3 = 60,$$

$$f_\times^1 = f_P^1 f_Q^0 + f_P^0 f_Q^1 = 30\cdot 3 + 20\cdot 3 = 150,.$$

$$f_\times^2 = f_P^2(F_{P5}^2) f_Q^0(F_{Q1}^0) + f_P^1(F_{P2}^1) f_Q^1(F_{Q2}^1) + f_P^0(F_{P1}^0) f_Q^2(F_{Q3}^2)$$
$$= 12\cdot 3 + 30\cdot 3 + 20\cdot 1 = 146 = 36F_{\times 5}^2 + 90F_{\times 4}^2 + 20F_{\times 3}^2,$$

$$f_\times^3 = f_P^3(F_{P20}^3)f_Q^0(F_{Q1}^0) + f_P^2(F_{P5}^2)f_Q^1(F_{Q2}^1) + f_P^1(F_{P2}^1)f_Q^2(F_{Q3}^2)$$
$$= 1\cdot 3 + 12\cdot 3 + 30\cdot 1 = 69 = 3F_{\times 20}^3 + 36F_{\times 10}^3 + 30F_{\times 6}^3,$$

$$f_\times^4 = f_P^3(F_{P20}^3)f_Q^1(F_{Q2}^1) + f_P^2(F_{P5}^2)f_Q^2(F_{Q3}^2) = 1\cdot 3 + 12\cdot 1 = 15 = 3F_{\times 40}^4 + 12F_{\times 15}^4,$$

$$f_\times^5 = f_P^3(F_{P20}^3)f_Q^2(F_{Q3}^2) = 1\cdot 1 = 1 = P_{\times 60}^5.$$

Thus, the product of a dodecahedron by a triangle is a 5 - polytope including 3 the dodecahedral prisms (5.12) and 12 the 3*5 - angular prismahedrons (5.17).

The structural formula of the polytope P_{60}^5 has the form

$$P_{60}^5\{3F_{40}^4[12F_{10}^3(5F_4^2,2F_5^2),2F_{20}^3(12F_5^2)],12F_{15}^4[3F_{10}^3(5F_4^2,2F_5^2),5F_6^3(3F_4^2,2F_3^2)]\}. \quad (5.21)$$

One can to call the polytope (5.21) a dodeca - polytopic prismahedron.

The Product of an Icosahedron by a Triangle

According to theorem 5.1 and ratios (5.5) for the numbers of elements of different dimension in the product $P_{12}^3(20F_5^2)\times Q_3^2$ there are

$$f_\times^0 = f_P^0 f_Q^0 = 12 \cdot 3 = 36,$$

$$f_\times^1 = f_P^1 f_Q^0 + f_P^0 f_Q^1 = 30\cdot 3 + 12\cdot 3 = 126,.$$

$$f_\times^2 = f_P^2(F_{P3}^2)f_Q^0(F_{Q1}^0) + f_P^1(F_{P2}^1)f_Q^1(F_{Q2}^1) + f_P^0(F_{P1}^0)f_Q^2(F_{Q3}^2)$$
$$= 20\cdot 3 + 30\cdot 3 + 12\cdot 1 = 162 = 90F_{\times 4}^2 + 72F_{\times 3}^2,$$

$$f_\times^3 = f_P^3(F_{P12}^3)f_Q^0(F_{Q1}^0) + f_P^2(F_{P3}^2)f_Q^1(F_{Q2}^1) + f_P^1(F_{P2}^1)f_Q^2(F_{Q3}^2)$$
$$= 1\cdot 3 + 20\cdot 3 + 30\cdot 1 = 93 = 3F_{\times 12}^3 + 90F_{\times 6}^3,$$

$$f_\times^4 = f_P^3(F_{P12}^3)f_Q^1(F_{Q2}^1) + f_P^2(F_{P3}^2)f_Q^2(F_{Q3}^2) = 1\cdot 3 + 20\cdot 1 = 23 = 3F_{\times 24}^4 + 20F_{\times 9}^4,$$

$$f_\times^5 = f_P^3(F_{P12}^3)f_Q^2(F_{Q3}^2) = 1\cdot 1 = 1 = P_{\times 36}^5.$$

Thus, the product of an icosahedron by a triangle is a 5 – polytope. It has 60 vertices, 126 edges, 72 triangles, 90 quadrangles, 3 dodecahedrons, 90 triangular prisms. Facets of it including 3 the icosahedral prisms (5.11) and 20 the 3*3 - angular prismahedrons (5.17).

The structural formula of the polytope P^5_{36} has the form

$$P^5_{36}\{20F^4_9[6F^3_6(3F^2_4,2F^2_3)],3F^4_{24}[20F^3_6(3F^2_4,2F^2_3),2F^3_{12}(20F^2_3)]\}. \quad (5.22)$$

One can call the polytope (5.22) an icosa - polytopic prismahedron.

The Product of a 4 - Simplex by a Triangle

According to theorem 5.1 and ratios (5.5) for the numbers of elements of different dimension in the product $P^4_5[5F^3_4(4F^2_3)]\times Q^2_3$ there are

$$f^0_\times = f^0_P f^0_Q = 5\cdot 3 = 15,$$

$$f^1_\times = f^1_P f^0_Q + f^0_P f^1_Q = 10\cdot 3+5\cdot 3 = 45,.$$

$$f^2_\times = f^2_P(F^2_{P3})f^0_Q(F^0_{Q1}) + f^1_P(F^1_{P2})f^1_Q(F^1_{Q2}) + f^0_P(F^0_{P1})f^2_Q(F^2_{Q3})$$
$$=10\cdot 3+10\cdot 3+5\cdot 1 = 65 = 30F^2_{\times 4}+35F^2_{\times 3},$$

$$f^3_\times = f^3_P(F^3_{P4})f^0_Q(F^0_{Q1}) + f^2_P(F^2_{P3})f^1_Q(F^1_{Q2}) + f^1_P(F^1_{P2})f^2_Q(F^2_{Q3})$$
$$=5\cdot 3+10\cdot 3+10\cdot 1 = 55 = 15F^3_{\times 4}+40F^3_{\times 6},$$

$$f^4_\times = f^4_P(F^4_{P5})f^0_Q(F^0_{Q1}) + f^3_P(F^3_{P4})f^1_Q(F^1_{Q2}) + f^2_{P3}(F^2_P)f^2_Q(F^2_{Q3})$$
$$=1\cdot 3+5\cdot 3+10\cdot 1 = 28 = 3F^4_{\times 5}+15F^4_{\times 8}+10F^4_{\times 9},$$

$$f^5_\times = f^4_P(F^4_{P5})f^1_Q(F^1_{Q2}) + f^3_P(F^3_{P4})f^2_Q(F^2_{Q3}) = 1\cdot 3+5\cdot 1 = 8 = 3F^5_{\times 10}+5F^5_{\times 12},$$

$$f^6_\times = f^4_P(F^4_{P5})f^2_Q(F^2_{Q3}) = 1\cdot 1 = 1 = P^6_{\times 15}.$$

Thus, the product of a 4 - simplex by a triangle is a 6 – polytope. It has 15 vertices, 45 edges, 35 triangles, 30 quadrangles, 15 tetrahedrons, 40 triangular prisms, 3 polytopes 4 – simplex, 15 tetrahedron prisms, 10 triangular prismahedrons. Facets of it including 3 the 5 – simplex - polytopic prisms (5.13) and 5 the tetrahedral prismahedrons (5.18).

The structural formula of the polytope P^6_{15} has the form

$$P_{15}^{6}\{3F_{10}^{5}[5F_{8}^{4}[4F_{4}^{3}(3F_{4}^{2},2F_{3}^{2}),2F_{4}^{3}(4F_{3}^{2})],2F_{5}^{4}[5F_{4}^{3}(4F_{3}^{2})]],$$
$$5F_{12}^{5}[4F_{9}^{4}[6F_{6}^{3}(3F_{4}^{2},2F_{3}^{2}),3F_{8}^{4}[4F_{6}^{3}(3F_{4}^{2},2F_{3}^{2}),2F_{4}^{3}(4F_{3}^{2})]]\}. \tag{5.23}$$

One can call the polytope (5.23) a 6 – simplex - polytopic prismahedron. The image of this polytope is shown in Figure 12.

Figure 12. The 6 – simplex - polytopic prismahedron

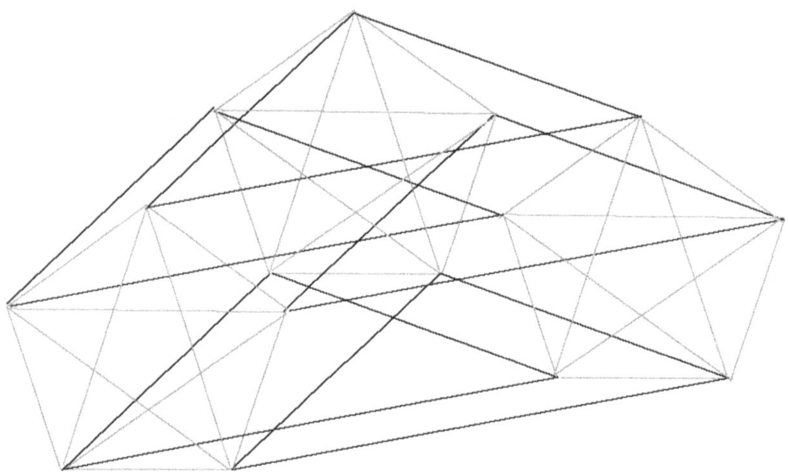

The Product of a 4 – Cross - Polytope by a Triangle

According to theorem 5.1 and ratios (5.5) for the numbers of elements of different dimension in the product $P_{8}^{4}[16F_{4}^{3}(4F_{3}^{2})]\times Q_{3}^{2}$ there are

$$f_{\times}^{0}=f_{P}^{0}f_{Q}^{0}=8\cdot 3=24,$$

$$f_{\times}^{1}=f_{P}^{1}f_{Q}^{0}+f_{P}^{0}f_{Q}^{1}=24\cdot 3+8\cdot 3=96,$$

$$f_{\times}^{2}=f_{P}^{2}(F_{P3}^{2})f_{Q}^{0}(F_{Q1}^{0})+f_{P}^{1}(F_{P2}^{1})f_{Q}^{1}(F_{Q2}^{1})+f_{P}^{0}(F_{P1}^{0})f_{Q}^{2}(F_{Q3}^{2})$$
$$=32\cdot 3+24\cdot 3+8\cdot 1=176=72F_{\times 4}^{2}+104F_{\times 3}^{2},$$

$$f_{\times}^{3}=f_{P}^{3}(F_{P4}^{3})f_{Q}^{0}(F_{Q1}^{0})+f_{P}^{2}(F_{P3}^{2})f_{Q}^{1}(F_{Q2}^{1})+f_{P}^{1}(F_{P2}^{1})f_{Q}^{2}(F_{Q3}^{2})$$
$$=16\cdot 3+32\cdot 3+24\cdot 1=168=48F_{\times 4}^{3}+120F_{\times 6}^{3},$$

$$f_{\times}^{4} = f_{P}^{4}(F_{P8}^{4}) f_{Q}^{0}(F_{Q1}^{0}) + f_{P}^{3}(F_{P4}^{3}) f_{Q}^{1}(F_{Q2}^{1}) + f_{P3}^{2}(F_{P}^{2}) f_{Q}^{2}(F_{Q3}^{2})$$
$$= 1\cdot 3 + 16\cdot 3 + 32\cdot 1 = 83 = 51F_{\times 8}^{4} + 32F_{\times 9}^{4},$$

$$f_{\times}^{5} = f_{P}^{4}(F_{P8}^{4}) f_{Q}^{1}(F_{Q2}^{1}) + f_{P}^{3}(F_{P4}^{3}) f_{Q}^{2}(F_{Q3}^{2}) = 1\cdot 3 + 16\cdot 1 = 19 = 3F_{\times 16}^{5} + 16F_{\times 12}^{5},$$

$$f_{\times}^{6} = f_{P}^{4}(F_{P8}^{4}) f_{Q}^{2}(F_{Q3}^{2}) = 1\cdot 1 = 1 = P_{\times 24}^{6}.$$

Thus, the product of a 4 – cross - polytope by a triangle is a 6-polytope. It has 24 vertices, 96 edges, 104 triangles, 72 quadrangles, 48 tetrahedrons, 120 triangular prisms, 51 tetrahedral prisms, 32 triangular prismahedrons. Facets of it including 3 the 5 – cross -polytopic prisms (5.10) and 16 the tetrahedral prismahedrons (5.3).

The structural formula of the polytope P_{24}^{6} has the form

$$P_{24}^{6}\{3F_{16}^{5}[16F_{8}^{4}[4F_{6}^{3}(3F_{4}^{2},2F_{3}^{2}),2F_{4}^{3}(4F_{3}^{2})],2F_{8}^{4}[16F_{4}^{3}(4F_{3}^{2})]],$$
$$16F_{12}^{5}[4F_{9}^{4}[6F_{6}^{3}(3F_{4}^{2},2F_{3}^{2}),3F_{8}^{4}[4F_{6}^{3}(3F_{4}^{2},2F_{3}^{2}),2F_{4}^{3}(4F_{3}^{2})]]\}.$$

One can call the polytope a 6 - cross - polytopic prismahedron.

The General Structural Formula for the Product of Polytopes by a Triangle

If we pay attention to a polytope taking into account its dimensions and the number of vertices and the number of its facets and the number of vertices of every facet, we can prove the general rule of changing these parameters of a polytope for its multiplication by a triangle.

Theorem 5.3: Let's assume that a convex polytope of dimension *n,* with the number of vertices *m,* has a number of facets ν with vertex number *k* for every facet. Then its product by a triangle is defined by the equation

$$P_{m}^{n}(\nu F_{k}^{n-1}) \times Q_{3}^{2} = P_{3m}^{n+2}(\nu F_{3k}^{n+1}, 3F_{2m}^{n+1}). \qquad (5.24)$$

Proof. When the polytope P_{m}^{n} is multiplied by the triangle Q_{3}^{2}, the number of vertices of the polytope is multiplied by 3 since the triangle has 3 edges and each edge at this multiplication translates polytope vertices in its direction. At that the dimension of the polytope naturally, according to the definition (Ziegler, 1995), is increased by two. Thus, one will get a polytope P_{3m}^{n+2}. If one multiply the polytope by the triangle, its facets F_{k}^{n-1} are multiplied by the triangle also. The number of these facets of the product is equal to the number of facets of the polytope P_{m}^{n}, i.e. it is equal to ν. In addition, at the multiplication by the triangle the polytope P_{m}^{n} is multiplied by each edge (i.e. multiplication by Q_{2}^{1} 3 times). Therefore, 3 facets F_{2m}^{n+1} more appear as facets of the polytope P_{3m}^{n+2}. It is easy to make sure that all the considered above products of the polytopes by the triangle satisfy the equality (5.24).

Corollary: If the polygon Q_t^2 with the number of vertices t is the polytope factor in (5.24), the equation (5.24) can be rewritten as

$$P_m^n(\nu F_k^{n-1})\times Q_t^2 = P_{tm}^{n+2}(\nu F_{tk}^{n+1}, \mathrm{t} F_{2m}^{n+1}). \tag{5.25}$$

As an example, the result of the product of a tetrahedron by a square is given in Figure 13.

Figure 13. The product of a tetrahedron by a square

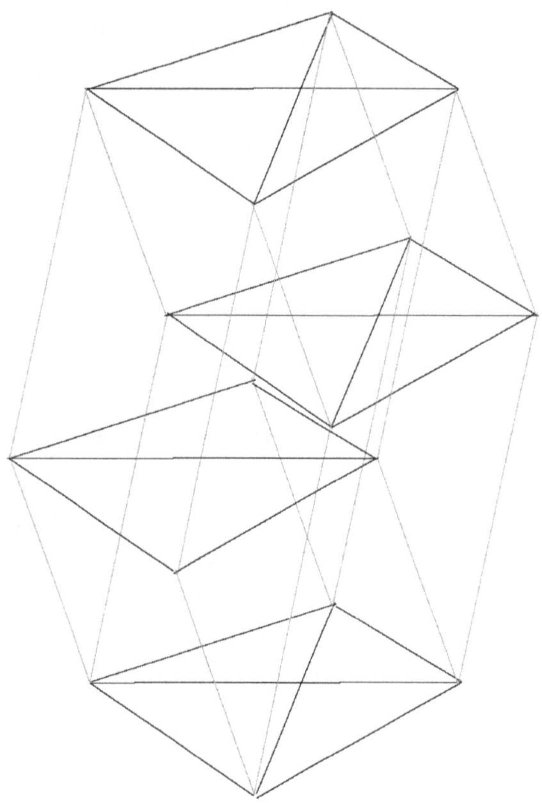

THE PRODUCTS OF POLYTOPES BY A TETRAHEDRON

The Product of a Tetrahedron by a Tetrahedron

According to theorem 5.1 and ratios (5.5) for the numbers of elements of different dimension in the product $P_4^3(4F_3^2)\times Q_4^3(4F_3^2)$ there are

$f_\times^0 = f_P^0 f_Q^0 = 4\cdot 4 = 16,$

$$f_P^0 f_Q^1 = 6\cdot 4 + 4\cdot 6 = 48,$$

$$f_\times^2 = f_P^2(F_{P3}^2)f_Q^0(F_{Q1}^0) + f_P^1(F_{P2}^1)f_Q^1(F_{Q2}^1) + f_P^0(F_{P1}^0)f_Q^2(F_{Q3}^2)$$
$$= 4\cdot 4 + 6\cdot 4 + 4\cdot 4 = 68 = 36F_{\times 4}^2 + 32F_{\times 3}^2,$$

$$f_\times^3 = f_P^3(F_{P4}^3)f_Q^0(F_{Q1}^0) + f_P^2(F_{P3}^2)f_Q^1(F_{Q2}^1) + f_P^1(F_{P2}^1)f_Q^2(F_{Q3}^2) + f_P^0(F_{P1}^0)f_Q^3(F_{Q4}^3)$$
$$= 1\cdot 4 + 4\cdot 6 + 6\cdot 4 + 4\cdot 1 = 56 = 8F_{\times 4}^3 + 48F_{\times 6}^3,$$

$$f_\times^4 = f_P^3(F_{P4}^3)f_Q^1(F_{Q1}^1) + f_P^2(F_{P3}^2)f_Q^2(F_{Q3}^2) + f_P^1(F_{P2}^1)f_Q^3(F_{Q4}^3)$$
$$= 1\cdot 6 + 4\cdot 4 + 6\cdot 1 = 28 = 12F_{\times 8}^4 + 16F_{\times 9}^4,$$

$$f_\times^5 = f_P^3(F_{P4}^3)f_Q^2(F_{Q3}^2) + f_P^2(F_{P3}^2)f_Q^3(F_{Q4}^3) = 1\cdot 4 + 4\cdot 1 = 8 = 8F_{\times 12}^5,$$

$$f_\times^6 = f_P^3(F_{P4}^3)f_Q^2(F_{Q4}^3) = 1\cdot 1 = 1 = P_{\times 16}^6.$$

Thus, the product of a tetrahedron by a tetrahedron is a 6 – polytope. It has 16 vertices, 48 edges, 32 triangles, 36 quadrangles, 8 tetrahedrons, 48 triangular prisms, 12 tetrahedral prisms, 16 triangular prismahedrons. Facets of it including 8 the tetrahedral prismahedrons (5.18).

The short structural formula of the polytope P_{16}^6 has the form

$$P_{16}^6(8F_{12}^5) \tag{5.26}$$

The image of this polytope is shown in Figure 14.

One can call the polytope (5.26) a 6 - complex of tetrahedral prismahedrons.

The Product of an Octahedron by a Tetrahedron

According to theorem 5.1 and ratio (5.5) for the numbers of elements of different dimension in the product $P_6^3(8F_{P3}^2)\times Q_4^3(4F_{Q3}^2)$ there are

$$f_\times^0 = f_P^0 f_Q^0 = 6\cdot 4 = 24,$$

$$f_P^0 f_Q^1 = 12\cdot 4 + 6\cdot 6 = 84,$$

$$f_\times^2 = f_P^2(F_{P3}^2)f_Q^0(F_{Q1}^0) + f_P^1(F_{P2}^1)f_Q^1(F_{Q2}^1) + f_P^0(F_{P1}^0)f_Q^2(F_{Q3}^2)$$
$$= 8\cdot 4 + 12\cdot 6 + 6\cdot 4 = 128 = 72F_{\times 4}^2 + 56F_{\times 3}^2,$$

Figure 14. The product of a tetrahedron by a tetrahedron

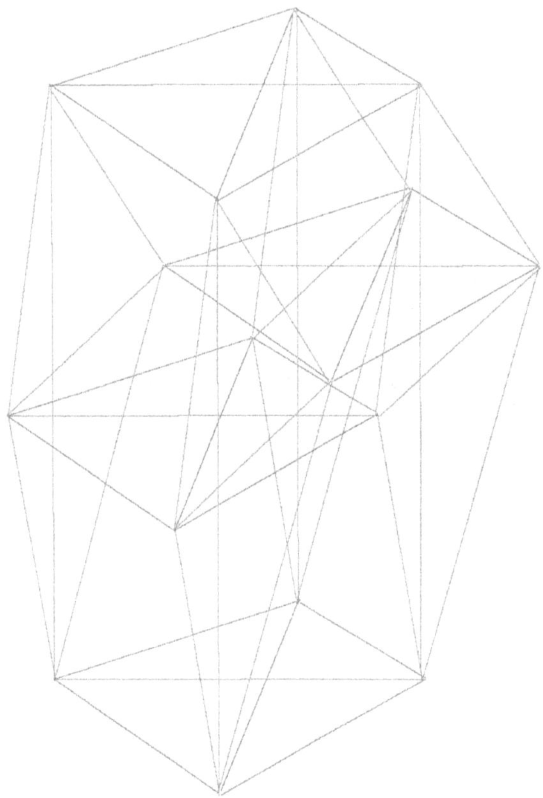

$$f_{\times}^{3} = f_{P}^{3}(F_{P6}^{3}) f_{Q}^{0}(F_{Q1}^{0}) + f_{P}^{2}(F_{P3}^{2}) f_{Q}^{1}(F_{Q2}^{1}) + f_{P}^{1}(F_{P2}^{1}) f_{Q}^{2}(F_{Q3}^{2}) + f_{P}^{0}(F_{P1}^{0}) f_{Q}^{3}(F_{Q4}^{3})$$
$$= 1\cdot 4 + 8\cdot 6 + 12\cdot 4 + 6\cdot 1 = 106 = 6F_{\times 4}^{3} + 100F_{\times 6}^{3},$$

$$f_{\times}^{4} = f_{P}^{3}(F_{P6}^{3}) f_{Q}^{1}(F_{Q1}^{1}) + f_{P}^{2}(F_{P3}^{2}) f_{Q}^{2}(F_{Q3}^{2}) + f_{P}^{1}(F_{P2}^{1}) f_{Q}^{3}(F_{Q4}^{3})$$
$$= 1\cdot 6 + 8\cdot 4 + 12\cdot 1 = 50 = 12F_{\times 8}^{4} + 32F_{\times 9}^{4} + 6F_{\times 12}^{4},$$

$$f_{\times}^{5} = f_{P}^{3}(F_{P6}^{3}) f_{Q}^{2}(F_{Q3}^{2}) + f_{P}^{2}(F_{P3}^{2}) f_{Q}^{3}(F_{Q4}^{3}) = 1\cdot 4 + 8\cdot 1 = 12 = 8F_{\times 12}^{5} + 4F_{\times 18}^{5},$$

$$f_{\times}^{6} = f_{P}^{3}(F_{P6}^{3}) f_{Q}^{2}(F_{Q4}^{3}) = 1\cdot 1 = 1 = P_{\times 24}^{6}.$$

Thus, the product of an octahedron by a tetrahedron is a 6 – polytope. It has 24 vertices, 84 edges, 56 triangles, 72 quadrangles, 6 tetrahedrons, 100 triangular prisms, 12 tetrahedral prisms, 32 triangular prismahedrons, 6 the 3*4 - angular prismahedrons (5.17). Facets of it including 8 the tetrahedral prismahedrons (5.18) and 4 the octahedral polytopic prisms (5.19).

The short structural formula of the polytope P_{24}^{6} has the form

$$P_{24}^{6}(4F_{18}^{5}, 8F_{12}^{5}). \tag{5.27}$$

One can call the polytope P_{24}^{6} a 6 - complex of the tetrahedral prismahedrons and the octahedral polytopic prisms.

THE PRODUCNT OF POLYTOPES WITH UNIFORM FACETS

Comparing (5.26) and (5.27) with equalities defined by theorems 1, 2 we note their general structure. This allows us to formulate and prove a general theorem on the product of polytopes with uniform facets.

Theorem 5.4: Let's assume that there is a convex polytope P of dimension n, with the number of vertices m, which has ν facets with the number of vertices k for each facet. Let's also there is set a convex polytope Q of dimension π, with the number of vertices l, having μ facets with the number of vertices λ for every facet. Then the product of these polytopes is defined by equation

$$P_{m}^{n}(\nu F_{k}^{n-1}) \times Q_{l}^{\pi}(\mu F_{\lambda}^{\pi-1}) = P_{lm}^{n+\pi}(\nu F_{lk}^{n+\pi-1}, \mu F_{\lambda m}^{n+\pi-1}). \tag{5.28}$$

Proof. For multiplication of the polytope P_{m}^{n} by the polytope Q_{l}^{π}, the number of the vertices of polytope P is multiplied by the number of the vertices of polytope Q since each vertex of the polytope P at multiplication will correspond to all vertices of the polytope Q. The dimension of the product polytope according to the definition (Ziegler, 1995) is the sum of the dimensions of polytopes-factors, i.e. we receive the polytope $P_{lm}^{n+\pi}$ as the result of the product. Multiplication of the polytope P by the polytope Q is accompanied also by multiplication of the polytope P facets by the polytope Q. That is why the result of this multiplication will be the facets of the polytopes product defined according to the same rules, i.e. $F_{lk}^{n+\pi-1}$ .The number of these facets is equal to the number of facets of the polytope P, i.e. ν. Besides, in the product of the polytopes P and Q there are facets formed by the product of the facets of polytope Q by polytope P. These are facets $F_{lk}^{n+\pi-1}$, their number is equal to the number of facets of the polytope Q, i.e. μ. The theorem 5.4 is proved.

Let's use theorem 5.4 for the product of an octahedron by a tetrahedron

$$P_{6}^{3}(8F_{P3}^{2}) \times Q_{4}^{3}(4F_{Q3}^{2}) = P_{24}^{6}(4F_{18}^{5}, 8F_{12}^{5}).$$

The received formula coincides with (5.27) got after detailed derivation.

It's easy to verify that all the previous formulas about polytopes product are particular cases of the formula (5.28). Theorem 5.4 allows to write down at once the result of the product of any convex polytopes with uniform faces without detailed derivation of this result. Let's build the multiplication table of existing polytopes of dimensions 3 and 4. For comfort of presentation the table is divided into a few tables (tables 5.1 – 5.4) to cover the products of all above - mentioned polytopes. The structure of new formed facets can be defined according to already mentioned formulas (Zhizhin, 2019a).

BOUNDARY COMPLEXES AND INTERIOR POINTS OF THE PRODUCT OF POLYTOPES

Boundary Complexes of the Polytopic Prisms

In the monograph "The Geometry of High-Dimensional Polytopes" (Zhizhin, 2019 a), two theorems were proved regarding the incidence coefficients of elements of different dimensions in an n - simplex, n - cube, n - cross - polytope. These polytopes are often used in the products of polytopes. Therefore, it is of interest to determine the value of these theorems for polytopic prismahedrons. To state these theorems, we introduce additional notation. There is denoted $k_{b(i)}$ the incidence value of the element of the boundary complex b with dimension i to the elements of the boundary complex having dimension $i+1$. This value is equal to the number of elements of the boundary complex which the given element belongs having dimension one greater than the given element of the boundary complex .

Theorem 5.5. The incidence of an element of dimension i of the boundary complex of an n - cube and n – simplex is equal to the difference of the dimension of the polytope and the dimension of this element, i.e.

$$k_{b(i)}(n) = n - i. \tag{5.29}$$

Theorem 5.6: The incidence coefficients of the elements of the boundary complex of n – cross -polytopes of dimension i if $i < n - 1$ are defined by the equality $k_{b(i)}(n) = 2(n-1-i)$. If $i = n$ - 1, so

$$k_{b(n-1)}(n) = 1. \tag{5.30}$$

Now we consider in more detail the geometric properties of polytopic prisms. Let's turn to the simplest of polytopic prisms - the tetrahedral prism (Figure 5.1) and decide whether the tetrahedral prism has internal points, i.e. points not belonging to its boundary complex. For this purpose, in Figure 5.1, coordinate systems are introduced, as well as in Figure 2.6 of Chapter 2 (Zhizhin, 2019 a), when the tetrahedron is considered. Taking into account that in this case we are dealing with a space of dimension 4, a fourth coordinate t is added to the coordinate system parallel to the generating lines of the tetrahedral prism. Assuming the length of all edges of the tetrahedral prism per unit, one can establish the coordinates of all the vertices of the tetrahedral prism (Figure 5.1) in the given coordinate system (x, y, z, t)

$$b_1\left(0,\frac{1}{\sqrt{3}},0,0\right), b_2\left(\frac{1}{2},-\frac{1}{2\sqrt{3}},0,0\right), b_3\left(-\frac{1}{2},-\frac{1}{2\sqrt{3}},0,0\right), b_4\left(0,0,\sqrt{\frac{2}{3}},0\right),$$
$$b_5\left(0,\frac{1}{\sqrt{3}},0,1\right), b_6\left(\frac{1}{2},-\frac{1}{2\sqrt{3}},0,1\right), b_7\left(-\frac{1}{2},-\frac{1}{2\sqrt{3}},0,1\right), b_8\left(0,0,\sqrt{\frac{2}{3}},1\right).$$

Then the set of interior points of the tetrahedral prism is determined by the system of inequalities (see Chapter 2 (Zhizhin, 2019 a))

$z>0,\ 0<t<1;$

$$z<2\sqrt{\frac{2}{3}}x-\sqrt{2}y+\sqrt{\frac{2}{3}},-\frac{1}{2}<x<0,-\frac{1}{2\sqrt{3}}<y<\frac{1}{\sqrt{3}},0<\mathrm{t}<1;$$

$$z<2\sqrt{2}y+\sqrt{\frac{2}{3}},-\frac{1}{2}<x<\frac{1}{2};-\frac{1}{2\sqrt{3}}<y<0,0<t<1;$$

$$z<-\sqrt{6}x-\sqrt{2}y+\sqrt{\frac{2}{3}},0<x<\frac{1}{2},-\frac{1}{2\sqrt{3}}<y<\frac{1}{\sqrt{3}},0<t<1.$$

Similarly, in principle, one can define a set of internal points and other polytopic prisms.

Following from Figure 5.1 of this Chapter, we find the incidence of the elements of a boundary complex of different dimensions of a tetrahedral prism

$$k_{b(i=0)}(4)=4,k_{b(i=1)}(4)=3,k_{b(i=2)}(4)=2,k_{b(i=3)}(4)=1.$$

Thus, the incidence values of the elements of the boundary complex of the tetrahedral prism obey the formula (5.29) and Theorem 5.5.

The same incidence values have elements of the boundary complex of the pyramidal prism (Figure 5.2) with one exception: the tops of the Egyptian pyramids have an incidence value

$$k_{b(i=0)}(4)=5.$$

The polytopic prisms with increasing dimension, if they do not include cross - polytopes, also obey theorem 5.5 and formula (5.29). For example, the 5 – simplex - prism (Figure 5.3) has the following incidence of elements of its boundary complex

$$k_{b(i=0)}(5)=5,k_{b(i=1)}(5)=4,k_{b(i=2)}(5)=3,k_{b(i=3)}(5)=2,k_{b(i=4)}(5)=1.$$

This agrees with formula (5.29) and theorem 5.5.

However, if the cross - polytope is included in the boundary polytope prism complex, the situation changes. For example, Figure 5.4 shows an image of a 5 – cross - prism, which includes two the 4 – cross - polytopes. From this image it follows that in this case the incidence of the elements of the boundary complex have the values

$$k_{b(i=0)}(5)=7,k_{b(i=1)}(5)=\begin{cases}6\\5\end{cases},k_{b(i=2)}(5)=4,k_{b(i=3)}(5)=2,k_{b(i=4)}(5)=1.$$

The maximum value of incidence of edge 6 refers to the forming lines of the 5 – cross - prism, the minimum incidence value of edge 4 refers to the edges of the 4 – cross - polytopes.

The Incidence Values in the Polytopic Prismahedrons

The incidence of elements of the boundary complexes of polytopic prisms is also preserved for polytopic prismahedrons. For many of them Theorem 5.5 and formula (5.29) are satisfied. For example, for a polytopic prismahedron that is a product of a square and a tetrahedron (Figure 5.13), the incidence values of elements of a boundary complex of different dimensions have the form

$$k_{b(i=0)}(5)=5, k_{b(i=1)}(5)=4, k_{b(i=2)}(5)=3, k_{b(i=3)}(5)=2, k_{b(i=4)}(5)=1.$$

This completely coincides with Theorem 5.5 and formula (5.29).

For the polytopic prismahedron, which is the product of a tetrahedron on a tetrahedron (Figure 5.14), the incidence values of elements of a boundary complex of different dimensions have the form

$$k_{b(i=0)}(6)=6, k_{b(i=1)}(6)=5, k_{b(i=2)}(6)=4, k_{b(i=3)}(6)=3, k_{b(i=4)}(6)=2, k_{b(i=5)}(6)=1.$$

This also agrees with theorem 5.5 and formula (5.29).

However, for polytopic prismahedron in which a pyramid participates (or an octahedron that can be considered as a double pyramid), there is a difference from theorem 5.5 and the formula (5.29). This difference was already observed when considering the product of the Egyptian pyramid on a one - dimensional segment (Figure 5.3). The incidence of vertices in this case is greater than the value determined by theorem 5.5 and formula (5.29). For example, for a polytopic prismahedron that is the product of an octahedron per triangle (Figure 5.10), the incidence values of the elements of the boundary complex have the form

$$k_{b(i=0)}(5)=6, k_{b(i=1)}(5)=4, k_{b(i=2)}(5)=3, k_{b(i=3)}(5)=2, k_{b(i=4)}(5)=1.$$

The incidence of vertices in this case is greater than the dimension of the polytopic prismahedron.

The participation of a cross - polytopes as elements of the boundary complex of the polytopic prismahedrons significantly changes the incidence of the elements of the boundary complex, as this has already been noted when considering polytopic prisms. The influence of a cross - polytopes on the incidence of elements of the boundary complex in complicated polytopic prismahedrons has yet to be investigated.

CONCLUSION

A study of the geometric structure of the work of polytopes showed that, having the highest dimension, they have in their stream a series of families of edges parallel to each other. The number of these families increases with the dimensionality of the work of political scientists. This can serve as the basis for a possible normal difference in the n - dimensional space with their participation. It should be noted that the works of polytopes (polytopic prismahedrons) have an important feature. They have polytopes of different dimensions of different types as facets (in the same polytopic prismahedron). It is important that polytopic prismahedrons can have n - cross - polytopes as faces, which is impossible in simplicial polytopes. It is interesting to note that the incidence coefficients of elements of different dimensions in polytopic prismahedron grow linearly with increasing dimension of a polytopic prismahedron and decreases linearly with increasing dimension of the element.

The polytopic prismahedrons are the products of polytopes with different dimension. By their geometry, they will represent a complex of the polytopic prisms, each of which is a higher - dimensional prism with bases in the form of polytopes of higher dimension. As Zigler was noted (Zigler, 1995) the product of polytopes is not a simplex even if factors are simplexes. In this regard, the developed theory of simplicial polytopes (Alexandrov, 1975, 1979; Robertson, 1984; Fomenko, 1992) for analysis polytopic prismshedrons becomes inapplicable. The polytopic prismahedrons can be considered as a result of the chemical interaction of molecules, which, from among which there is a polytope of a certain dimension. The possibility of the multivaluedness of the incidence coefficients of geometric elements of polytopic prisms is revealed. It is shown that polytopic prismahedrons, due to their nature, as products of polytopes, ensure the filling of n - dimensional spaces during their translation. Thus, the polytopic prismahedron is the sterehedron, the existence of which was suggested by B.N. Delaunay (Delaunay, 1961; Delaunay & Sandakova, 1961). However, Delaunay never succeeded in constructing any particular stereohedron in a space with a dimension greater than three. In this chapter, tables of polytopic prismahedrons are given, their analytical expressions and geometric images are given also.

REFERENCES

Alexandrov, P. S. (1975). *Introduction to homological dimension theory and general combinatorial topology*. Science.

Alexandrov, P. S. (1979). *Selected work. The general theory of homology*. Science.

Delaunay, B. (1961). Proof of the main theorem of the theory of stereohedrons. *Reports of the Academy of Sciences of the USSR, 138*(6), 1270–1272.

Delaunay, B., & Sandakova, N. (1961). Theory of stereohedrons. *Proceedings of the Mathematical Institute. V.A. Steklov, 64*, 28 – 51.

Fomenko, A. T. (1992). *Visual geometry and topology. Mathematical images in the real world*. Moscow State University.

Ilyushin, G. D. (2003). *Modeling of self - organization processes in crystal-forming systems*. URSS.

Lord, E. A., Mackay, A. L., & Ranganathan, S. (2006). *New geometry for new materials*. Cambridge University Press.

Mukhopadhyay, N.K. (1993). Diffraction studies of icosahedral phases in $Al_{70}Fe_{20}W_{10}$. *Journal of Non-Crystalline Solids, 153 – 154*, 1193 – 1197.

Panina, G. Y. (2006). *The Algebra of Polyhedrons.* Academic Press.

Poincaré, A. (1895). Analysis situs. *J. de é Ecole Polytechnique, 1,* 1 – 121.

Pontryagin, L. S. (1976). *The foundations of combinatorial topology*. Science.

Robertson, S. A. (1984). *Polytopes and Symmetry*. Cambridge University Press.

Shechtman, D., Blech, I., Gratias, D., & Cahn, J. W. (1984). Metallic phase with longerange orientational order and no translational symmetry. *Physical Review Letters, 53*(20), 1951–1953. doi:10.1103/PhysRevLett.53.1951

Shevchenko, V. Ya., Zhizhin, G. V., & Mackay, A. (2013). On the structure of the quasi - crystals in the high dimension space. In M. V. Diudea (Ed.), *Diamonds and related nanostructures* (pp. 311–320). Springer. doi:10.1007/978-94-007-6371-5_17

Silk, J. (1980). *The big bang. The creation and evolution of the Universe*. W.H. Freeman and Company.

Snegirev, I. L., & Frantsev, Yu. P. (1938). *Ancient Egypt*. OGIZ.

Zang, X., & Kelton, K. (1993). High-order crystal approximant alloys $Ti_{54}Zr_{26}Ni_{20}$. *Journal of Non-Crystalline Solids, 153-154*, 114–118. doi:10.1016/0022-3093(93)90325-R

Zhizhin, G. V. (2012, October). *Hierarchical filling of spaces with polytopes.* Paper presented at "St. Petersburg Scientific Forum: Science and Human Progress". 7th St.- Petersburg meeting of Nobel Prize laureates, St. Petersburg, Russia.

Zhizhin, G. V. (2014a). *World – 4D*. Polytechnic Service.

Zhizhin, G. V. (2014b). Fractal nature of disproportionate phases. *Materials (Basel)*, 10.

Zhizhin, G. V. (2015, November). *Polytopic prismahedrons – fundamental regions of the n-dimension nanostructures.* Paper presented at The International conference "Nanoscience in Chemistry, Physics, Biology and Mathematics", Cluj-Napoca, Romania.

Zhizhin, G.V. (2017a). N – prismahedrons and their dual polytopes. *International Journal Chemical Modeling, 5*(4).

Zhizhin, G. V. (2017b). *Chemical Compounds Structures and the Higher Dimension of Molecules: Emerging Research and Opportunities*. IGI Global.

Zhizhin, G. V., & Diudea, M. V. (2016). Space of nanoworld. In M. V. Putz & C. M. Marius (Eds.), *Sustainable nanosystems, development, properties, and applications* (pp. 221–245). IGI Dlobal.

Ziegler, G. (1995). *Lectures on polytopes*. Springer. doi:10.1007/978-1-4613-8431-1

KEY TERMS AND DEFINITIONS

5-Cross-Prism: The product of 4-cross-polytope by one dimension segment.
5-Simplex-Prism: The product of 4-simplex by one dimension segment.
6-Simplex-Polytopic Prismahedron: The product of 4-simplex by the triangle.

6-Complex-Polytopic Prismahedrons: The product of tetrahedron by the tetrahedron.

6-Complex-Tetrahedral Prismahedrons and Octahedral Prismahedrons: The product octahedral by the tetrahedral.

Cube-Polytopic Prismahedron: The product of cube by the triangle.

Dodeca-Polytopic Prismahedron: The product of dodecahedron by the triangle.

Dodecahedral Prism: The product of dodecahedral by one dimension segment.

Hierarchical and Translation Filling Spaces: At the same time hierarchical and translation filling.

Icosa-Polytopic Prismahedron: The product of icosahedron by the triangle.

Icosahedral Prism: The product of icosahedral by one dimension segment.

N*3-Angular Prismahedron: The product of n-angle by the triangle.

Octahedral Prism: The product of octahedral by one dimension segment.

Octahedral Prismahedron: The product of octahedral by the triangle.

Polytopic Prismahedron: The product of polytope by one dimension segment.

Tetrahedral Prism: The product of tetrahedral by one dimension segment.

Tetrahedral Prismahedron: The product of tetrahedral by the triangle.

Triangular Prismahedron: The product of triangle by the triangle.

Chapter 6
The Products of Molecules and Clusters on Polytopes of Higher Dimension

ABSTRACT

The molecules and clusters given in the previous chapters are multiplied by polytopes of different dimensions and types: from a one-dimensional segment to polytopes of arbitrary dimension n. The structure of their products is analytically determined. Due to the presence in the products of polytopes of several families of parallel edges, these polytopes can become the basis of parallelohedrons, dividing the space of higher dimension face to face. The products of polytopes can simulate continuous areas of living matter of finite sizes.

INTRODUCTION

After considering the geometric structure of the product of polytopes of higher dimension in the previous chapter, in accordance with the main purpose of this monograph, we turn to an analysis of the geometric structure of the product of molecules and clusters on the polytopes of higher dimension. Moreover, according to Chapters 1 - 4, molecules and clusters are also considered as polytopes of the highest dimensionality of the species characteristic of chemical compounds. In this monograph, passing from simple cases of molecules of chemical compounds to polytopes, general expressions for such works in cases of n - dimensional polytopes will be obtained. The results of these studies can be used in the future to build n - dimensional spaces of matter.

DOI: 10.4018/978-1-7998-6768-5.ch006

THE PRODUCT OF A TETRAHEDRON WITH CENTER ON ONE – DIMENSIONAL SEGMENT

Molecules of chemical compounds (many acids, hydrocarbons, oxides) often exhibit tetrahedral coordination (Gillespie, 1972; Gillespie, & Hargittai, 1991; Zhizhin, 2018). In these compounds, atoms are located not only at the vertices of the tetrahedron, but also in the center of the tetrahedron. One example may be chromium oxide $Cr\,O_3$. In this case, the chromium atom is located in the center of the tetrahedron, and the oxygen atoms are located at the vertices of the tetrahedron (Chapter 1). In accordance with Chapter 5, we denote the tetrahedron with the center by *P*, and the one - dimensional segment by *Q*. The number of vertices in the tetrahedron with center is 5, the number of edges is 10, the number of two - dimensional faces is 10, the number of three - dimensional faces is 5 (one large tetrahedron and four tetrahedra inside it). Thus, for a tetrahedron with a center, in accordance with the notation of Chapter 5, we have $f_P^o = 5, f_P^1 = 10, f_P^2 = 10, f_P^3 = 5$. The equation of the Euler - Poincaré (Poincaré, 1895) in this case has view

$$\sum_{i=0}^{d-1} (-1)^i f^i(P) = 1 + (-1)^{d-1}. \tag{1}$$

In this equation $f^i(P)$ is the number of faces with dimension *i* in polytope *P* with dimension *d*.

Substitute obtained numbers $f^i(P)$ in the equation of the Euler - Poincaré it is easy to verify that this equation for the polytope (8) is satisfied for $d = 4$; $5 - 10 + 10 - 5 = 0$.

Thus, polytope *P* is simplex of dimension 4 (4 – simplex). For one – dimensional segment we have $f_Q^o = 2, f_Q^1 = 1$.

For the product of tetrahedron with center on a one – dimensional segment in accordance of Theorem 1 of Chapter 5 we have

$$f_\times^0 = f_P^0 f_Q^0 = 5 \cdot 2 = 10,$$

$$f_\times^1 = f_P^1 f_Q^0 + f_P^0 f_Q^1 = 10 \cdot 2 + 6 \cdot 1 = 25,.$$

$$f_\times^2 = f_P^2 \cdot f_Q^0 + f_P^1 \cdot f_Q^1 = 10 \cdot 2 + 10 \cdot 1 = 30,$$

$$f_\times^3 = f_P^3 \cdot f_Q^0 + f_P^2 \cdot f_Q^1 = 5 \cdot 2 + 10 \cdot 1 = 20,$$

$$f_\times^4 = f_P^4 \cdot f_Q^0 + f_P^3 \cdot f_Q^1 = 1 \cdot 2 + 5 \cdot 1 = 7,$$

$$f_\times^5 = f_P^4 \cdot f_Q^1 = 1 \cdot 1 = 1.$$

Consequently, the product of a tetrahedron centered on a one - dimensional segment has 10 vertices, 25 edges, 30 two – dimensional faces, 20 three - dimensional faces, 7 four - dimensional faces. This product itself has dimension 5. This can be easily verified by substituting the obtained values of the numbers of faces of various dimensions into the Euler – Poincare equation 10 – 25 + 30 – 20 + 7 = 2. Thus, the Euler - Poincaré equation (1) is satisfied for dimension $d = 5$.

THE PRODUCT OF A TETRAHEDRON WITH CENTER ON TRIANGLE

In this case, as well as the previous paragraph for polytope P, we have $f_P^o = 5, f_P^1 = 10, f_P^2 = 10, f_P^3 = 5$.

For triangle the numbers of elements with different dimension are $f_Q^o = 3, f_Q^1 = 3, f_Q^2 = 1$.

For the product of tetrahedron with center on a triangle in accordance of Theorem 1 of Chapter 5 we have

$$f_\times^0 = f_P^0 f_Q^0 = 5 \cdot 3 = 15,$$

$$f_\times^1 = f_P^1 f_Q^0 + f_P^0 f_Q^1 = 10 \cdot 3 + 5 \cdot 3 = 45,.$$

$$f_\times^2 = f_P^2 \cdot f_Q^0 + f_P^1 \cdot f_Q^1 + f_P^0 \cdot f_Q^2 = 10 \cdot 3 + 10 \cdot 3 + 5 \cdot 1 = 65,$$

$$f_\times^3 = f_P^3 \cdot f_Q^0 + f_P^2 \cdot f_Q^1 + f_P^1 \cdot f_Q^2 = 5 \cdot 3 + 10 \cdot 3 + 10 \cdot 1 = 55,$$

$$f_\times^4 = f_P^4 \cdot f_Q^0 + f_P^3 \cdot f_Q^1 + f_P^2 \cdot f_Q^2 = 1 \cdot 3 + 5 \cdot 3 + 10 \cdot 1 = 28,$$

$$f_\times^5 = f_P^4 f_Q^1 + f_P^3 \cdot f_Q^2 = 1 \cdot 3 + 5 \cdot 1 = 8,$$

$$f_\times^6 = f_P^4 \cdot f_Q^2 = 1 \cdot 1 = 1.$$

Consequently, the product of a tetrahedron centered on a triangle has 15 vertices, 45 edges, 65 two – dimensional faces, 55 three - dimensional faces, 28 four - dimensional faces and 8 five – dimensional faces. This product itself has dimension 6. This can be easily verified by substituting the obtained values of the numbers of faces of various dimensions into the Euler – Poincare equation 15 – 45 + 65 – 55 + 28 - 8 = 0. Thus, the Euler - Poincaré equation (1) is satisfied for dimension $d = 6$.

THE PRODUCT OF A TETRAHEDRON WITH CENTER ON A TETRAHEDRON

In this case, as well as the previous paragraph for polytope P, we have $f_P^o = 5, f_P^1 = 10, f_P^2 = 10, f_P^3 = 5$.

For a tetrahedron the numbers of elements with different dimension are $f_Q^o = 4, f_Q^1 = 6, f_Q^2 = 4, f_Q^3 = 1$.

For the product of tetrahedron with center on a tetrahedron in accordance of Theorem 1 of Chapter 5 we have

$$f_{\times}^{0} = f_{P}^{0} f_{Q}^{0} = 5 \cdot 4 = 20,$$

$$f_{\times}^{1} = f_{P}^{1} f_{Q}^{0} + f_{P}^{0} f_{Q}^{1} = 10 \cdot 4 + 5 \cdot 6 = 70, .$$

$$f_{\times}^{2} = f_{P}^{2} \cdot f_{Q}^{0} + f_{P}^{1} \cdot f_{Q}^{1} + f_{P}^{0} \cdot f_{Q}^{2} = 10 \cdot 4 + 10 \cdot 6 + 5 \cdot 4 = 120,$$

$$f_{\times}^{3} = f_{P}^{3} \cdot f_{Q}^{0} + f_{P}^{2} \cdot f_{Q}^{1} + f_{P}^{1} \cdot f_{Q}^{2} + f_{P}^{0} \cdot f_{Q}^{3} = 5 \cdot 4 + 10 \cdot 6 + 10 \cdot 4 + 5 \cdot 1 = 125,$$

$$f_{\times}^{4} = f_{P}^{4} \cdot f_{Q}^{0} + f_{P}^{3} \cdot f_{Q}^{1} + f_{P}^{2} \cdot f_{Q}^{2} + f_{P}^{1} \cdot f_{Q}^{3} = 1 \cdot 4 + 5 \cdot 6 + 10 \cdot 4 + 10 \cdot 1 = 84,$$

$$f_{\times}^{5} = f_{P}^{4} \cdot f_{Q}^{1} + f_{P}^{3} \cdot f_{Q}^{2} + f_{P}^{2} \cdot f_{Q}^{3} = 1 \cdot 6 + 5 \cdot 4 + 10 \cdot 1 = 36,$$

$$f_{\times}^{6} = f_{P}^{4} f_{Q}^{2} + f_{P}^{3} \cdot f_{Q}^{3} = 1 \cdot 4 + 5 \cdot 1 = 9,$$

$$f_{\times}^{7} = f_{P}^{4} \cdot f_{Q}^{3} = 1 \cdot 1 = 1.$$

Consequently, the product of a tetrahedron centered on a tetrahedron has 20 vertices, 70 edges, 120 two – dimensional faces, 125 three - dimensional faces, 84 four - dimensional faces, 36 five – dimensional faces, and 9 six – dimensional faces. This product itself has dimension 7. This can be easily verified by substituting the obtained values of the numbers of faces of various dimensions into the Euler – Poincare equation 20 – 70 + 120 – 125 + 84 – 36 + 9 = 2. Thus, the Euler - Poincaré equation (1) is satisfied for dimension $d = 7$.

THE PRODUCT OF A TETREHEDRON WITH CENTER ON A M - SIMPLEX

In this case, as well as the previous paragraph for polytope *P*, we have $f_{P}^{o} = 5, f_{P}^{1} = 10, f_{P}^{2} = 10, f_{P}^{3} = 5$.

For a m – simplex the numbers of elements with different dimension are (Zhizhin, 2019 a) $f_{Q}^{i}(m) = C_{m+1}^{i+1}, i = 0,1,...,m.$

For the product of tetrahedron with center on a m - simplex in accordance of Theorem 1 of Chapter 5 we have

$$f_{\times}^{0} = f_{P}^{0} f_{Q}^{0} = 5 \cdot C_{m+1}^{1},$$

$$f_{\times}^{1} = f_P^1 f_Q^0 + f_P^0 f_Q^1 = 10\cdot C_{m+1}^{1} + 5\cdot C_{m+1}^{2},.$$

$$f_{\times}^{2} = f_P^2\cdot f_Q^0 + f_P^1\cdot f_Q^1 + f_P^0\cdot f_Q^2 = 10\cdot C_{m+1}^{1} + 10\cdot C_{m+1}^{2} + 5\cdot C_{m+1}^{3},$$

$$f_{\times}^{3} = f_P^3\cdot f_Q^0 + f_P^2\cdot f_Q^1 + f_P^1\cdot f_Q^2 + f_P^0\cdot f_Q^3 = 5\cdot C_{m+1}^{1} + 10\cdot C_{m+1}^{2} + 10\cdot C_{m+1}^{3} + 5\cdot C_{m+1}^{4},$$

$$f_{\times}^{4} = f_P^4\cdot f_Q^0 + f_P^3\cdot f_Q^1 + f_P^2\cdot f_Q^2 + f_P^1\cdot f_Q^3 + f_P^0 f_Q^4 = 1\cdot C_{m+1}^{1} + 5\cdot C_{m+1}^{2} + 10\cdot C_{m+1}^{3} + 10\cdot C_{m+1}^{4} + 5\cdot C_{m+1}^{5},$$

$$f_{\times}^{5} = f_P^4\cdot f_Q^1 + f_P^3\cdot f_Q^2 + f_P^2\cdot f_Q^3 + f_P^1\cdot f_Q^4 + f_P^0 f_Q^5 = 1\cdot C_{m+1}^{2} + 5\cdot C_{m+1}^{3} + 10\cdot C_{m+1}^{4} + 10\cdot C_{m+1}^{5} + 5\cdot C_{m+1}^{6},$$

$$f_{\times}^{m+2} = f_P^4\cdot f_Q^{m-2} + f_P^3\cdot f_Q^{m-1} + f_P^2\cdot f_Q^{m} = 1\cdot C_{m+1}^{m-1} + 5\cdot C_{m+1}^{m} + 10,$$

$$f_{\times}^{m+3} = f_P^4\cdot f_Q^{m-1} + f_P^3\cdot f_Q^{m} = 1\cdot C_{m+1}^{m} + 5,$$

$$f_{\times}^{m+4} = f_P^4\cdot f_Q^{m} = 1\cdot 1 = 1.$$

Consequently, the product of a tetrahedron centered on a tetrahedron has

$5\cdot C_{m+1}^{1}$ vertices,

$10\cdot C_{m+1}^{1} + 5\cdot C_{m+1}^{2}$ edges,

$10\cdot C_{m+1}^{1} + 10\cdot C_{m+1}^{2} + 5\cdot C_{m+1}^{3}$ two – dimensional faces,

$5\cdot C_{m+1}^{1} + 10\cdot C_{m+1}^{2} + 10\cdot C_{m+1}^{3} + 5\cdot C_{m+1}^{4}$ three - dimensional faces,

$1\cdot C_{m+1}^{1} + 5\cdot C_{m+1}^{2} + 10\cdot C_{m+1}^{3} + 10\cdot C_{m+1}^{4} + 5\cdot C_{m+1}^{5}$ four - dimensional faces,

$1\cdot C_{m+1}^{2} + 5\cdot C_{m+1}^{3} + 10\cdot C_{m+1}^{4} + 10\cdot C_{m+1}^{5} + 5\cdot C_{m+1}^{6}$ five – dimensional faces, …,

$1\cdot C_{m+1}^{m-1} + 5\cdot C_{m+1}^{m} + 10\,(m+2)$ – dimensional faces, and

$1\cdot C_{m+1}^{m} + 5\,(m+3)$ – dimensional faces.

This product itself has dimension $m + 4$.

THE PRODUCT OF A TETRAHEDRON WITH CENTER ON A M - CUBE

In this case, as well as the previous paragraph for polytope P, we have $f_P^o = 5, f_P^1 = 10, f_P^2 = 10, f_P^3 = 5$.

For a m – cube the numbers of elements with different dimension are (Zhizhin, 2019 a) $f_Q^i(m) = 2^{m-i}C_m^i, i = 0,1,...,m.$

For the product of tetrahedron with center on a tetrahedron in accordance of Theorem 1 of Chapter 5 we have

$$f_\times^0 = f_P^0 f_Q^0 = 5\cdot 2^m,$$

$$f_\times^1 = f_P^1 f_Q^0 + f_P^0 f_Q^1 = 10\cdot 2^m + 5\cdot 2^{m-1}C_{m+1}^1,.$$

$$f_\times^2 = f_P^2\cdot f_Q^0 + f_P^1\cdot f_Q^1 + f_P^0\cdot f_Q^2 = 10\cdot 2^m + 10\cdot 2^{m-1}C_m^1 + 5\cdot 2^{m-2}C_m^2,$$

$$f_\times^3 = f_P^3\cdot f_Q^0 + f_P^2\cdot f_Q^1 + f_P^1\cdot f_Q^2 + f_P^0\cdot f_Q^3 = 5\cdot 2^m + 10\cdot 2^{m-1}C_m^1 + 10\cdot 2^{m-2}C_m^2 + 5\cdot 2^{m-3}C_m^3,$$

$$f_\times^4 = f_P^4\cdot f_Q^0 + f_P^3\cdot f_Q^1 + f_P^2\cdot f_Q^2 + f_P^1\cdot f_Q^3 + f_P^0 f_Q^4$$
$$= 1\cdot 2^m + 5\cdot 2^{m-1}C_m^1 + 10\cdot 2^{m-2}C_m^2 + 10\cdot 2^{m-3}C_m^3 + 5\cdot 2^{m-4}C_m^4,$$

$$f_\times^5 = f_P^4\cdot f_Q^1 + f_P^3\cdot f_Q^2 + f_P^2\cdot f_Q^3 + f_P^1\cdot f_Q^4 + f_P^0 f_Q^5$$
$$= 1\cdot 2^{m-1}C_m^1 + 5\cdot 2^{m-2}C_m^2 + 10\cdot 2^{m-3}C_m^3 + 10\cdot 2^{m-4}C_m^4 + 5\cdot 2^{m-5}C_m^5,$$

$$f_\times^{m+2} = f_P^4\cdot f_Q^{m-2} + f_P^3\cdot f_Q^{m-1} + f_P^2\cdot f_Q^m = 1\cdot 2^2 C_m^{m-2} + 5\cdot 2C_m^{m-1} + 10,$$

$$f_\times^{m+3} = f_P^4\cdot f_Q^{m-1} + f_P^3\cdot f_Q^m = 1\cdot 2C_m^{m-1} + 5,$$

$$f_\times^{m+4} = f_P^4\cdot f_Q^m = 1\cdot 1 = 1.$$

Consequently, the product of a tetrahedron centered on a tetrahedron has

$5\bullet 2^m$ vertices,

$10\cdot 2^m + 5\cdot 2^{m-1}C_m^1$ edges,

$10\cdot 2^m + 10\cdot 2^{m-1}C_m^1 + 5\cdot 2^{m-2}C_m^2$ two – dimensional faces,

$5\cdot 2^{m} + 10\cdot 2^{m-1}C_m^1 + 10\cdot 2^{m-2}C_m^2 + 5\cdot 2^{m-3}C_m^3$ three - dimensional faces,

$1\cdot 2^{m} + 5\cdot 2^{m-1}C_m^1 + 10\cdot 2^{m-2}C_{m+1}^2 + 10\cdot 2^{m-3}C_m^3 + 5\cdot 2^{m-4}C_m^4$ four - dimensional faces,

$1\cdot 2^{m-1}C_m^1 + 5\cdot 2^{m-2}C_m^2 + 10\cdot 2^{m-3}C_m^3 + 10\cdot 2^{m-4}C_m^4 + 5\cdot 2^{m-5}C_m^5$ five – dimensional faces, …,

$1\cdot 2^2 C_m^{m-2} + 5\cdot 2C_m^{m-1} + 10\,(m+2)$ – dimensional faces, and

$1\cdot 2C_m^{m-1} + 5\,(m+3)$ – dimensional faces.

This product itself has dimension $m + 4$.

THE PRODUCT OF A TETRAHEDRON WITH CENTER ON A *M* – CROSS - POLYTOPE

In this case, as well as the previous paragraph for polytope P, we have $f_P^o = 5, f_P^1 = 10, f_P^2 = 10, f_P^3 = 5$.

For a m – cross - polytope the numbers of elements with different dimension are (Zhizhin, 2019 a) $f_Q^i(m) = 2^{1+i}C_m^{m-1-i}, i = 0,1,...,m-1$.

For the product of tetrahedron with center on a m – cross - polytope in accordance of Theorem 1 of Chapter 5 we have

$$f_\times^0 = f_P^0 f_Q^0 = 5\cdot 2C_m^{m-1},$$

$$f_\times^1 = f_P^1 f_Q^0 + f_P^0 f_Q^1 = 10\cdot 2C_m^{m-1} + 5\cdot 2^2 C_m^{m-2},.$$

$$f_\times^2 = f_P^2\cdot f_Q^0 + f_P^1\cdot f_Q^1 + f_P^0\cdot f_Q^2 = 10\cdot 2C_m^{m-1} + 10\cdot 2^2 C_m^{m-2} + 5\cdot 2^3 C_m^{m-3},$$

$$f_\times^3 = f_P^3\cdot f_Q^0 + f_P^2\cdot f_Q^1 + f_P^1\cdot f_Q^2 + f_P^0\cdot f_Q^3 = 5\cdot 2C_m^{m-1} + 10\cdot 2^2 C_m^{m-2} + 10\cdot 2^3 C_m^{m-3} + 5\cdot 2^4 C_m^{m-4},$$

$$f_\times^4 = f_P^4\cdot f_Q^0 + f_P^3\cdot f_Q^1 + f_P^2\cdot f_Q^2 + f_P^1\cdot f_Q^3 + f_P^0 f_Q^4$$
$$= 1\cdot 2C_m^{m-1} + 5\cdot 2^2 C_m^{m-2} + 10\cdot 2^3 C_m^{m-3} + 10\cdot 2^4 C_m^{m-4} + 5\cdot 2^5 C_m^{m-5},$$

$$f_\times^5 = f_P^4\cdot f_Q^1 + f_P^3\cdot f_Q^2 + f_P^2\cdot f_Q^3 + f_P^1\cdot f_Q^4 + f_P^0 f_Q^5$$
$$= 1\cdot 2^2 C_m^{m-2} + 5\cdot 2^3 C_m^{m-3} + 10\cdot 2^4 C_m^{m-4} + 10\cdot 2^5 C_m^{m-5} + 5\cdot 2^6 C_m^{m-6},$$

$$f_{\times}^{m+2} = f_P^4 \cdot f_Q^{m-2} + f_P^3 \cdot f_Q^{m-1} + f_P^2 \cdot f_Q^m = 1 \cdot 2^{m-1} C_m^1 + 5 \cdot 2^m C_m^0 + 10,$$

$$f_{\times}^{m+3} = f_P^4 \cdot f_Q^{m-1} + f_P^3 \cdot f_Q^m = 1 \cdot 2^m + 5,$$

$$f_{\times}^{m+4} = f_P^4 \cdot f_Q^m = 1 \cdot 1 = 1.$$

Consequently, the product of a tetrahedron centered on a m – cross - polytope has

$5 \cdot 2C_m^{m-1}$ vertices,

$10 \cdot 2C_m^{m-1} + 5 \cdot 2^2 C_m^{m-2}$ edges,

$10 \cdot 2C_m^{m-1} + 10 \cdot 2^2 C_m^{m-2} + 5 \cdot 2^3 C_m^{m-3}$ two – dimensional faces,

$5 \cdot 2C_m^{m-1} + 10 \cdot 2^2 C_m^{m-2} + 10 \cdot 2^3 C_m^{m-3} + 5 \cdot 2^4 C_m^{m-4}$ three - dimensional faces,

$1 \cdot 2C_m^{m-1} + 5 \cdot 2^2 C_m^{m-2} + 10 \cdot 2^3 C_m^{m-3} + 10 \cdot 2^4 C_m^{m-4} + 5 \cdot 2^5 C_m^{m-5}$ four - dimensional faces,

$1 \cdot 2^2 C_m^{m-2} + 5 \cdot 2^3 C_m^{m-3} + 10 \cdot 2^4 C_m^{m-4} + 10 \cdot 2^5 C_m^{m-5} + 5 \cdot 2^6 C_m^{m-6}$ five – dimensional faces, …,

$1 \cdot 2^{m-1} C_m^1 + 5 \cdot 2^m + 10\,(m + 2)$ – dimensional faces, and

$1 \cdot C_{m+1}^m + 5\,(m + 3)$ – dimensional faces.

This product itself has dimension $m + 4$.

THE PRODUCT OF A MOLECULE OF WURTZITE ON ONE – DIMENSIONAL SEGMENT

Molecules of chemical compounds (in particular cluster compounds) often exhibit octahedral coordination (Gillespie, 1972; Gillespie, & Hargittai, 1991; Gubin, 2019). In these compounds, atoms are located not only at the vertices of the octahedron, but also in the center of the tetrahedron. In accordance with Chapter 5, we denote the octahedron with the center by P, and the one - dimensional segment by Q. The number of vertices in the octahedron with center is 7, the number of edges is 18, the number of two - dimensional faces is 20, the number of three -dimensional faces is 9 (one large octahedron and eight tetrahedrons inside it). Thus, for an octahedron with a center, in accordance with the notation of Chapter 5, we have $f_P^o = 7, f_P^1 = 18, f_P^2 = 20, f_P^3 = 9$.

Substitute this numbers $f^i(P)$ in the equation of the Euler – Poincaré (1) it is easy to verify that this equation for the polytope is satisfied for $d = 4$; $7 - 18 + 20 - 9 = 0$.

For one – dimensional segment we have $f_Q^o = 2, f_Q^1 = 1$.

For the product of an octahedron with center on one – dimensional segment in accordance of Theorem 1 of Chapter 5 we have

$$f_\times^0 = f_P^0 f_Q^0 = 7 \cdot 2 = 14,$$

$$f_\times^1 = f_P^1 f_Q^0 + f_P^0 f_Q^1 = 18 \cdot 2 + 7 \cdot 1 = 43,.$$

$$f_\times^2 = f_P^2 \cdot f_Q^0 + f_P^1 \cdot f_Q^1 = 20 \cdot 2 + 18 \cdot 1 = 58,$$

$$f_\times^3 = f_P^3 \cdot f_Q^0 + f_P^2 \cdot f_Q^1 = 9 \cdot 2 + 20 \cdot 1 = 38,$$

$$f_\times^4 = f_P^4 \cdot f_Q^0 + f_P^3 \cdot f_Q^1 = 1 \cdot 2 + 9 \cdot 1 = 11,$$

$$f_\times^5 = f_P^4 \cdot f_Q^1 = 1 \cdot 1 = 1.$$

Consequently, the product of an octahedron centered on one - dimensional segment has 14 vertices, 43 edges, 58 two – dimensional faces, 38 three - dimensional faces, 11 four - dimensional faces. This product itself has dimension 5. This can be easily verified by substituting the obtained values of the numbers of faces of various dimensions into the Euler – Poincare equation $14 - 43 + 58 - 38 + 11 = 2$. Thus, the Euler - Poincaré equation (1) is satisfied for dimension $d = 5$.

THE PRODUCT OF AN OCTAHEDRON WITH CENTER ON TRIANGLE

In this case, as well as the previous paragraph for polytope P, we have $f_P^o = 7, f_P^1 = 18, f_P^2 = 20, f_P^3 = 9$.

For triangle the numbers of elements with different dimension are $f_Q^o = 3, f_Q^1 = 3, f_Q^2 = 1$.

For the product of an octahedron with center on a triangle in accordance of Theorem 1 of Chapter 5 we have

$$f_\times^0 = f_P^0 f_Q^0 = 7 \cdot 3 = 21,$$

$$f_\times^1 = f_P^1 f_Q^0 + f_P^0 f_Q^1 = 18 \cdot 3 + 7 \cdot 3 = 75,.$$

$$f_\times^2 = f_P^2 \cdot f_Q^0 + f_P^1 \cdot f_Q^1 + f_P^0 \cdot f_Q^2 = 20 \cdot 3 + 18 \cdot 3 + 7 \cdot 1 = 121,$$

$$f_\times^3 = f_P^3 \cdot f_Q^0 + f_P^2 \cdot f_Q^1 + f_P^1 \cdot f_Q^2 = 9 \cdot 3 + 20 \cdot 3 + 18 \cdot 1 = 105,$$

$$f_\times^4 = f_P^4 \cdot f_Q^0 + f_P^3 \cdot f_Q^1 + f_P^2 \cdot f_Q^2 = 3 \cdot 1 + 3 \cdot 9 + 20 \cdot 1 = 50,$$

$$f_\times^5 = f_P^4 f_Q^1 + f_P^3 \cdot f_Q^2 = 1 \cdot 3 + 9 \cdot 1 = 12,$$

$$f_\times^6 = f_P^4 \cdot f_Q^2 = 1 \cdot 1 = 1.$$

Consequently, the product of a tetrahedron centered on a triangle has 21 vertices, 75 edges, 121 two – dimensional faces, 105 three - dimensional faces, 50 four - dimensional faces and 12 five – dimensional faces. This product itself has dimension 6. This can be easily verified by substituting the obtained values of the numbers of faces of various dimensions into the Euler – Poincare equation 21 – 75 + 121 – 105 + 50 - 12 = 0. Thus, the Euler - Poincaré equation (1) is satisfied for dimension $d = 6$.

THE PRODUCT OF AN OCTAHEDRON WITH CENTER ON A TETRAHEDRON

In this case, as well as the previous paragraph for polytope *P*, we have $f_P^o = 7, f_P^1 = 18, f_P^2 = 20, f_P^3 = 9$.

For a tetrahedron the numbers of elements with different dimension are $f_Q^o = 4, f_Q^1 = 6, f_Q^2 = 4, f_Q^3 = 1$.

For the product of an octahedron with center on a tetrahedron in accordance of Theorem 1 of Chapter 5 we have

$$f_\times^0 = f_P^0 f_Q^0 = 7 \cdot 4 = 28,$$

$$f_\times^1 = f_P^1 f_Q^0 + f_P^0 f_Q^1 = 18 \cdot 4 + 7 \cdot 6 = 114,.$$

$$f_\times^2 = f_P^2 \cdot f_Q^0 + f_P^1 \cdot f_Q^1 + f_P^0 \cdot f_Q^2 = 20 \cdot 4 + 18 \cdot 6 + 7 \cdot 4 = 216,$$

$$f_\times^3 = f_P^3 \cdot f_Q^0 + f_P^2 \cdot f_Q^1 + f_P^1 \cdot f_Q^2 + f_P^0 \cdot f_Q^3 = 9 \cdot 4 + 20 \cdot 6 + 18 \cdot 4 + 7 \cdot 1 = 235,$$

$$f_\times^4 = f_P^4 \cdot f_Q^0 + f_P^3 \cdot f_Q^1 + f_P^2 \cdot f_Q^2 + f_P^1 \cdot f_Q^3 = 1 \cdot 4 + 9 \cdot 6 + 20 \cdot 4 + 18 \cdot 1 = 156,$$

$$f_\times^5 = f_P^4 \cdot f_Q^1 + f_P^3 \cdot f_Q^2 + f_P^2 \cdot f_Q^3 = 1 \cdot 6 + 9 \cdot 4 + 20 \cdot 1 = 62,$$

$$f_\times^6 = f_P^4 f_Q^2 + f_P^3 \cdot f_Q^3 = 1 \cdot 4 + 9 \cdot 1 = 13,$$

$f_{\times}^{7} = f_{P}^{4} \cdot f_{Q}^{3} = 1 \cdot 1 = 1.$

Consequently, the product of a tetrahedron centered on a tetrahedron has 28 vertices, 114 edges, 216 two – dimensional faces, 235 three - dimensional faces, 156 four - dimensional faces, 62 five – dimensional faces, and 13 six – dimensional faces. This product itself has dimension 7. This can be easily verified by substituting the obtained values of the numbers of faces of various dimensions into the Euler – Poincare equation 28 – 114 + 216 – 235 + 156 – 62 + 13 = 2. Thus, the Euler - Poincaré equation (1) is satisfied for dimension $d = 7$.

THE PRODUCT OF AN OCTAHEDRON WITH CENTER ON AN OCTAHEDRON

In this case, as well as the previous paragraph for polytope P, we have $f_{P}^{o} = 7, f_{P}^{1} = 18, f_{P}^{2} = 20, f_{P}^{3} = 9.$

For an octahedron the numbers of elements with different dimension are $f_{Q}^{o} = 6, f_{Q}^{1} = 12, f_{Q}^{2} = 8, f_{Q}^{3} = 1.$

For the product of an octahedron with center on a tetrahedron in accordance of Theorem 1 of Chapter 5 we have

$$f_{\times}^{0} = f_{P}^{0} f_{Q}^{0} = 7 \cdot 6 = 42,$$

$$f_{\times}^{1} = f_{P}^{1} f_{Q}^{0} + f_{P}^{0} f_{Q}^{1} = 18 \cdot 6 + 7 \cdot 12 = 192,.$$

$$f_{\times}^{2} = f_{P}^{2} \cdot f_{Q}^{0} + f_{P}^{1} \cdot f_{Q}^{1} + f_{P}^{0} \cdot f_{Q}^{2} = 20 \cdot 6 + 18 \cdot 12 + 7 \cdot 8 = 392,$$

$$f_{\times}^{3} = f_{P}^{3} \cdot f_{Q}^{0} + f_{P}^{2} \cdot f_{Q}^{1} + f_{P}^{1} \cdot f_{Q}^{2} + f_{P}^{0} \cdot f_{Q}^{3} = 9 \cdot 6 + 20 \cdot 12 + 18 \cdot 8 + 7 \cdot 1 = 445,$$

$$f_{\times}^{4} = f_{P}^{4} \cdot f_{Q}^{0} + f_{P}^{3} \cdot f_{Q}^{1} + f_{P}^{2} \cdot f_{Q}^{2} + f_{P}^{1} \cdot f_{Q}^{3} = 1 \cdot 6 + 9 \cdot 12 + 20 \cdot 8 + 18 \cdot 1 = 292,$$

$$f_{\times}^{5} = f_{P}^{4} \cdot f_{Q}^{1} + f_{P}^{3} \cdot f_{Q}^{2} + f_{P}^{2} \cdot f_{Q}^{3} = 1 \cdot 12 + 9 \cdot 8 + 20 \cdot 1 = 104,$$

$$f_{\times}^{6} = f_{P}^{4} f_{Q}^{2} + f_{P}^{3} \cdot f_{Q}^{3} = 1 \cdot 8 + 9 \cdot 1 = 17,$$

$$f_{\times}^{7} = f_{P}^{4} \cdot f_{Q}^{3} = 1 \cdot 1 = 1.$$

Consequently, the product of a tetrahedron centered on a tetrahedron has 42 vertices, 192 edges, 392 two – dimensional faces, 445 three - dimensional faces, 292 four - dimensional faces, 104 five – dimensional faces, and 17 six – dimensional faces. This product itself has dimension 7. This can be easily verified by substituting the obtained values of the numbers of faces of various dimensions into the Euler – Poincare equation 42 – 192 + 392 – 445 + 292 – 104 + 18 = 2. Thus, the Euler - Poincaré equation (1) is satisfied for dimension $d = 7$.

THE PRODUCT OF AN OCTAHEDRON WITH CENTER ON A *M* - SIMPLEX

In this case, as well as the previous paragraph for polytope *P*, we have $f_P^o = 7, f_P^1 = 18, f_P^2 = 20, f_P^3 = 9.$

For a m – simplex the numbers of elements with different dimension are (Zhizhin, 2019 a) $f_Q^i(m) = C_{m+1}^{i+1}, i = 0,1,...,m.$

For the product of an octahedron with center on a m - simplex in accordance of Theorem 1 of Chapter 5 we have

$$f_\times^0 = f_P^0 f_Q^0 = 7 \cdot C_{m+1}^1,$$

$$f_\times^1 = f_P^1 f_Q^0 + f_P^0 f_Q^1 = 18 \cdot C_{m+1}^1 + 7 \cdot C_{m+1}^2,.$$

$$f_\times^2 = f_P^2 \cdot f_Q^0 + f_P^1 \cdot f_Q^1 + f_P^0 \cdot f_Q^2 = 20 \cdot C_{m+1}^1 + 18 \cdot C_{m+1}^2 + 7 \cdot C_{m+1}^3,$$

$$f_\times^3 = f_P^3 \cdot f_Q^0 + f_P^2 \cdot f_Q^1 + f_P^1 \cdot f_Q^2 + f_P^0 \cdot f_Q^3 = 9 \cdot C_{m+1}^1 + 20 \cdot C_{m+1}^2 + 18 \cdot C_{m+1}^3 + 7 \cdot C_{m+1}^4,$$

$$f_\times^4 = f_P^4 \cdot f_Q^0 + f_P^3 \cdot f_Q^1 + f_P^2 \cdot f_Q^2 + f_P^1 \cdot f_Q^3 + f_P^0 f_Q^4 = 1 \cdot C_{m+1}^1 + 9 \cdot C_{m+1}^2 + 20 \cdot C_{m+1}^3 + 18 \cdot C_{m+1}^4 + 7 \cdot C_{m+1}^5,$$

$$f_\times^5 = f_P^4 \cdot f_Q^1 + f_P^3 \cdot f_Q^2 + f_P^2 \cdot f_Q^3 + f_P^1 \cdot f_Q^4 + f_P^0 f_Q^5 = 1 \cdot C_{m+1}^2 + 9 \cdot C_{m+1}^3 + 20 \cdot C_{m+1}^4 + 18 \cdot C_{m+1}^5 + 7 \cdot C_{m+1}^6,$$

$$f_\times^{m+2} = f_P^4 \cdot f_Q^{m-2} + f_P^3 \cdot f_Q^{m-1} + f_P^2 \cdot f_Q^m = 1 \cdot C_{m+1}^{m-1} + 9 \cdot C_{m+1}^m + 20,$$

$$f_\times^{m+3} = f_P^4 \cdot f_Q^{m-1} + f_P^3 \cdot f_Q^m = 1 \cdot C_{m+1}^m + 9,$$

$$f_\times^{m+4} = f_P^4 \cdot f_Q^m = 1 \cdot 1 = 1.$$

Consequently, the product of a tetrahedron centered on a tetrahedron has

$7 \cdot C_{m+1}^1$ vertices,

$18 \cdot C_{m+1}^1 + 7 \cdot C_{m+1}^2$ edges,

$20 \cdot C^1_{m+1} + 18 \cdot C^2_{m+1} + 7 \cdot C^3_{m+1}$ two – dimensional faces,

$9 \cdot C^1_{m+1} + 20 \cdot C^2_{m+1} + 18 \cdot C^3_{m+1} + 7 \cdot C^4_{m+1}$ three - dimensional faces,

$1 \cdot C^1_{m+1} + 9 \cdot C^2_{m+1} + 20 \cdot C^3_{m+1} + 18 \cdot C^4_{m+1} + 7 \cdot C^5_{m+1}$ four - dimensional faces,

$1 \cdot C^2_{m+1} + 9 \cdot C^3_{m+1} + 20 \cdot C^4_{m+1} + 18 \cdot C^5_{m+1} + 7 \cdot C^6_{m+1}$ five – dimensional faces, …,

$1 \cdot C^{m-1}_{m+1} + 9 \cdot C^m_{m+1} + 20$ $(m + 2)$ – dimensional faces, and

$1 \cdot C^m_{m+1} + 9$ $(m + 3)$ – dimensional faces.

This product itself has dimension $m + 4$.

THE PRODUCT OF AN OCTAHEDRON WITH CENTER ON A *M* - CUBE

In this case, as well as the previous paragraph for polytope *P*, we have $f_P^o = 7, f_P^1 = 18, f_P^2 = 20, f_P^3 = 9$.

For a m – cube the numbers of elements with different dimension are (Zhizhin, 2019 a) $f_Q^i(m) = 2^{m-i} C_m^i, i = 0,1,\ldots,m.$

For the product of tetrahedron with center on a tetrahedron in accordance of Theorem 1 of Chapter 5 we have

$$f_\times^0 = f_P^0 f_Q^0 = 7 \cdot 2^m,$$

$$f_\times^1 = f_P^1 f_Q^0 + f_P^0 f_Q^1 = 18 \cdot 2^m + 7 \cdot 2^{m-1} C^1_{m+1},.$$

$$f_\times^2 = f_P^2 \cdot f_Q^0 + f_P^1 \cdot f_Q^1 + f_P^0 \cdot f_Q^2 = 20 \cdot 2^m + 18 \cdot 2^{m-1} C_m^1 + 7 \cdot 2^{m-2} C_m^2,$$

$$f_\times^3 = f_P^3 \cdot f_Q^0 + f_P^2 \cdot f_Q^1 + f_P^1 \cdot f_Q^2 + f_P^0 \cdot f_Q^3 = 9 \cdot 2^m + 20 \cdot 2^{m-1} C_m^1 + 18 \cdot 2^{m-2} C_m^2 + 7 \cdot 2^{m-3} C_m^3,$$

$$f_\times^4 = f_P^4 \cdot f_Q^0 + f_P^3 \cdot f_Q^1 + f_P^2 \cdot f_Q^2 + f_P^1 \cdot f_Q^3 + f_P^0 f_Q^4$$
$$= 1 \cdot 2^m + 9 \cdot 2^{m-1} C_m^1 + 20 \cdot 2^{m-2} C_m^2 + 18 \cdot 2^{m-3} C_m^3 + 7 \cdot 2^{m-4} C_m^4,$$

$$f_\times^5 = f_P^4 \cdot f_Q^1 + f_P^3 \cdot f_Q^2 + f_P^2 \cdot f_Q^3 + f_P^1 \cdot f_Q^4 + f_P^0 f_Q^5$$
$$= 1 \cdot 2^{m-1} C_m^1 + 9 \cdot 2^{m-2} C_m^2 + 20 \cdot 2^{m-3} C_m^3 + 18 \cdot 2^{m-4} C_m^4 + 7 \cdot 2^{m-5} C_m^5,$$

$$f_{\times}^{m+2} = f_P^4 \cdot f_Q^{m-2} + f_P^3 \cdot f_Q^{m-1} + f_P^2 \cdot f_Q^m = 1 \cdot 2^2 C_m^{m-2} + 9 \cdot 2C_m^{m-1} + 20,$$

$$f_{\times}^{m+3} = f_P^4 \cdot f_Q^{m-1} + f_P^3 \cdot f_Q^m = 1 \cdot 2C_m^{m-1} + 9,$$

$$f_{\times}^{m+4} = f_P^4 \cdot f_Q^m = 1 \cdot 1 = 1.$$

Consequently, the product of a tetrahedron centered on a tetrahedron has

$7 \bullet 2^m$ vertices,

$18 \cdot 2^m + 7 \cdot 2^{m-1} C_m^1$ edges,

$20 \cdot 2^m + 18 \cdot 2^{m-1} C_m^1 + 7 \cdot 2^{m-2} C_m^2$ two – dimensional faces,

$9 \cdot 2^m + 20 \cdot 2^{m-1} C_m^1 + 18 \cdot 2^{m-2} C_m^2 + 7 \cdot 2^{m-3} C_m^3$ three - dimensional faces,

$1 \cdot 2^m + 9 \cdot 2^{m-1} C_m^1 + 20 \cdot 2^{m-2} C_{m+1}^2 + 18 \cdot 2^{m-3} C_m^3 + 7 \cdot 2^{m-4} C_m^4$ four - dimensional faces,

$1 \cdot 2^{m-1} C_m^1 + 9 \cdot 2^{m-2} C_m^2 + 20 \cdot 2^{m-3} C_m^3 + 18 \cdot 2^{m-4} C_m^4 + 7 \cdot 2^{m-5} C_m^5$ five – dimensional faces, …,

$1 \cdot 2^2 C_m^{m-2} + 9 \cdot 2C_m^{m-1} + 20$ $(m + 2)$ – dimensional faces, and

$1 \cdot 2C_m^{m-1} + 9$ $(m + 3)$ – dimensional faces.

This product itself has dimension $m + 4$.

The Product of an Octahedron with Center on a *M* – Cross - Polytope

In this case, as well as the previous paragraph for polytope *P*, we have $f_P^o = 7, f_P^1 = 18, f_P^2 = 20, f_P^3 = 9$.

For a m – cross - polytope the numbers of elements with different dimension are (Zhizhin, 2019 a) $f_Q^i(m) = 2^{1+i} C_m^{m-1-i}, i = 0,1,\dots,m-1$.

For the product of tetrahedron with center on a m – cross - polytope in accordance of Theorem 1 of Chapter 5 we have

$$f_{\times}^0 = f_P^0 f_Q^0 = 7 \cdot 2C_m^{m-1},$$

$$f_{\times}^1 = f_P^1 f_Q^0 + f_P^0 f_Q^1 = 18 \cdot 2C_m^{m-1} + 7 \cdot 2^2 C_m^{m-2},.$$

$$f_{\times}^{2}=f_{P}^{2}\cdot f_{Q}^{0}+f_{P}^{1}\cdot f_{Q}^{1}+f_{P}^{0}\cdot f_{Q}^{2}=20\cdot 2C_{m}^{m-1}+18\cdot 2^{2}C_{m}^{m-2}+7\cdot 2^{3}C_{m}^{m-3},$$

$$f_{\times}^{3}=f_{P}^{3}\cdot f_{Q}^{0}+f_{P}^{2}\cdot f_{Q}^{1}+f_{P}^{1}\cdot f_{Q}^{2}+f_{P}^{0}\cdot f_{Q}^{3}=9\cdot 2C_{m}^{m-1}+20\cdot 2^{2}C_{m}^{m-2}+18\cdot 2^{3}C_{m}^{m-3}+7\cdot 2^{4}C_{m}^{m-4},$$

$$f_{\times}^{4}=f_{P}^{4}\cdot f_{Q}^{0}+f_{P}^{3}\cdot f_{Q}^{1}+f_{P}^{2}\cdot f_{Q}^{2}+f_{P}^{1}\cdot f_{Q}^{3}+f_{P}^{0}f_{Q}^{4}$$
$$=1\cdot 2C_{m}^{m-1}+9\cdot 2^{2}C_{m}^{m-2}+20\cdot 2^{3}C_{m}^{m-3}+18\cdot 2^{4}C_{m}^{m-4}+7\cdot 2^{5}C_{m}^{m-5},$$

$$f_{\times}^{5}=f_{P}^{4}\cdot f_{Q}^{1}+f_{P}^{3}\cdot f_{Q}^{2}+f_{P}^{2}\cdot f_{Q}^{3}+f_{P}^{1}\cdot f_{Q}^{4}+f_{P}^{0}f_{Q}^{5}$$
$$=1\cdot 2^{2}C_{m}^{m-2}+9\cdot 2^{3}C_{m}^{m-3}+20\cdot 2^{4}C_{m}^{m-4}+18\cdot 2^{5}C_{m}^{m-5}+7\cdot 2^{6}C_{m}^{m-6},$$

$$f_{\times}^{m+2}=f_{P}^{4}\cdot f_{Q}^{m-2}+f_{P}^{3}\cdot f_{Q}^{m-1}+f_{P}^{2}\cdot f_{Q}^{m}=1\cdot 2^{m-1}C_{m}^{1}+9\cdot 2^{m}C_{m}^{0}+20,$$

$$f_{\times}^{m+3}=f_{P}^{4}\cdot f_{Q}^{m-1}+f_{P}^{3}\cdot f_{Q}^{m}=1\cdot 2^{m}+9,$$

$$f_{\times}^{m+4}=f_{P}^{4}\cdot f_{Q}^{m}=1\cdot 1=1.$$

Consequently, the product of a tetrahedron centered on a m – cross - polytope has

$7\cdot 2C_{m}^{m-1}$ vertices,

$18\cdot 2C_{m}^{m-1}+7\cdot 2^{2}C_{m}^{m-2}$ edges,

$20\cdot 2C_{m}^{m-1}+18\cdot 2^{2}C_{m}^{m-2}+7\cdot 2^{3}C_{m}^{m-3}$ two – dimensional faces,

$9\cdot 2C_{m}^{m-1}+20\cdot 2^{2}C_{m}^{m-2}+18\cdot 2^{3}C_{m}^{m-3}+7\cdot 2^{4}C_{m}^{m-4}$ three - dimensional faces,

$1\cdot 2C_{m}^{m-1}+9\cdot 2^{2}C_{m}^{m-2}+20\cdot 2^{3}C_{m}^{m-3}+18\cdot 2^{4}C_{m}^{m-4}+7\cdot 2^{5}C_{m}^{m-5}$ four - dimensional faces,

$1\cdot 2^{2}C_{m}^{m-2}+9\cdot 2^{3}C_{m}^{m-3}+20\cdot 2^{4}C_{m}^{m-4}+18\cdot 2^{5}C_{m}^{m-5}+7\cdot 2^{6}C_{m}^{m-6}$ five – dimensional faces, …,

$1\cdot 2^{m-1}C_{m}^{1}+9\cdot 2^{m}+20$ $(m+2)$ – dimensional faces, and

$1\bullet 2^{m}+9$ $(m+3)$ – dimensional faces.

This product itself has dimension $m+4$.

The Product of a Molecule of Wurtzite on One – Dimensional Segment

A series of binary compounds have structure of wurtzite - mineral ZnS, in which from compound ZnS with the structure of the adamantane zinc atom and sulfur atom have the tetrahedron coordination (Gillespie, 1972). The centrum each tetrahedron is vertex of another tetrahedron. The wurtzite structure has compounds ZnO, CdS, ZnS, BeO, AlN. The unit cell the lattice is a convex shape it is shown in Figure 12 of Chapter 1. In accordance with Chapter 5, we denote the molecule of wurzite by *P*, and the one - dimensional segment by *Q*. The number of vertices in the octahedron with center is 14, the number of edges is 29, the number of two - dimensional faces is 21, the number of three -dimensional faces is 6. Thus, for an octahedron with a center, in accordance with the notation of Chapter 5, we have $f_P^o = 14, f_P^1 = 29, f_P^2 = 21, f_P^3 = 6.$

Substitute this numbers $f^i(P)$ in the equation of the Euler – Poincaré (1) it is easy to verify that this equation for the polytope is satisfied for $d = 4$; 14 – 29 + 21 – 6 = 0.

For one – dimensional segment we have $f_Q^o = 2, f_Q^1 = 1.$

For the product of a molecule of wurzite on one – dimensional segment in accordance of Theorem 1 of Chapter 5 we have

$$f_\times^0 = f_P^0 f_Q^0 = 14 \cdot 2 = 28,$$

$$f_\times^1 = f_P^1 f_Q^0 + f_P^0 f_Q^1 = 29 \cdot 2 + 14 \cdot 1 = 72,.$$

$$f_\times^2 = f_P^2 \cdot f_Q^0 + f_P^1 \cdot f_Q^1 = 21 \cdot 2 + 29 \cdot 1 = 71,$$

$$f_\times^3 = f_P^3 \cdot f_Q^0 + f_P^2 \cdot f_Q^1 = 6 \cdot 2 + 21 \cdot 1 = 33,$$

$$f_\times^4 = f_P^4 \cdot f_Q^0 + f_P^3 \cdot f_Q^1 = 1 \cdot 2 + 6 \cdot 1 = 8,$$

$$f_\times^5 = f_P^4 \cdot f_Q^1 = 1 \cdot 1 = 1.$$

Consequently, the product of an octahedron centered on one - dimensional segment has 28 vertices, 72 edges, 71 two – dimensional faces, 33 three - dimensional faces, 8 four - dimensional faces. This product itself has dimension 5. This can be easily verified by substituting the obtained values of the numbers of faces of various dimensions into the Euler – Poincare equation 28 – 72 + 71 – 33 + 8 = 2. Thus, the Euler - Poincaré equation (1) is satisfied for dimension $d = 5$.

The Product of a Molecule of Wurzite on Triangle

In this case, as well as the previous paragraph for polytope *P*, we have $f_P^o = 14, f_P^1 = 28, f_P^2 = 21, f_P^3 = 6.$

For triangle the numbers of elements with different dimension are $f_Q^o = 3, f_Q^1 = 3, f_Q^2 = 1.$

For the product of an octahedron with center on a triangle in accordance of Theorem 1 of Chapter 5 we have

$$f_{\times}^{0} = f_{P}^{0} f_{Q}^{0} = 7 \cdot 3 = 21,$$

$$f_{\times}^{1} = f_{P}^{1} f_{Q}^{0} + f_{P}^{0} f_{Q}^{1} = 29 \cdot 3 + 14 \cdot 3 = 129,.$$

$$f_{\times}^{2} = f_{P}^{2} \cdot f_{Q}^{0} + f_{P}^{1} \cdot f_{Q}^{1} + f_{P}^{0} \cdot f_{Q}^{2} = 21 \cdot 3 + 29 \cdot 3 + 14 \cdot 1 = 164,$$

$$f_{\times}^{3} = f_{P}^{3} \cdot f_{Q}^{0} + f_{P}^{2} \cdot f_{Q}^{1} + f_{P}^{1} \cdot f_{Q}^{2} = 6 \cdot 3 + 21 \cdot 3 + 29 \cdot 1 = 110,$$

$$f_{\times}^{4} = f_{P}^{4} \cdot f_{Q}^{0} + f_{P}^{3} \cdot f_{Q}^{1} + f_{P}^{2} \cdot f_{Q}^{2} = 3 \cdot 1 + 3 \cdot 6 + 21 \cdot 1 = 42,$$

$$f_{\times}^{5} = f_{P}^{4} f_{Q}^{1} + f_{P}^{3} \cdot f_{Q}^{2} = 1 \cdot 3 + 6 \cdot 1 = 9,$$

$$f_{\times}^{6} = f_{P}^{4} \cdot f_{Q}^{2} = 1 \cdot 1 = 1.$$

Consequently, the product of a molecule of wurzite on a triangle has 21 vertices, 129 edges, 164 two – dimensional faces, 110 three - dimensional faces, 42 four - dimensional faces and 9 five – dimensional faces. This product itself has dimension 6. This can be easily verified by substituting the obtained values of the numbers of faces of various dimensions into the Euler – Poincare equation 42 – 129 + 164 – 110 + 42 - 9 = 0. Thus, the Euler - Poincaré equation (1) is satisfied for dimension $d = 6$.

The Product of a Molecule of Wurzite on a Tetrahedron

In this case, as well as the previous paragraph for polytope P, we have $f_{P}^{o} = 14, f_{P}^{1} = 298, f_{P}^{2} = 21, f_{P}^{3} = 6$.

For a tetrahedron the numbers of elements with different dimension are

$$f_{Q}^{o} = 4, f_{Q}^{1} = 6, f_{Q}^{2} = 4, f_{Q}^{3} = 1.$$

For the product of an octahedron with center on a tetrahedron in accordance of Theorem 1 of Chapter 5 we have

$$f_{\times}^{0} = f_{P}^{0} f_{Q}^{0} = 14 \cdot 4 = 56,$$

$$f_{\times}^{1} = f_{P}^{1} f_{Q}^{0} + f_{P}^{0} f_{Q}^{1} = 29 \cdot 4 + 14 \cdot 6 = 200,.$$

$$f_{\times}^{2} = f_P^2 \cdot f_Q^0 + f_P^1 \cdot f_Q^1 + f_P^0 \cdot f_Q^2 = 21\cdot 4 + 29\cdot 6 + 14\cdot 4 = 314,$$

$$f_{\times}^{3} = f_P^3 \cdot f_Q^0 + f_P^2 \cdot f_Q^1 + f_P^1 \cdot f_Q^2 + f_P^0 \cdot f_Q^3 = 6\cdot 4 + 21\cdot 6 + 29\cdot 4 + 14\cdot 1 = 280,$$

$$f_{\times}^{4} = f_P^4 \cdot f_Q^0 + f_P^3 \cdot f_Q^1 + f_P^2 \cdot f_Q^2 + f_P^1 \cdot f_Q^3 = 1\cdot 4 + 6\cdot 6 + 21\cdot 4 + 29\cdot 1 = 153,$$

$$f_{\times}^{5} = f_P^4 \cdot f_Q^1 + f_P^3 \cdot f_Q^2 + f_P^2 \cdot f_Q^3 = 1\cdot 6 + 6\cdot 4 + 21\cdot 1 = 51,$$

$$f_{\times}^{6} = f_P^4 f_Q^2 + f_P^3 \cdot f_Q^3 = 1\cdot 4 + 6\cdot 1 = 10,$$

$$f_{\times}^{7} = f_P^4 \cdot f_Q^3 = 1\cdot 1 = 1.$$

Consequently, the product of a tetrahedron centered on a tetrahedron has 56 vertices, 200 edges, 314 two – dimensional faces, 280 three - dimensional faces, 153 four - dimensional faces, 51 five – dimensional faces, and 10 six – dimensional faces. This product itself has dimension 7. This can be easily verified by substituting the obtained values of the numbers of faces of various dimensions into the Euler – Poincare equation 56 – 200 + 314 – 280 + 153 – 51 + 10 = 2. Thus, the Euler - Poincaré equation (1) is satisfied for dimension $d = 7$.

The Product of a Molecule of Wurzite on a *M* - Simplex

In this case, as well as the previous paragraph for polytope P, we have $f_P^o = 14, f_P^1 = 29, f_P^2 = 21, f_P^3 = 6$.

For a m – simplex the numbers of elements with different dimension are (Zhizhin, 2019 a) $f_Q^i(m) = C_{m+1}^{i+1}, i = 0,1,\ldots,m$.

For the product of a molecule of wurzite on a m - simplex in accordance of Theorem 1 of Chapter 5 we have

$$f_{\times}^{0} = f_P^0 f_Q^0 = 14\cdot C_{m+1}^{1},$$

$$f_{\times}^{1} = f_P^1 f_Q^0 + f_P^0 f_Q^1 = 29\cdot C_{m+1}^{1} + 14\cdot C_{m+1}^{2},.$$

$$f_{\times}^{2} = f_P^2 \cdot f_Q^0 + f_P^1 \cdot f_Q^1 + f_P^0 \cdot f_Q^2 = 21\cdot C_{m+1}^{1} + 29\cdot C_{m+1}^{2} + 14\cdot C_{m+1}^{3},$$

$$f_{\times}^{3} = f_P^3 \cdot f_Q^0 + f_P^2 \cdot f_Q^1 + f_P^1 \cdot f_Q^2 + f_P^0 \cdot f_Q^3 = 6\cdot C_{m+1}^{1} + 21\cdot C_{m+1}^{2} + 29\cdot C_{m+1}^{3} + 14\cdot C_{m+1}^{4},$$

$$f_{\times}^{4} = f_P^4 \cdot f_Q^0 + f_P^3 \cdot f_Q^1 + f_P^2 \cdot f_Q^2 + f_P^1 \cdot f_Q^3 + f_P^0 f_Q^4 = 1\cdot C_{m+1}^{1} + 6\cdot C_{m+1}^{2} + 21\cdot C_{m+1}^{3} + 29\cdot C_{m+1}^{4} + 14\cdot C_{m+1}^{5},$$

$$f_\times^5 = f_P^4 \cdot f_Q^1 + f_P^3 \cdot f_Q^2 + f_P^2 \cdot f_Q^3 + f_P^1 \cdot f_Q^4 + f_P^0 f_Q^5 = 1 \cdot C_{m+1}^2 + 6 \cdot C_{m+1}^3 + 21 \cdot C_{m+1}^4 + 29 \cdot C_{m+1}^5 + 14 \cdot C_{m+1}^6,$$

$$f_\times^{m+2} = f_P^4 \cdot f_Q^{m-2} + f_P^3 \cdot f_Q^{m-1} + f_P^2 \cdot f_Q^m = 1 \cdot C_{m+1}^{m-1} + 6 \cdot C_{m+1}^m + 21,$$

$$f_\times^{m+3} = f_P^4 \cdot f_Q^{m-1} + f_P^3 \cdot f_Q^m = 1 \cdot C_{m+1}^m + 6,$$

$$f_\times^{m+4} = f_P^4 \cdot f_Q^m = 1 \cdot 1 = 1.$$

Consequently, the product of a tetrahedron centered on a tetrahedron has

$14 \cdot C_{m+1}^1$ vertices,

$29 \cdot C_{m+1}^1 + 14 \cdot C_{m+1}^2$ edges,

$21 \cdot C_{m+1}^1 + 29 \cdot C_{m+1}^2 + 14 \cdot C_{m+1}^3$ two – dimensional faces,

$6 \cdot C_{m+1}^1 + 21 \cdot C_{m+1}^2 + 29 \cdot C_{m+1}^3 + 14 \cdot C_{m+1}^4$ three - dimensional faces,

$1 \cdot C_{m+1}^1 + 6 \cdot C_{m+1}^2 + 21 \cdot C_{m+1}^3 + 29 \cdot C_{m+1}^4 + 14 \cdot C_{m+1}^5$ four - dimensional faces,

$1 \cdot C_{m+1}^2 + 6 \cdot C_{m+1}^3 + 21 \cdot C_{m+1}^4 + 29 \cdot C_{m+1}^5 + 14 \cdot C_{m+1}^6$ five – dimensional faces, …,

$1 \cdot C_{m+1}^{m-1} + 6 \cdot C_{m+1}^m + 21$ $(m + 2)$ – dimensional faces, and

$1 \cdot C_{m+1}^m + 6$ $(m + 3)$ – dimensional faces.

This product itself has dimension $m + 4$.

The Product of a Molecules of Wurzite on a *M* - Cube

In this case, as well as the previous paragraph for polytope *P*, we have $f_P^o = 14, f_P^1 = 29, f_P^2 = 21, f_P^3 = 6$.

For a m – cube the numbers of elements with different dimension are (Zhizhin, 2019 a) $f_Q^i(m) = 2^{m-i} C_m^i, i = 0,1,...,m$.

For the product of tetrahedron with center on a tetrahedron in accordance of Theorem 1 of Chapter 5 we have

$$f_{\times}^{0} = f_{P}^{0} f_{Q}^{0} = 14 \cdot 2^{m},$$

$$f_{\times}^{1} = f_{P}^{1} f_{Q}^{0} + f_{P}^{0} f_{Q}^{1} = 29 \cdot 2^{m} + 14 \cdot 2^{m-1} C_{m+1}^{1},.$$

$$f_{\times}^{2} = f_{P}^{2} \cdot f_{Q}^{0} + f_{P}^{1} \cdot f_{Q}^{1} + f_{P}^{0} \cdot f_{Q}^{2} = 21 \cdot 2^{m} + 29 \cdot 2^{m-1} C_{m}^{1} + 14 \cdot 2^{m-2} C_{m}^{2},$$

$$f_{\times}^{3} = f_{P}^{3} \cdot f_{Q}^{0} + f_{P}^{2} \cdot f_{Q}^{1} + f_{P}^{1} \cdot f_{Q}^{2} + f_{P}^{0} \cdot f_{Q}^{3} = 6 \cdot 2^{m} + 21 \cdot 2^{m-1} C_{m}^{1} + 29 \cdot 2^{m-2} C_{m}^{2} + 14 \cdot 2^{m-3} C_{m}^{3},$$

$$f_{\times}^{4} = f_{P}^{4} \cdot f_{Q}^{0} + f_{P}^{3} \cdot f_{Q}^{1} + f_{P}^{2} \cdot f_{Q}^{2} + f_{P}^{1} \cdot f_{Q}^{3} + f_{P}^{0} f_{Q}^{4}$$
$$= 1 \cdot 2^{m} + 6 \cdot 2^{m-1} C_{m}^{1} + 21 \cdot 2^{m-2} C_{m}^{2} + 29 \cdot 2^{m-3} C_{m}^{3} + 14 \cdot 2^{m-4} C_{m}^{4},$$

$$f_{\times}^{5} = f_{P}^{4} \cdot f_{Q}^{1} + f_{P}^{3} \cdot f_{Q}^{2} + f_{P}^{2} \cdot f_{Q}^{3} + f_{P}^{1} \cdot f_{Q}^{4} + f_{P}^{0} f_{Q}^{5}$$
$$= 1 \cdot 2^{m-1} C_{m}^{1} + 6 \cdot 2^{m-2} C_{m}^{2} + 21 \cdot 2^{m-3} C_{m}^{3} + 29 \cdot 2^{m-4} C_{m}^{4} + 14 \cdot 2^{m-5} C_{m}^{5},$$

$$f_{\times}^{m+2} = f_{P}^{4} \cdot f_{Q}^{m-2} + f_{P}^{3} \cdot f_{Q}^{m-1} + f_{P}^{2} \cdot f_{Q}^{m} = 1 \cdot 2^{2} C_{m}^{m-2} + 6 \cdot 2 C_{m}^{m-1} + 21,$$

$$f_{\times}^{m+3} = f_{P}^{4} \cdot f_{Q}^{m-1} + f_{P}^{3} \cdot f_{Q}^{m} = 1 \cdot 2 C_{m}^{m-1} + 6,$$

$$f_{\times}^{m+4} = f_{P}^{4} \cdot f_{Q}^{m} = 1 \cdot 1 = 1.$$

Consequently, the product of a tetrahedron centered on a tetrahedron has

$14 \cdot 2^{m}$ vertices,

$29 \cdot 2^{m} + 14 \cdot 2^{m-1} C_{m}^{1}$ edges,

$21 \cdot 2^{m} + 29 \cdot 2^{m-1} C_{m}^{1} + 14 \cdot 2^{m-2} C_{m}^{2}$ two – dimensional faces,

$6 \cdot 2^{m} + 21 \cdot 2^{m-1} C_{m}^{1} + 29 \cdot 2^{m-2} C_{m}^{2} + 14 \cdot 2^{m-3} C_{m}^{3}$ three - dimensional faces,

$1 \cdot 2^{m} + 6 \cdot 2^{m-1} C_{m}^{1} + 21 \cdot 2^{m-2} C_{m+1}^{2} + 29 \cdot 2^{m-3} C_{m}^{3} + 14 \cdot 2^{m-4} C_{m}^{4}$ four - dimensional faces,

$1 \cdot 2^{m-1} C_{m}^{1} + 6 \cdot 2^{m-2} C_{m}^{2} + 21 \cdot 2^{m-3} C_{m}^{3} + 29 \cdot 2^{m-4} C_{m}^{4} + 14 \cdot 2^{m-5} C_{m}^{5}$ five – dimensional faces, …,

$1 \cdot 2^{2} C_{m}^{m-2} + 6 \cdot 2 C_{m}^{m-1} + 21$ $(m + 2)$ – dimensional faces, and

$1 \cdot 2C_m^{m-1} + 6\,(m+3)$ – dimensional faces.

This product itself has dimension $m + 4$.

The Product of a Molecule of Wurzite on a *M* – Cross - Polytope

In this case, as well as the previous paragraph for polytope *P*, we have $f_P^o = 14, f_P^1 = 29, f_P^2 = 21, f_P^3 = 6.$

For a m – cross - polytope the numbers of elements with different dimension are (Zhizhin, 2019 a) $f_Q^i(m) = 2^{1+i} C_m^{m-1-i}, i = 0,1,...,m-1.$

For the product of tetrahedron with center on a m – cross - polytope in accordance of Theorem 1 of Chapter 5 we have

$$f_\times^0 = f_P^0 f_Q^0 = 14 \cdot 2C_m^{m-1},$$

$$f_\times^1 = f_P^1 f_Q^0 + f_P^0 f_Q^1 = 29 \cdot 2C_m^{m-1} + 14 \cdot 2^2 C_m^{m-2},.$$

$$f_\times^2 = f_P^2 \cdot f_Q^0 + f_P^1 \cdot f_Q^1 + f_P^0 \cdot f_Q^2 = 21 \cdot 2C_m^{m-1} + 29 \cdot 2^2 C_m^{m-2} + 14 \cdot 2^3 C_m^{m-3},$$

$$f_\times^3 = f_P^3 \cdot f_Q^0 + f_P^2 \cdot f_Q^1 + f_P^1 \cdot f_Q^2 + f_P^0 \cdot f_Q^3 = 6 \cdot 2C_m^{m-1} + 21 \cdot 2^2 C_m^{m-2} + 29 \cdot 2^3 C_m^{m-3} + 14 \cdot 2^4 C_m^{m-4},$$

$$f_\times^4 = f_P^4 \cdot f_Q^0 + f_P^3 \cdot f_Q^1 + f_P^2 \cdot f_Q^2 + f_P^1 \cdot f_Q^3 + f_P^0 f_Q^4$$
$$= 1 \cdot 2C_m^{m-1} + 6 \cdot 2^2 C_m^{m-2} + 21 \cdot 2^3 C_m^{m-3} + 29 \cdot 2^4 C_m^{m-4} + 14 \cdot 2^5 C_m^{m-5},$$

$$f_\times^5 = f_P^4 \cdot f_Q^1 + f_P^3 \cdot f_Q^2 + f_P^2 \cdot f_Q^3 + f_P^1 \cdot f_Q^4 + f_P^0 f_Q^5$$
$$= 1 \cdot 2^2 C_m^{m-2} + 6 \cdot 2^3 C_m^{m-3} + 21 \cdot 2^4 C_m^{m-4} + 29 \cdot 2^5 C_m^{m-5} + 14 \cdot 2^6 C_m^{m-6},$$

$$f_\times^{m+2} = f_P^4 \cdot f_Q^{m-2} + f_P^3 \cdot f_Q^{m-1} + f_P^2 \cdot f_Q^m = 1 \cdot 2^{m-1} C_m^1 + 6 \cdot 2^m C_m^0 + 21,$$

$$f_\times^{m+3} = f_P^4 \cdot f_Q^{m-1} + f_P^3 \cdot f_Q^m = 1 \cdot 2^m + 6,$$

$$f_\times^{m+4} = f_P^4 \cdot f_Q^m = 1 \cdot 1 = 1.$$

Consequently, the product of a tetrahedron centered on a m – cross - polytope has

$14 \cdot 2C_m^{m-1}$ vertices,

$29\cdot 2C_m^{m-1}+14\cdot 2^2C_m^{m-2}$ edges,

$21\cdot 2C_m^{m-1}+29\cdot 2^2C_m^{m-2}+14\cdot 2^3C_m^{m-3}$ two – dimensional faces,

$6\cdot 2C_m^{m-1}+21\cdot 2^2C_m^{m-2}+29\cdot 2^3C_m^{m-3}+14\cdot 2^4C_m^{m-4}$ three - dimensional faces,

$1\cdot 2C_m^{m-1}+6\cdot 2^2C_m^{m-2}+21\cdot 2^3C_m^{m-3}+29\cdot 2^4C_m^{m-4}+14\cdot 2^5C_m^{m-5}$ four - dimensional faces,

$1\cdot 2^2C_m^{m-2}+6\cdot 2^3C_m^{m-3}+21\cdot 2^4C_m^{m-4}+29\cdot 2^5C_m^{m-5}+14\cdot 2^6C_m^{m-6}$ five – dimensional faces, …,

$1\cdot 2^{m-1}C_m^1+6\cdot 2^m+21\,(m+2)$ – dimensional faces, and

$1\bullet 2^m+6\ (m+3)$ – dimensional faces.

This product itself has dimension $m+4$.

The Product of a Molecule of Fluorite on One – Dimensional Segment

A series of binary compound have a structure of the mineral fluorite CdF_2, $MnCl_2$, $FeCl_2$, $CoCl_2$, $NiCl_2$, $ZnCl_2$, $CdCl_2$, Cu_2S, CaF_2, SrF_2, BaF_2, RbF_2, AlF_8, $MgCl_2$, $CaCl_2$, $SrCl_2$, $BaCl_2$, $PbCl_2$, $AlCl_3$, $SiCl_4$, Li_2O, Na_2O, K_2O, Rb_2O, Cs_2O, Li_2S, Na_2S, K_2S, Rb_2S. Example of compound $MnCl_2$ is presented on Figure 13 in Chapter 1. From Figure 13 of Chapter 1 it follows that the number of vertices is 14, the number of edges is 36, the number of flat faces is 30 (sum from number of triangles 24 and number of rectangles 6). The number three - dimension shape is 8. This is a poly - incident polytope, since six of its vertices are incident to four edges, and eight vertices are incident to six edges.

In accordance with Chapter 5, we denote the molecule of fluorite by P, and the one - dimensional segment by Q. Thus, for a molecule fluorite, in accordance with the notation of Chapter 5, we have $f_P^o=14, f_P^1=36, f_P^2=30, f_P^3=8$.

Substitute this numbers $f^i(P)$ in the equation of the Euler – Poincaré (1) it is easy to verify that this equation for the polytope is satisfied for $d=4$; $14-36+30-8=0$.

For one – dimensional segment we have $f_Q^o=2, f_Q^1=1$.

For the product of a molecule of fluorite on one – dimensional segment in accordance of Theorem 1 of Chapter 5 we have

$$f_\times^0=f_P^0f_Q^0=14\cdot 2=28,$$

$$f_\times^1=f_P^1f_Q^0+f_P^0f_Q^1=36\cdot 2+14\cdot 1=86,$$

$$f_{\times}^{2} = f_{P}^{2} \cdot f_{Q}^{0} + f_{P}^{1} \cdot f_{Q}^{1} = 30 \cdot 2 + 36 \cdot 1 = 96,$$

$$f_{\times}^{3} = f_{P}^{3} \cdot f_{Q}^{0} + f_{P}^{2} \cdot f_{Q}^{1} = 8 \cdot 2 + 30 \cdot 1 = 46,$$

$$f_{\times}^{4} = f_{P}^{4} \cdot f_{Q}^{0} + f_{P}^{3} \cdot f_{Q}^{1} = 1 \cdot 2 + 8 \cdot 1 = 10,$$

$$f_{\times}^{5} = f_{P}^{4} \cdot f_{Q}^{1} = 1 \cdot 1 = 1.$$

Consequently, the product of an octahedron centered on one - dimensional segment has 28 vertices, 86 edges, 96 two – dimensional faces, 46 three - dimensional faces, 10 four - dimensional faces. This product itself has dimension 5. This can be easily verified by substituting the obtained values of the numbers of faces of various dimensions into the Euler – Poincare equation 28 – 86 + 96 – 46 + 10 = 2. Thus, the Euler - Poincaré equation (1) is satisfied for dimension $d = 5$.

The Product of a Molecule of Fluorite on Triangle

In this case, as well as the previous paragraph for polytope P, we have $f_{P}^{o} = 14, f_{P}^{1} = 36, f_{P}^{2} = 30, f_{P}^{3} = 8$.

For triangle the numbers of elements with different dimension are $f_{Q}^{o} = 3, f_{Q}^{1} = 3, f_{Q}^{2} = 1$.

For the product of an octahedron with center on a triangle in accordance of Theorem 1 of Chapter 5 we have

$$f_{\times}^{0} = f_{P}^{0} f_{Q}^{0} = 14 \cdot 3 = 42,$$

$$f_{\times}^{1} = f_{P}^{1} f_{Q}^{0} + f_{P}^{0} f_{Q}^{1} = 36 \cdot 3 + 14 \cdot 3 = 150, .$$

$$f_{\times}^{2} = f_{P}^{2} \cdot f_{Q}^{0} + f_{P}^{1} \cdot f_{Q}^{1} + f_{P}^{0} \cdot f_{Q}^{2} = 30 \cdot 3 + 36 \cdot 3 + 14 \cdot 1 = 212,$$

$$f_{\times}^{3} = f_{P}^{3} \cdot f_{Q}^{0} + f_{P}^{2} \cdot f_{Q}^{1} + f_{P}^{1} \cdot f_{Q}^{2} = 8 \cdot 3 + 30 \cdot 3 + 36 \cdot 1 = 150,$$

$$f_{\times}^{4} = f_{P}^{4} \cdot f_{Q}^{0} + f_{P}^{3} \cdot f_{Q}^{1} + f_{P}^{2} \cdot f_{Q}^{2} = 3 \cdot 1 + 3 \cdot 8 + 30 \cdot 1 = 57,$$

$$f_{\times}^{5} = f_{P}^{4} f_{Q}^{1} + f_{P}^{3} \cdot f_{Q}^{2} = 1 \cdot 3 + 8 \cdot 1 = 11,$$

$$f_{\times}^{6} = f_{P}^{4} \cdot f_{Q}^{2} = 1 \cdot 1 = 1.$$

Consequently, the product of a molecule of wurzite on a triangle has 42 vertices, 150 edges, 212 two – dimensional faces, 150 three - dimensional faces, 57 four - dimensional faces and 11 five – dimensional faces. This product itself has dimension 6. This can be easily verified by substituting the obtained values of the numbers of faces of various dimensions into the Euler – Poincare equation 42 – 150 + 212 – 150 + 57 - 11 = 0. Thus, the Euler - Poincaré equation (1) is satisfied for dimension $d = 6$.

The Product of a Molecule of Fluorite on a Tetrahedron

In this case, as well as the previous paragraph for polytope P, we have $f_P^o = 14, f_P^1 = 36, f_P^2 = 30, f_P^3 = 8$.

For a tetrahedron the numbers of elements with different dimension are $f_Q^o = 4, f_Q^1 = 6, f_Q^2 = 4, f_Q^3 = 1$.

For the product of an octahedron with center on a tetrahedron in accordance of Theorem 1 of Chapter 5 we have

$$f_\times^0 = f_P^0 f_Q^0 = 14 \cdot 4 = 56,$$

$$f_\times^1 = f_P^1 f_Q^0 + f_P^0 f_Q^1 = 36 \cdot 4 + 14 \cdot 6 = 228,.$$

$$f_\times^2 = f_P^2 \cdot f_Q^0 + f_P^1 \cdot f_Q^1 + f_P^0 \cdot f_Q^2 = 30 \cdot 4 + 36 \cdot 6 + 14 \cdot 4 = 392,$$

$$f_\times^3 = f_P^3 \cdot f_Q^0 + f_P^2 \cdot f_Q^1 + f_P^1 \cdot f_Q^2 + f_P^0 \cdot f_Q^3 = 8 \cdot 4 + 30 \cdot 6 + 36 \cdot 4 + 14 \cdot 1 = 370,$$

$$f_\times^4 = f_P^4 \cdot f_Q^0 + f_P^3 \cdot f_Q^1 + f_P^2 \cdot f_Q^2 + f_P^1 \cdot f_Q^3 = 1 \cdot 4 + 8 \cdot 6 + 30 \cdot 4 + 36 \cdot 1 = 208,$$

$$f_\times^5 = f_P^4 \cdot f_Q^1 + f_P^3 \cdot f_Q^2 + f_P^2 \cdot f_Q^3 = 1 \cdot 6 + 8 \cdot 4 + 30 \cdot 1 = 68,$$

$$f_\times^6 = f_P^4 f_Q^2 + f_P^3 \cdot f_Q^3 = 1 \cdot 4 + 8 \cdot 1 = 12,$$

$$f_\times^7 = f_P^4 \cdot f_Q^3 = 1 \cdot 1 = 1.$$

Consequently, the product of a tetrahedron centered on a tetrahedron has 56 vertices, 228 edges, 392 two – dimensional faces, 370 three - dimensional faces, 208 four - dimensional faces, 68 five – dimensional faces, and 12 six – dimensional faces. This product itself has dimension 7. This can be easily verified by substituting the obtained values of the numbers of faces of various dimensions into the Euler – Poincare equation 56 – 228 + 392– 370 + 208 – 681 + 12 = 2. Thus, the Euler - Poincaré equation (1) is satisfied for dimension $d = 7$.

The Product of a Molecule of Fluorite on a *M* - Simplex

In this case, as well as the previous paragraph for polytope P, we have $f_P^o = 14, f_P^1 = 36, f_P^2 = 30, f_P^3 = 8$.

For a m – simplex the numbers of elements with different dimension are (Zhizhin, 2019 a) $f_Q^i(m) = C_{m+1}^{i+1}, i = 0,1,...,m.$

For the product of a molecule of fluorite on a m - simplex in accordance of Theorem 1 of Chapter 5 we have

$$f_\times^0 = f_P^0 f_Q^0 = 14 \cdot C_{m+1}^1,$$

$$f_\times^1 = f_P^1 f_Q^0 + f_P^0 f_Q^1 = 36 \cdot C_{m+1}^1 + 14 \cdot C_{m+1}^2,.$$

$$f_\times^2 = f_P^2 \cdot f_Q^0 + f_P^1 \cdot f_Q^1 + f_P^0 \cdot f_Q^2 = 30 \cdot C_{m+1}^1 + 36 \cdot C_{m+1}^2 + 14 \cdot C_{m+1}^3,$$

$$f_\times^3 = f_P^3 \cdot f_Q^0 + f_P^2 \cdot f_Q^1 + f_P^1 \cdot f_Q^2 + f_P^0 \cdot f_Q^3 = 8 \cdot C_{m+1}^1 + 30 \cdot C_{m+1}^2 + 36 \cdot C_{m+1}^3 + 14 \cdot C_{m+1}^4,$$

$$f_\times^4 = f_P^4 \cdot f_Q^0 + f_P^3 \cdot f_Q^1 + f_P^2 \cdot f_Q^2 + f_P^1 \cdot f_Q^3 + f_P^0 f_Q^4 = 1 \cdot C_{m+1}^1 + 8 \cdot C_{m+1}^2 + 30 \cdot C_{m+1}^3 + 36 \cdot C_{m+1}^4 + 14 \cdot C_{m+1}^5,$$

$$f_\times^5 = f_P^4 \cdot f_Q^1 + f_P^3 \cdot f_Q^2 + f_P^2 \cdot f_Q^3 + f_P^1 \cdot f_Q^4 + f_P^0 f_Q^5 = 1 \cdot C_{m+1}^2 + 8 \cdot C_{m+1}^3 + 30 \cdot C_{m+1}^4 + 36 \cdot C_{m+1}^5 + 14 \cdot C_{m+1}^6,$$

$$f_\times^{m+2} = f_P^4 \cdot f_Q^{m-2} + f_P^3 \cdot f_Q^{m-1} + f_P^2 \cdot f_Q^m = 1 \cdot C_{m+1}^{m-1} + 8 \cdot C_{m+1}^m + 30,$$

$$f_\times^{m+3} = f_P^4 \cdot f_Q^{m-1} + f_P^3 \cdot f_Q^m = 1 \cdot C_{m+1}^m + 8,$$

$$f_\times^{m+4} = f_P^4 \cdot f_Q^m = 1 \cdot 1 = 1.$$

Consequently, the product of a tetrahedron centered on a tetrahedron has

$14 \cdot C_{m+1}^1$ vertices,

$36 \cdot C_{m+1}^1 + 14 \cdot C_{m+1}^2$ edges,

$30 \cdot C_{m+1}^1 + 36 \cdot C_{m+1}^2 + 14 \cdot C_{m+1}^3$ two – dimensional faces,

$8 \cdot C_{m+1}^1 + 30 \cdot C_{m+1}^2 + 36 \cdot C_{m+1}^3 + 14 \cdot C_{m+1}^4$ three - dimensional faces,

$1 \cdot C_{m+1}^1 + 8 \cdot C_{m+1}^2 + 30 \cdot C_{m+1}^3 + 36 \cdot C_{m+1}^4 + 14 \cdot C_{m+1}^5$ four - dimensional faces,

$1\cdot C_{m+1}^{2}+8\cdot C_{m+1}^{3}+30\cdot C_{m+1}^{4}+36\cdot C_{m+1}^{5}+14\cdot C_{m+1}^{6}$ five – dimensional faces, …,

$1\cdot C_{m+1}^{m-1}+8\cdot C_{m+1}^{m}+30$ $(m+2)$ – dimensional faces, and

$1\cdot C_{m+1}^{m}+8$ $(m+3)$ – dimensional faces.

This product itself has dimension $m+4$.

The Product of a Molecules of Fluorite on a *M* - Cube

In this case, as well as the previous paragraph for polytope *P*, we have $f_P^o=14, f_P^1=36, f_P^2=30, f_P^3=8$.

For a m – cube the numbers of elements with different dimension are (Zhizhin, 2019 a) $f_Q^i(m)=2^{m-i}C_m^i, i=0,1,\dots,m.$

For the product of tetrahedron with center on a tetrahedron in accordance of Theorem 1 of Chapter 5 we have

$$f_\times^0=f_P^0 f_Q^0=14\cdot 2^m,$$

$$f_\times^1=f_P^1 f_Q^0+f_P^0 f_Q^1=36\cdot 2^m+14\cdot 2^{m-1}C_m^1,.$$

$$f_\times^2=f_P^2\cdot f_Q^0+f_P^1\cdot f_Q^1+f_P^0\cdot f_Q^2=30\cdot 2^m+36\cdot 2^{m-1}C_m^1+14\cdot 2^{m-2}C_m^2,$$

$$f_\times^3=f_P^3\cdot f_Q^0+f_P^2\cdot f_Q^1+f_P^1\cdot f_Q^2+f_P^0\cdot f_Q^3=8\cdot 2^m+30\cdot 2^{m-1}C_m^1+36\cdot 2^{m-2}C_m^2+14\cdot 2^{m-3}C_m^3,$$

$$f_\times^4=f_P^4\cdot f_Q^0+f_P^3\cdot f_Q^1+f_P^2\cdot f_Q^2+f_P^1\cdot f_Q^3+f_P^0 f_Q^4$$
$$=1\cdot 2^m+8\cdot 2^{m-1}C_m^1+30\cdot 2^{m-2}C_m^2+36\cdot 2^{m-3}C_m^3+14\cdot 2^{m-4}C_m^4,$$

$$f_\times^5=f_P^4\cdot f_Q^1+f_P^3\cdot f_Q^2+f_P^2\cdot f_Q^3+f_P^1\cdot f_Q^4+f_P^0 f_Q^5$$
$$=1\cdot 2^{m-1}C_m^1+8\cdot 2^{m-2}C_m^2+30\cdot 2^{m-3}C_m^3+36\cdot 2^{m-4}C_m^4+14\cdot 2^{m-5}C_m^5,$$

$$f_\times^{m+2}=f_P^4\cdot f_Q^{m-2}+f_P^3\cdot f_Q^{m-1}+f_P^2\cdot f_Q^m=1\cdot 2^2C_m^{m-2}+8\cdot 2C_m^{m-1}+30,$$

$$f_\times^{m+3}=f_P^4\cdot f_Q^{m-1}+f_P^3\cdot f_Q^m=1\cdot 2C_m^{m-1}+8,$$

$$f_\times^{m+4}=f_P^4\cdot f_Q^m=1\cdot 1=1.$$

Consequently, the product of a tetrahedron centered on a tetrahedron has

$14 \cdot 2^m$ vertices,

$36 \cdot 2^m + 14 \cdot 2^{m-1} C_m^1$ edges,

$30 \cdot 2^m + 36 \cdot 2^{m-1} C_m^1 + 14 \cdot 2^{m-2} C_m^2$ two – dimensional faces,

$8 \cdot 2^m + 30 \cdot 2^{m-1} C_m^1 + 36 \cdot 2^{m-2} C_m^2 + 14 \cdot 2^{m-3} C_m^3$ three - dimensional faces,

$1 \cdot 2^m + 8 \cdot 2^{m-1} C_m^1 + 30 \cdot 2^{m-2} C_{m+1}^2 + 36 \cdot 2^{m-3} C_m^3 + 14 \cdot 2^{m-4} C_m^4$ four - dimensional faces,

$1 \cdot 2^{m-1} C_m^1 + 8 \cdot 2^{m-2} C_m^2 + 30 \cdot 2^{m-3} C_m^3 + 36 \cdot 2^{m-4} C_m^4 + 14 \cdot 2^{m-5} C_m^5$ five – dimensional faces, …,

$1 \cdot 2^2 C_m^{m-2} + 8 \cdot 2 C_m^{m-1} + 30$ $(m + 2)$ – dimensional faces, and

$1 \cdot 2 C_m^{m-1} + 8$ $(m + 3)$ – dimensional faces.

This product itself has dimension $m + 4$.

The Product of a Molecule of Fluorite on a *M* – Cross - Polytope

In this case, as well as the previous paragraph for polytope P, we have $f_P^o = 14, f_P^1 = 36, f_P^2 = 30, f_P^3 = 8$.

For a m – cross - polytope the numbers of elements with different dimension are (Zhizhin, 2019 a) $f_Q^i(m) = 2^{1+i} C_m^{m-1-i}, i = 0, 1, ..., m-1$.

For the product of tetrahedron with center on a m – cross - polytope in accordance of Theorem 1 of Chapter 5 we have

$$f_\times^0 = f_P^0 f_Q^0 = 14 \cdot 2 C_m^{m-1},$$

$$f_\times^1 = f_P^1 f_Q^0 + f_P^0 f_Q^1 = 36 \cdot 2 C_m^{m-1} + 14 \cdot 2^2 C_m^{m-2},.$$

$$f_\times^2 = f_P^2 \cdot f_Q^0 + f_P^1 \cdot f_Q^1 + f_P^0 \cdot f_Q^2 = 30 \cdot 2 C_m^{m-1} + 36 \cdot 2^2 C_m^{m-2} + 14 \cdot 2^3 C_m^{m-3},$$

$$f_\times^3 = f_P^3 \cdot f_Q^0 + f_P^2 \cdot f_Q^1 + f_P^1 \cdot f_Q^2 + f_P^0 \cdot f_Q^3 = 8 \cdot 2 C_m^{m-1} + 30 \cdot 2^2 C_m^{m-2} + 36 \cdot 2^3 C_m^{m-3} + 14 \cdot 2^4 C_m^{m-4},$$

$$f_\times^4 = f_P^4 \cdot f_Q^0 + f_P^3 \cdot f_Q^1 + f_P^2 \cdot f_Q^2 + f_P^1 \cdot f_Q^3 + f_P^0 f_Q^4$$
$$= 1 \cdot 2C_m^{m-1} + 8 \cdot 2^2 C_m^{m-2} + 30 \cdot 2^3 C_m^{m-3} + 36 \cdot 2^4 C_m^{m-4} + 14 \cdot 2^5 C_m^{m-5},$$

$$f_\times^5 = f_P^4 \cdot f_Q^1 + f_P^3 \cdot f_Q^2 + f_P^2 \cdot f_Q^3 + f_P^1 \cdot f_Q^4 + f_P^0 f_Q^5$$
$$= 1 \cdot 2^2 C_m^{m-2} + 8 \cdot 2^3 C_m^{m-3} + 30 \cdot 2^4 C_m^{m-4} + 36 \cdot 2^5 C_m^{m-5} + 14 \cdot 2^6 C_m^{m-6},$$

$$f_\times^{m+2} = f_P^4 \cdot f_Q^{m-2} + f_P^3 \cdot f_Q^{m-1} + f_P^2 \cdot f_Q^m = 1 \cdot 2^{m-1} C_m^1 + 8 \cdot 2^m C_m^0 + 30,$$

$$f_\times^{m+3} = f_P^4 \cdot f_Q^{m-1} + f_P^3 \cdot f_Q^m = 1 \cdot 2^m + 8,$$

$$f_\times^{m+4} = f_P^4 \cdot f_Q^m = 1 \cdot 1 = 1.$$

Consequently, the product of a tetrahedron centered on a m – cross - polytope has

$14 \cdot 2C_m^{m-1}$ vertices,

$36 \cdot 2C_m^{m-1} + 14 \cdot 2^2 C_m^{m-2}$ edges,

$30 \cdot 2C_m^{m-1} + 36 \cdot 2^2 C_m^{m-2} + 14 \cdot 2^3 C_m^{m-3}$ two – dimensional faces,

$8 \cdot 2C_m^{m-1} + 30 \cdot 2^2 C_m^{m-2} + 36 \cdot 2^3 C_m^{m-3} + 14 \cdot 2^4 C_m^{m-4}$ three - dimensional faces,

$1 \cdot 2C_m^{m-1} + 8 \cdot 2^2 C_m^{m-2} + 30 \cdot 2^3 C_m^{m-3} + 36 \cdot 2^4 C_m^{m-4} + 14 \cdot 2^5 C_m^{m-5}$ four - dimensional faces,

$1 \cdot 2^2 C_m^{m-2} + 8 \cdot 2^3 C_m^{m-3} + 30 \cdot 2^4 C_m^{m-4} + 36 \cdot 2^5 C_m^{m-5} + 14 \cdot 2^6 C_m^{m-6}$ five – dimensional faces, …,

$1 \cdot 2^{m-1} C_m^1 + 8 \cdot 2^m + 30$ $(m + 2)$ – dimensional faces, and

$1 \bullet 2^m + 8$ $(m + 3)$ – dimensional faces.

This product itself has dimension $m + 4$.

The Product of a Cluster of Similar Sells (Plato Bodies) with the Number of Faces j with a Common Center on a One - Dimensional Segment

In the literature, various intermetallic clusters are known (Mackay, 1962; Bergman, et al.,1952, 1957; Samson, 1972; Pauling, 1960; and others). Most attention was paid to clusters of icosahedrons. The second chapter of this monograph defines the dimension of a cluster of *N* free shells in the form of Plato bodies with a common center when there is an atom in the common center and when it is not there. In the framework of this chapter, we will consider the results of the product of such clusters on polytopes of higher dimension. For the convenience of writing analytical expressions, we use the following notation. The cluster, in which there is no atom in the center of the cluster, will be denoted by the subscript *0*. The cluster, which has an atom in the center, will be denoted by the subscript *a*. In Theorem 4 of Chapter 2 of this book, formulas are established for the numbers of faces of various dimensions in a cluster of *N* similar shells in the form of Plato bodies with the number of faces *j* in each shell with a common center. If there is no atom in the common center, then these formulas, taking into account additional designation conditions, have the form

$$f_{P_0}^0 = Nn_j, f_{P_0}^1 = N\frac{n_j t_j}{2} + (N-1)n_j, f_{P_0}^2 = Nj + (N-1)\frac{n_j t_j}{2}, f_{P_0}^3 = (N-1)\,\text{j} + N,$$
$$f_{P_0}^4 = C_N^2, f_{P_0}^5 = C_N^3, \ldots, f_{P_0}^{N+1} = C_N^{N-1}. \quad (2)$$

For one – dimensional segment we have $f_Q^o = 2, f_Q^1 = 1$.

For the product of a cluster of *N* similar shells in the form of Plato bodies with the number of faces *j* in each shell with a common center (if there is no atom in the common center) on one – dimensional segment in accordance of Theorem 1 of Chapter 5 we have

$$f_\times^0 = f_{P_0}^0 f_Q^0 = f_{P_0}^0 2;$$

$$f_\times^1 = f_{P_0}^1 f_Q^0 + f_{P_0}^0 f_Q^1 = f_{P_0}^1 2 + f_{P_0}^0;$$

$$f_\times^2 = f_{P_0}^2 \cdot f_Q^0 + f_{P_0}^1 \cdot f_Q^1 = f_{P_0}^2 \cdot 2 + f_{P_0}^1 \cdot 1,$$

$$f_\times^3 = f_{P_0}^3 \cdot f_Q^0 + f_{P_0}^2 \cdot f_Q^1 = f_{P_0}^3 \cdot 2 + f_{P_0}^2 \cdot 1,$$

$$f_\times^4 = f_{P_0}^4 \cdot f_Q^0 + f_{P_0}^3 \cdot f_Q^1 = f_{P_0}^4 \cdot 2 + f_{P_0}^3 \cdot 1,$$

$$f_\times^{N+3} = f_{P_0}^{N+2} \cdot f_Q^1 = 1 \cdot 1 = 1..$$

Consequently, the product of a cluster of *N* similar shells in the form of Plato bodies with the number of faces *j* in each shell with a common center (if there is no atom in the common center) on one - dimensional segment has $f_{P_0}^{0}2$ vertices, $f_{P_0}^{1}2+f_{P_0}^{0}$ edges, $f_{P_0}^{2}\cdot 2+f_{P_0}^{1}\cdot 1$ two – dimensional faces, $f_{P_0}^{3}\cdot 2+f_{P_0}^{2}\cdot 1$ three - dimensional faces, $f_{P_0}^{4}\cdot 2+f_{P_0}^{3}\cdot 1$ four - dimensional faces and go on. This product itself has dimension N + 3. In these relations, the quantities $f_{P_0}^{i}, i=0\div(N+2)$ are given by equalities (2).

The Product of a Cluster of Similar Sells (Plato Bodies) with the Number of Faces j with a Common Center on a Triangle

For a triangle we have $f_Q^{o}=3, f_Q^{1}=3, f_Q^{2}=1.$

For the product of a cluster of N similar shells in the form of Plato bodies with the number of faces j in each shell with a common center (if there is no atom in the common center) on triangle in accordance of Theorem 1 of Chapter 5 we have

$$f_{\times}^{0}=f_{P_0}^{0}f_Q^{0}=f_{P_0}^{0}3;$$

$$f_{\times}^{1}=f_{P_0}^{1}f_Q^{0}+f_{P_0}^{0}f_Q^{1}=f_{P_0}^{1}3+f_{P_0}^{0}3;$$

$$f_{\times}^{2}=f_{P_0}^{2}\cdot f_Q^{0}+f_{P_0}^{1}\cdot f_Q^{1}+f_{P_0}^{0}\cdot f_Q^{2}=f_{P_0}^{2}\cdot 3+f_{P_0}^{1}\cdot 3+f_{P_0}^{0}\cdot 1,$$

$$f_{\times}^{3}=f_{P_0}^{3}\cdot f_Q^{0}+f_{P_0}^{2}\cdot f_Q^{1}+f_{P_0}^{1}\cdot f_Q^{2}=f_{P_0}^{3}\cdot 3+f_{P_0}^{2}\cdot 3+f_{P_0}^{1}\cdot 1,$$

$$f_{\times}^{4}=f_{P_0}^{4}\cdot f_Q^{0}+f_{P_0}^{3}\cdot f_Q^{1}+f_{P_0}^{2}\cdot f_Q^{2}=f_{P_0}^{4}\cdot 3+f_{P_0}^{3}\cdot 3+f_{P_0}^{2}\cdot 1,$$

$$f_{\times}^{N+3}=f_{P_0}^{N+2}\cdot f_Q^{1}+f_{P_0}^{N+1}\cdot f_Q^{2}=f_{P_0}^{N+2}\cdot 3+f_{P_0}^{N+1}\cdot 1;$$
$$f_{\times}^{N+4}=f_{P_0}^{N+2}\cdot f_Q^{2}=1\cdot 1=1.$$

Consequently, the product of a cluster of N similar shells in the form of Plato bodies with the number of faces j in each shell with a common center (if there is no atom in the common center) on triangle $f_{P_0}^{0}2$ vertices, $f_{P_0}^{1}2+f_{P_0}^{0}$ edges, $f_{P_0}^{2}\cdot 2+f_{P_0}^{1}\cdot 1$ two – dimensional faces, $f_{P_0}^{3}\cdot 2+f_{P_0}^{2}\cdot 1$ three - dimensional faces, $f_{P_0}^{4}\cdot 2+f_{P_0}^{3}\cdot 1$ four - dimensional faces and go on. This product itself has dimension N + 4. In these relations, the quantities $f_{P_0}^{i}, i=0\div(N+2)$ are given by equalities (2).

The Product of a Cluster of Similar Sells (Plato Bodies) with the Number of Faces j with a Common Center on a Tetrahedron

For a tetrahedron we have $f_Q^{o}=4, f_Q^{1}=6, f_Q^{2}=4, f_Q^{3}=1.$

For the product of a cluster of N similar shells in the form of Plato bodies with the number of faces j in each shell with a common center (if there is no atom in the common center) on tetrahedron in accordance of Theorem 1 of Chapter 5 we have

$$f_{\times}^{0}=f_{P_0}^{0}f_{Q}^{0}=f_{P_0}^{0}\cdot 4;$$

$$f_{\times}^{1}=f_{P_0}^{1}f_{Q}^{0}+f_{P_0}^{0}f_{Q}^{1}=f_{P_0}^{1}\cdot 4+f_{P_0}^{0}\cdot 6;$$

$$f_{\times}^{2}=f_{P_0}^{2}\cdot f_{Q}^{0}+f_{P_0}^{1}\cdot f_{Q}^{1}+f_{P_0}^{0}\cdot f_{Q}^{2}=f_{P_0}^{2}\cdot 4+f_{P_0}^{1}\cdot 6+f_{P_0}^{0}\cdot 4,$$

$$f_{\times}^{3}=f_{P_0}^{3}\cdot f_{Q}^{0}+f_{P_0}^{2}\cdot f_{Q}^{1}+f_{P_0}^{1}\cdot f_{Q}^{2}+f_{P_0}^{0}\cdot f_{Q}^{3}=f_{P_0}^{3}\cdot 4+f_{P_0}^{2}\cdot 6+f_{P_0}^{1}\cdot 4+f_{P_0}^{0}\cdot 1,$$

$$f_{\times}^{4}=f_{P_0}^{4}\cdot f_{Q}^{0}+f_{P_0}^{3}\cdot f_{Q}^{1}+f_{P_0}^{2}\cdot f_{Q}^{2}+f_{P_0}^{1}\cdot f_{Q}^{3}=f_{P_0}^{4}\cdot 4+f_{P_0}^{3}\cdot 6+f_{P_0}^{2}\cdot 4+f_{P_0}^{1}\cdot 1,$$

$$f_{\times}^{N+4}=f_{P_0}^{N+2}\cdot f_{Q}^{2}+f_{P_0}^{N+1}\cdot f_{Q}^{3}=f_{P_0}^{N+2}\cdot 4+f_{P_0}^{N+1}\cdot 1;$$
$$f_{\times}^{N+5}=f_{P_0}^{N+2}\cdot f_{Q}^{3}=1\cdot 1=1.$$

Consequently, the product of a cluster of N similar shells in the form of Plato bodies with the number of faces j in each shell with a common center (if there is no atom in the common center) on tetrahedron has $f_{P_0}^{0}4$ vertices, $f_{P_0}^{1}\cdot 4+f_{P_0}^{0}\cdot 6$ edges, $f_{P_0}^{2}\cdot 4+f_{P_0}^{1}\cdot 6+f_{P_0}^{0}\cdot 4$ two – dimensional faces, $f_{P_0}^{3}\cdot 4+f_{P_0}^{2}\cdot 6+f_{P_0}^{1}\cdot 4+f_{P_0}^{0}\cdot 1$ three - dimensional faces, $f_{P_0}^{4}\cdot 4+f_{P_0}^{3}\cdot 6+f_{P_0}^{2}\cdot 4+f_{P_0}^{1}\cdot 1$ four - dimensional faces and go on. This product itself has dimension N + 5. In these relations, the quantities $f_{P_0}^{i}, i=0\div(N+2)$ are given by equalities (2).

The Product of a Cluster of Similar Sells (Plato Bodies) with the Number of Fces j with a Common Center on a *D* - Simplex

For a d - simplex we have $f_{Q}^{i}(d)=C_{d+1}^{i+1}, i=1,2,...,d.$

For the product of a cluster of N similar shells in the form of Plato bodies with the number of faces j in each shell with a common center (if there is no atom in the common center) on a d – simplex in accordance of Theorem 1 of Chapter 5 we have

$$f_{\times}^{0}=f_{P_0}^{0}f_{Q}^{0}=f_{P_0}^{0}\cdot C_{d+1}^{1};$$

$$f_{\times}^{1}=f_{P_0}^{1}f_{Q}^{0}+f_{P_0}^{0}f_{Q}^{1}=f_{P_0}^{1}\cdot C_{d+1}^{1}+f_{P_0}^{0}\cdot C_{d+1}^{2};$$

$$f_{\times}^{2}=f_{P_0}^{2}\cdot f_{Q}^{0}+f_{P_0}^{1}\cdot f_{Q}^{1}+f_{P_0}^{0}\cdot f_{Q}^{2}=f_{P_0}^{2}\cdot C_{d+1}^{1}+f_{P_0}^{1}\cdot C_{d+1}^{2}+f_{P_0}^{0}\cdot C_{d+1}^{3},$$

$$f_{\times}^{3} = f_{P_0}^{3} \cdot f_Q^{0} + f_{P_0}^{2} \cdot f_Q^{1} + f_{P_0}^{1} \cdot f_Q^{2} + f_{P_0}^{0} \cdot f_Q^{3} = f_{P_0}^{3} \cdot C_{d+1}^{1} + f_{P_0}^{2} \cdot C_{d+1}^{2} + f_{P_0}^{1} \cdot C_{d+1}^{3} + f_{P_0}^{0} \cdot C_{d+1}^{4},$$

$$f_{\times}^{4} = f_{P_0}^{4} \cdot f_Q^{0} + f_{P_0}^{3} \cdot f_Q^{1} + f_{P_0}^{2} \cdot f_Q^{2} + f_{P_0}^{1} \cdot f_Q^{3} + f_{P_0}^{0} \cdot f_Q^{4}$$
$$= f_{P_0}^{4} \cdot C_{d+1}^{1} + f_{P_0}^{3} \cdot C_{d+1}^{2} + f_{P_0}^{2} \cdot C_{d+1}^{3} + f_{P_0}^{1} \cdot C_{d+1}^{4} + f_{P_0}^{0} \cdot C_{d+1}^{5},$$

$$f_{\times}^{N+1+d} = f_{P_0}^{N+2} \cdot f_Q^{d-1} + f_{P_0}^{N+1} \cdot f_Q^{d} = f_{P_0}^{N+2} \cdot C_{d+1}^{d} + f_{P_0}^{N+1} \cdot 1;$$
$$f_{\times}^{N+2+d} = f_{P_0}^{N+2} \cdot f_Q^{d} = 1 \cdot 1 = 1.$$

Consequently, the product of a cluster of *N* similar shells in the form of Plato bodies with the number of faces *j* in each shell with a common center (if there is no atom in the common center) on d - simplex has

$f_{P_0}^{0} C_{d+1}^{1}$ vertices,

$f_{P_0}^{1} \cdot C_{d+1}^{1} + f_{P_0}^{0} \cdot C_{d+1}^{2}$ edges,

$f_{P_0}^{2} \cdot C_{d+1}^{1} + f_{P_0}^{1} \cdot C_{d+1}^{2} + f_{P_0}^{0} \cdot C_{d+1}^{3}$ two – dimensional faces,

$f_{P_0}^{3} \cdot C_{d+1}^{1} + f_{P_0}^{2} \cdot C_{d+1}^{2} + f_{P_0}^{1} \cdot C_{d+1}^{3} + f_{P_0}^{0} \cdot C_{d+1}^{4}$ three - dimensional faces,

$f_{P_0}^{4} \cdot C_{d+1}^{1} + f_{P_0}^{3} \cdot C_{d+1}^{2} + f_{P_0}^{2} \cdot C_{d+1}^{3} + f_{P_0}^{1} \cdot C_{d+1}^{4} + f_{P_0}^{0} \cdot C_{d+1}^{5}$ four - dimensional faces and go on.

This product itself has dimension *N* + 2 + *d*. In these relations, the quantities $f_{P_0}^{i}, i = 0 \div (N+2)$ are given by equalities (2).

The Product of a Cluster of Similar Sells (Plato Bodies) with the Number of Faces j with a Common Center on a *D* - Cube

For a *d* - cube we have $f_Q^{i}(d) = 2^{d-i} C_d^{i}, i = 0,1,...,d.$

For the product of a cluster of *N* similar shells in the form of Plato bodies with the number of faces *j* in each shell with a common center (if there is no atom in the common center) on a *d* – cube in accordance of Theorem 1 of Chapter 5 we have

$$f_{\times}^{0} = f_{P_0}^{0} f_Q^{0} = f_{P_0}^{0} \cdot 2^{d} C_d^{0};$$

$$f_{\times}^{1} = f_{P_0}^{1} f_Q^{0} + f_{P_0}^{0} f_Q^{1} = f_{P_0}^{1} \cdot 2^{d} C_d^{0} + f_{P_0}^{0} \cdot 2^{d-1} C_d^{1};$$

$$f_\times^2 = f_{P_0}^2 \cdot f_Q^0 + f_{P_0}^1 \cdot f_Q^1 + f_{P_0}^0 \cdot f_Q^2 = f_{P_0}^2 \cdot 2^d C_d^0 + f_{P_0}^1 \cdot 2^{d-1} C_d^1 + f_{P_0}^0 \cdot 2^{d-2} C_d^2,$$

$$f_\times^3 = f_{P_0}^3 \cdot f_Q^0 + f_{P_0}^2 \cdot f_Q^1 + f_{P_0}^1 \cdot f_Q^2 + f_{P_0}^0 \cdot f_Q^3 = f_{P_0}^3 \cdot 2^d C_d^0 + f_{P_0}^2 \cdot 2^{d-1} C_d^1 + f_{P_0}^1 \cdot 2^{d-2} C_d^2 + f_{P_0}^0 \cdot 2^{d-3} C_d^3,$$

$$f_\times^4 = f_{P_0}^4 \cdot f_Q^0 + f_{P_0}^3 \cdot f_Q^1 + f_{P_0}^2 \cdot f_Q^2 + f_{P_0}^1 \cdot f_Q^3 + f_{P_0}^0 \cdot f_Q^4$$
$$= f_{P_0}^4 \cdot 2^d C_d^0 + f_{P_0}^3 \cdot 2^{d-1} C_d^1 + f_{P_0}^2 \cdot 2^{d-2} C_d^2 + f_{P_0}^1 \cdot 2^{d-3} C_d^3 + f_{P_0}^0 \cdot 2^{d-4} C_d^4,$$

$$f_\times^{N+1+d} = f_{P_0}^{N+2} \cdot f_Q^{d-1} + f_{P_0}^{N+1} \cdot f_Q^d = f_{P_0}^{N+2} \cdot 2C_d^{d-1} + f_{P_0}^{N+1} \cdot 1;$$
$$f_\times^{N+2+d} = f_{P_0}^{N+2} \cdot f_Q^d = 1 \cdot 1 = 1.$$

Consequently, the product of a cluster of N similar shells in the form of Plato bodies with the number of faces j in each shell with a common center (if there is no atom in the common center) on d - cube has

$f_{P_0}^0 \; 2^d C_d^0$ vertices,

$f_{P_0}^1 \cdot 2^d C_d^0 + f_{P_0}^0 \cdot 2^{d-1} C_d^1$ edges,

$f_{P_0}^2 \cdot 2^d C_d^0 + f_{P_0}^1 \cdot 2^{d-1} C_d^1 + f_{P_0}^0 \cdot 2^{d-2} C_d^2$ two – dimensional faces,

$f_{P_0}^3 \cdot 2^d C_d^0 + f_{P_0}^2 \cdot 2^{d-1} C_d^1 + f_{P_0}^1 \cdot 2^{d-2} C_d^2 + f_{P_0}^0 \cdot 2^{d-3} C_d^3$ three - dimensional faces,

$f_{P_0}^4 \cdot 2^d C_d^0 + f_{P_0}^3 \cdot 2^{d-1} C_d^1 + f_{P_0}^2 \cdot 2^{d-2} C_d^2 + f_{P_0}^1 \cdot 2^{d-3} C_d^3 + f_{P_0}^0 \cdot 2^{d-4} C_d^4$, four - dimensional faces and go on.

This product itself has dimension $N + 2 + d$. In these relations, the quantities $f_{P_0}^i, i = 0 \div (N+2)$ are given by equalities (2).

The Product of a Cluster of Similar Sells (Plato Bodies) with the Number of Faces j with a Common Center on a *D* – Cross - Polytope

For a d – cross - polytope we have $f_Q^i(d) = 2^{1+i} C_d^{d-1-i}, i = 0,1,...,d-1.$

For the product of a cluster of N similar shells in the form of Plato bodies with the number of faces j in each shell with a common center (if there is no atom in the common center) on a d – cross - polytope in accordance of Theorem 1 of Chapter 5 we have

$$f_\times^0 = f_{P_0}^0 f_Q^0 = f_{P_0}^0 \cdot 2C_d^{d-1};$$

$$f_{\times}^{1}=f_{P_0}^{1}f_{Q}^{0}+f_{P_0}^{0}f_{Q}^{1}=f_{P_0}^{1}\cdot 2C_d^{d-1}+f_{P_0}^{0}\cdot 2^2C_d^{d-2};$$

$$f_{\times}^{2}=f_{P_0}^{2}\cdot f_{Q}^{0}+f_{P_0}^{1}\cdot f_{Q}^{1}+f_{P_0}^{0}\cdot f_{Q}^{2}=f_{P_0}^{2}\cdot 2C_d^{d-1}+f_{P_0}^{1}\cdot 2^2C_d^{d-2}+f_{P_0}^{0}\cdot 2^3C_d^{d-3},$$

$$f_{\times}^{3}=f_{P_0}^{3}\cdot f_{Q}^{0}+f_{P_0}^{2}\cdot f_{Q}^{1}+f_{P_0}^{1}\cdot f_{Q}^{2}+f_{P_0}^{0}\cdot f_{Q}^{3}=f_{P_0}^{3}\cdot 2C_d^{d-1}+f_{P_0}^{2}\cdot 2^2C_d^{d-2}+f_{P_0}^{1}\cdot 2^3C_d^{d-3}+f_{P_0}^{0}\cdot 2^4C_d^{d-4},$$

$$f_{\times}^{4}=f_{P_0}^{4}\cdot f_{Q}^{0}+f_{P_0}^{3}\cdot f_{Q}^{1}+f_{P_0}^{2}\cdot f_{Q}^{2}+f_{P_0}^{1}\cdot f_{Q}^{3}+f_{P_0}^{0}\cdot f_{Q}^{4}$$
$$=f_{P_0}^{4}\cdot 2C_d^{d-1}+f_{P_0}^{3}\cdot 2^2C_d^{d-2}+f_{P_0}^{2}\cdot 2^3C_d^{d-3}+f_{P_0}^{1}\cdot 2^4C_d^{d-4}+f_{P_0}^{0}\cdot 2^5C_d^{d-5},$$

$$f_{\times}^{N+1+d}=f_{P_0}^{N+2}\cdot f_{Q}^{d-1}+f_{P_0}^{N+1}\cdot f_{Q}^{d}=f_{P_0}^{N+2}\cdot 2^dC_d^{0}+f_{P_0}^{N+1}\cdot 1;$$
$$f_{\times}^{N+2+d}=f_{P_0}^{N+2}\cdot f_{Q}^{d}=1\cdot 1=1.$$

Consequently, the product of a cluster of N similar shells in the form of Plato bodies with the number of faces j in each shell with a common center (if there is no atom in the common center) on d – cross - polytope has

$f_{P_0}^{0}\cdot 2C_d^{d-1}$ vertices,

$f_{P_0}^{1}\cdot 2C_d^{d-1}+f_{P_0}^{0}\cdot 2^2C_d^{d-2}$ edges,

$f_{P_0}^{2}\cdot 2C_d^{d-1}+f_{P_0}^{1}\cdot 2^2C_d^{d-2}+f_{P_0}^{0}\cdot 2^3C_d^{d-3}$ two – dimensional faces,

$f_{P_0}^{3}\cdot 2C_d^{d-1}+f_{P_0}^{2}\cdot 2^2C_d^{d-2}+f_{P_0}^{1}\cdot 2^3C_d^{d-3}+f_{P_0}^{0}\cdot 2^4C_d^{d-4}$ three - dimensional faces,

$f_{P_0}^{4}\cdot 2C_d^{d-1}+f_{P_0}^{3}\cdot 2^2C_d^{d-2}+f_{P_0}^{2}\cdot 2^3C_d^{d-3}+f_{P_0}^{1}\cdot 2^4C_d^{d-4}+f_{P_0}^{0}\cdot 2^5C_d^{d-5}$, four - dimensional faces and go on.

This product itself has dimension $N + 2 + d$. In these relations, the quantities $f_{P_0}^{i}, i=0\div(N+2)$ are given by equalities (2).

The Product of a Cluster of Similar Sells (Plato Bodies) with the Number of Faces j with a Common Center and Atom in him on a One - dimensional Segment

In Theorem 4 of Chapter 2 of this book, formulas are established for the numbers of faces of various dimensions in a cluster of N similar shells in the form of Plato bodies with the number of faces j in each shell with a common center and atom in him. If there is no atom in the common center, then these formulas, taking into account additional designation conditions, have the form

$$f_{P_a}^0 = Nn_j + 1, f_{P_a}^1 = N\frac{n_j t_j}{2} + Nn_j, f_{P_a}^2 = Nj + N\frac{n_j t_j}{2}, f_{P_a}^3 = Nj + N,$$
$$f_{P_a}^4 = C_{N+1}^2, f_{P_a}^5 = C_{N+1}^3, \ldots, f_{P_a}^{N+2} = C_{N+1}^N. \tag{3}$$

For one – dimensional segment we have $f_Q^o = 2, f_Q^1 = 1$.

For the product of a cluster of *N* similar shells in the form of Plato bodies with the number of faces *j* in each shell with a common center (if there is no atom in the common center) on one – dimensional segment in accordance of Theorem 1 of Chapter 5 we have

$$f_\times^0 = f_{P_a}^0 f_Q^0 = f_{P_a}^0 2;$$

$$f_\times^1 = f_{P_a}^1 f_Q^0 + f_{P_a}^0 f_Q^1 = f_{P_a}^1 2 + f_{P_a}^0;$$

$$f_\times^2 = f_{P_a}^2 \cdot f_Q^0 + f_{P_a}^1 \cdot f_Q^1 = f_{P_a}^2 \cdot 2 + f_{P_a}^1 \cdot 1,$$

$$f_\times^3 = f_{P_a}^3 \cdot f_Q^0 + f_{P_a}^2 \cdot f_Q^1 = f_{P_a}^3 \cdot 2 + f_{P_a}^2 \cdot 1,$$

$$f_\times^4 = f_{P_a}^4 \cdot f_Q^0 + f_{P_a}^3 \cdot f_Q^1 = f_{P_a}^4 \cdot 2 + f_{P_a}^3 \cdot 1,$$

$$f_\times^{N+3} = f_{P_a}^{N+2} \cdot f_Q^1 = 1 \cdot 1 = 1.$$

Consequently, the product of a cluster of *N* similar shells in the form of Plato bodies with the number of faces *j* in each shell with a common center (if there is no atom in the common center) on one - dimensional segment has

$f_{P_a}^0 2$ vertices,

$f_{P_a}^1 2 + f_{P_a}^0$ edges,

$f_{P_a}^2 \cdot 2 + f_{P_a}^1 \cdot 1$ two – dimensional faces,

$f_{P_a}^3 \cdot 2 + f_{P_a}^2 \cdot 1$ three - dimensional faces,

$f_{P_a}^4 \cdot 2 + f_{P_a}^3 \cdot 1$ four - dimensional faces and go on.

This product itself has dimension $N + 3$. In these relations, the quantities $f_{P_a}^i, i = 0 \div (N+2)$ are given by equalities (3).

The Products of Polytopes and the Model of a Continuous Medium of Living Matter

As follows from the numerous examples examined in this chapter, the products of molecules and clusters on polytopes lead to the spread of these molecules and clusters in space. The greater the dimension of polytopes as factors, the greater the dimension of space is this distribution. Naturally, the dimension of the initial chemical compounds (molecules and clusters) remains unchanged. Since the number of vertices in a molecule or cluster, as well as a polytope, as a factor, is finite, of course, the number of vertices in their product is finite, i.e. a closed body with a finite number of vertices is formed. In the process of producing polytopes from each vertex of the initial chemical compound, as a polytope, vertices are put together that connect two copies of the initial figure along each coordinate of another factor (polytope). This geometric multiplication can acquire chemical or physical significance if the exposed edges (or at least part of these edges) correspond to the arising chemical or physical bonds. In this case, the remaining part of the exposed edges will have a purely geometric value, which will contribute to the creation of a closed convex figure. Vertices connecting chemical molecules or clusters corresponding to covalent bonds, hydrogen bonds or electrostatic attraction can be considered bridging compounds. Such compounds are widespread in biological objects. As is known, biomolecules are chains of monomer units (Lehninger, 1985; Metzler, 1980). Moreover, the monomer units in biomolecules are geometrically polytopes of the highest dimension (Zhizhin, 2016; 2019 b). Biomolecules like chains or fibers cannot form a continuous medium. For such a formation, chain linking with each other is necessary. This is carried out in nature using bridge connections. Connected by bonds are not only monomer units belonging to different chains, but also monomeric units belonging to the same chain. So - called binary, tertiary and quaternary structures are formed in various proteins, nucleic acids, and the ribosome. Quaternary structures are characterized by ordered packing of macromolecules (Metzler, 1980). In ribosomes, for example, stability is ensured by a certain concentration of divalent magnesium and calcium ions (Spirin, 2019). These ions combine with each other phosphoric acid residues associated with the monomeric units of nucleic acids. Moreover, this connection occurs both between different nucleic acid chains and between monomeric units of the same nucleic acid chain (ribosomal RNA). The phosphoric acid residue is geometrically a tetrahedron with a center, i.e. polytope of dimension 4. In this regard, the bridging compounds lead to the distribution of phosphoric acid residues in space, and the nucleic chains are stacked between the residues of phosphoric acid, like tetrahedra with a center. The question of the distribution of phosphoric acid residues in such structures has not been completely studied at present. But it can be assumed that there is some order in these distributions. In mathematical terms, this order is determined by the structure of another polytope, as a factor. This polytope is still unknown to us. In general, as a result of the product of polytopes, a certain structure of higher dimension is formed, which is a model of a continuous medium of a given living matter (ribosomal RNA or ribosome as a whole).

CONCLUSON

The products of molecules and clusters on polytopes of various dimensions and various types are analyzed analytically, starting from one - dimensional segments and ending with simplexes, cubes and cross - polytopes of arbitrary dimension *n*. The result of these products are polytopes with a dimension equal to the sum of the dimensions of the factors and with the number of vertices equal to the product of the numbers of vertices of the factors. In this case, the initial chemical compound is repeated in space as many times as there are vertices in the polytope of the factor. The distribution of these copies of chemical compounds over the space of the product of polytopes is uniform and the distribution order depends on the type of polytope of the factor. The edges resulting from the operation of a product have geometric meanings for creating a convex closed figure. In addition, part of these edges can correspond to bonds of various nature between chemical compounds: covalent chemical bonds, hydrogen bonds, electrostatic attraction. Thus, the products of polytopes correspond to larger chemical compounds. A distinctive feature of these compounds is the presence in their structure of several families of parallel edges. The edges resulting from the operation of the product can simulate the bridge joints characteristic of living matter. Thus, it becomes possible to construct a mathematical model of a continuous medium of living matter in ordered quaternary structures. The presence of several families of parallel edges in the work of polytopes can allow them to fill the space of higher dimension as a result of their translation.

REFERENCES

Bergman, G., Waugh, J. L. T., & Pauling, L. (1952). Crystal structure of the intermetallic compound. *Nature*, *169*(4312), 1057–1058. doi:10.1038/1691057a0

Bergman, G., Waugh, J. L. T., & Pauling, L. (1957). The crystal structure of the metallic phase. *Acta Crystallographica*, *10*(4), 254–259. doi:10.1107/S0365110X57000808

Gillespie, R. J. (1972). *Molecular Geometry*. Van Nostrand Reinhold Company.

Gillespie, R. J., & Hargittai, I. (1991). *The VSEPR Model of Molecular Geometry*. Allyn & Bacon.

Gubin, S. P. (2019). *Chemistry clusters. Basics of classification and structure*. Science.

Lehninger, A. L. (1982). *Principles of Biochemistry*. Worth Publishers, Inc.

Mackay, A. L. (1962). A dense non - crystalline packing of equal spheres. *Acta Crystallographica*, *15*(9), 1916–1918. doi:10.1107/S0365110X6200239X

Metzler, D. E. (1980). *Biochemistry. The Chemical Reactions in Living Cells*. Academic Press.

Poincaré, A. (1895). Analysis situs. *J de e'Ecole Polytechnique.*, *1*, 1–121.

Pouling, L. (1960). *The Nature of the Chemical Bond and the Structure of Molecules and Crystals: An Introduction to Modern Structural Chemistry* (3rd ed.). Cornell University Press.

Samson, S. (1972). Complex cubic A_6B compounds. II. The crystal structure of Mg_6Pd. *Acta Crystallographica. Section B, Structural Crystallography and Crystal Chemistry*, *28*(3), 936–945. doi:10.1107/S0567740872003437

Spirin, A. S. (2019). *Molecular biology. The ribosomes and protein biosynthesis.* Knowledge Lab.

Zhizhin, G. V. (2016). The Structure, Topological, and Functional Dimension of Biomolecules. *International Journal of Chemoinformatics and Chemical Engineering*, *5*(2), 44–58. doi:10.4018/IJCCE.2016070104

Zhizhin, G. V. (2018). Chemical Compound Structures and the Higher Dimension of Molecules: Emerging Research and Opportunities. Hershey, PA: IGI Global. doi:10.4018/978-1-5225-4108-0

Zhizhin, G. V. (2019a). *The Geometry of Higher – Dimensional Polytopes*. IGI Global. doi:10.4018/978-1-5225-6968-8

Zhizhin, G. V. (2019b). *Attractors and Higher Dimensions in Population and Molecular Biology: Emerging Research and Opportunities*. IGI Global. doi:10.4018/978-1-5225-9651-6

KEY TERMS AND DEFINITIONS

5-Cross-Prism: The product of 4-cross-polytope by one-dimension segment.

5-Simplex-Prism: The product of 4-simplex by one-dimension segment.

6-Complex-Polytopic Prismahedrons: The product of tetrahedron by the tetrahedron.

6-Complex-Tetrahedral Prismahedrons and Octahedral Prismahedrons: The product octahedral by the tetrahedral.

6-Simplex-Polytopic Prismahedron: The product of 4-simplex by the triangle.

Cube-Polytopic Prismahedron: The product of cube by the triangle.

Dodeca-Polytopic Prismahedron: The product of dodecahedron by the triangle.

Dodecahedral Prism: The product of dodecahedral by one-dimension segment.

Icosa-Polytopic Prismahedron: The product of icosahedron by the triangle.

Icosahedral Prism: The product of icosahedral by one-dimension segment.

N*3-Angular Prismahedron: The product of n-angle by the triangle.

Octahedral Prism: The product of octahedral by one-dimension segment.

Octahedral Prismahedron: The product of octahedral by the triangle.

Polytopic Prismahedron: The product of polytope by one-dimension segment.

Tetrahedral Prism: The product of tetrahedral by one-dimension segment.

Tetrahedral Prismahedron: The product of tetrahedral by the triangle.

Triangular Prismahedron: The product of triangle by the triangle.

Chapter 7
Parallelohedrons and Stereohedrons of Delaunay

ABSTRACT

The works of Delaunay and the followers of his ideas about the geometry of n-dimensional parallelohedrons and stereohedrons are considered. It is proved that these representations do not take into account the conditions for the existence of polytopes of higher dimension and the properties characteristic of figures of higher dimension. They are an attempt to extend the properties of three-dimensional figures to figures of higher dimension. A direct verification of the parallelohedrons from the Delaunay classification taking into account the Shtogrin parallelohedron showed that these figures do not satisfy the Euler-Poincaré equation and therefore the assertion that they are parallelohedrons with dimension 4 is erroneous.

INTRODUCTION

In 1885, the book of Efgraf Fedorov "Began the doctrine of figures" (Fedorov, 1885) went out of print. It classifies two - and three - dimensional figures that can fill the Euclidean plane and three - dimensional space adjacent to each other along whole edges and, accordingly, flat faces without gaps. These figures were called parallelogons and, correspondingly, parallelohedra, due to the fact that their sides are pairwise parallel. It gives the concept of stereohedrons as equal figures that make up a parallelohedron together. In 1889 and in 1891 the following works of Fedorov, "Symmetry of finite figures" and "Symmetry of regular systems of figures" (Fedorov, 1889, 1891), went out of print. In these works, Fedorov deduced 230 spatial groups, i.e. 230 unique geometric laws of arrangement of elementary particles of crystalline structures in three - dimensional Euclidean space and 17 two - dimensional groups of arrangement of elementary particles in the Euclidean plane. In 1891, Schönflis's book "Crystal Systems and the Structure of Crystals" was also published (Schönflis, 1891). Her author has repeatedly quoted Fedorov, pointing to his primacy in a number of questions on theoretical crystallography. Fedorov's extensive correspondence with Schönflis (Shafranovsky, 1963) has been preserved, from which the leading role of Fedorov in the derivation of 230 spatial symmetry groups follows. The studies were reflected in a report by David

DOI: 10.4018/978-1-7998-6768-5.ch007

Hilbert (Hilbert, 1901) at the International Congress of Mathematicians in Paris, in which a number of mathematical problems were formulated that determined the development of mathematics for the next century. One of these problems (the eighteenth) raised the question of constructing spaces from congruent polyhedrons. The report expressed the desire to investigate the question of constructing spaces of higher dimension from congruent figures (Aleksandrov, 1969). The answer to this request was a series of researchers' work in subsequent years. However, many of them tried to consider spaces of higher dimension and figures of higher dimension by mechanically extending to a space of higher dimension ideas about three - dimensional Euclidean space without taking into account the essential properties and features of spaces of higher dimension (Minkowski, 1911; Delaunay, 1929, 1937, 1961; Delaunay, & Sandakova,1961; Voronoi, 1952; Alexandrov, 1934, 1954; Venkov, 1954, 1959). Although back in 1880, the work of the American mathematician Stringham appeared on figures in n - dimensional space (Stringham, 1880). This work was apparently unknown to these researchers, as well as the work of Stott from Amsterdam (Stott, 1900, 1910). Much later, the works of Coxeter and Grünbaum, widely known at present, on figures of higher dimensions appeared (Coxeter, 1963; Grünbaum, 1967). From the analysis of geometric figures of higher dimensions, it would be necessary to begin the solution of the question of constructing spaces of higher dimensions with the help of these figures. That is exactly what the ingenious crystallographer Fedorov did when he studied the question of constructing three - dimensional space with the help of three - dimensional figures. However, this did not happen. As a result, mechanically extending Fedorov's teaching on parallelohedra in three - dimensional spaces to the four - dimensional space, Delaunay in 1929 systematized, as he believed, four - dimensional parallelohedons (Delaunay, 1929). Moreover, no evidence was given for any of these 51 parallelohedrons that they have dimension 4. Later, a student of Delaunay Shtogrin, as part of the teachings of Delaunay, discovered the existence of another figure, additionally included in the classification of Delaunay. It was believed that this figure is also a parallelogon with dimension 4 (Shtogrin, 1973). In this chapter, it will be shown that, as a result of neglecting the necessary conditions for the existence of polytopes of higher dimension, described in detail in the first chapter of this paper, none of the Delaunay classification figures, as well as the figure proposed by Shtogrin, are polytopes of higher dimension, i.e. due to this, they do not have, as it was supposed, a dimension equal to 4.

THE MAIN PROVISIONS OF THE DELAUNAY THEORY

Starting from 1929, Delaunay's theory has been stated more than once in various publications. Take as an example the presentation of this theory in the work of Shtogrin (1973). The paragraph "Basic concepts, definitions and tasks" says "The subject of our research will be the partitioning of n - dimensional Euclidean space into bodies - polyhedrons. The location of bodies in space at which they pairwise have no internal points is called packing. The location of bodies in space at where each point in space belongs to at least one of these bodies, is called a covering. A partition is a location of bodies in space that is both packing and covering. of interior points and cover the entire area. If the bodies in decomposition of convex polyhedrons that are. "

In these first lines, much remains uncertain. First of all, what are bodies in the space of higher dimension. If we follow Fedorov, who investigated the partition of three - dimensional space, then it would be necessary to first determine what kind of bodies, their properties and conditions of existence. This is nothing in the teaching of Delaunay.

The followers of Delaunay's teachings do not seem to know that a prerequisite for the existence of a polytope of higher dimension is the satisfaction of its Euler – Poincaré equation (equation (1) from the first chapter of this paper). Until today, the Euler - Poincaré equation has not been used in the presentation of Delaunay's theory by his followers. And this is understandable, since this equation would destroy the entire Delaunay theory.

Secondly, in Delaunay's theory, it remains unclear what to consider internal points of the body. This question is trivial when considering problems on the plane and in three - dimensional space. However, when moving into a space of higher dimension, this question becomes non - trivial. Therefore, back in 1967, Grünbaum introduced the concept of the boundary complex of a polytope of higher dimension, which includes many elements of different dimensions (Grünbaum, 1967). The ideas about the internal points of higher - dimensional polytopes are examined in detail in the author's monograph (Zhizhin, 2019 a). The issues of belonging of some elements of a certain dimension in a polytope of higher dimension to other elements of another dimension are also nontrivial. These issues are discussed in detail in this monograph (Zhizhin, 2019 a).

The division of space into polyhedrons Delaunay calls normal if these polyhedrons are adjacent along integer (n - 1) - dimensional faces (Shtogrin, 1973). It should be said here that this position again follows from representations of spaces of dimension 2 or 3. In these spaces, such a condition is equivalent to the absence of gaps between the elements of the "face to face" partition. However, in the space of higher dimension, the concept of "face to face" becomes uncertain, since a polytope of higher dimension has many faces of different dimensions. And the assertion that there will be no gaps between the elements of dividing a space of higher dimension is absolutely only necessary when they are adjacent along facets (i.e., (n - 1) - dimensional faces). The proof of Delaunay (1937) that polytopes in n - dimensional space either do not enter into each other or are adjacent along faces of dimension n -1 (facets) is not correct, since the concepts of three - dimensional space are used when considering n - dimensional space. The statement (Delaunay, 1937) that the vertices of the polytopes L_1 and L_2, if they lie on opposite sides of the chordal (n - 1) - dimensional "plane" of the balls (L_1) and (L_2) described around L_1 and L_2, has not been proved. We cite this statement from the work of Delaunay (1937).

"If the balls (L_1) and (L_2) intersect, then the vertices of the polytope L_1 lie on the same hat of the ball (L_1), which does not lie inside the ball (L_2), since the ball (L_2) is empty. Similarly, the vertices of the polytope L_2 lie on that cap of the ball (L_2), which does not lie inside the ball (L_1), since the ball (L_1) is empty, therefore, the vertices of the polytopes L_1 and L_2 lie on different sides of the chordal (n - 1) - dimensional "plane" of these balls, and therefore, the polytopes L_1 and L_2 themselves lie on opposite sides of this plane."

Obviously, if the polytopes L_1 and L_2 do not enter one into the other, but have common elements, then the remaining elements lie on opposite sides of the common elements. But why these common elements must have dimension n - 1 remained unproven. Polytopes in n - dimensional space can have common vertices, edges, common elements of any dimension from 0 to n - 1. Facets have no exclusivity in this regard.

We quote further the paragraph from the mentioned paragraph of Shtogrin's work. "Venkov, unlike Delaunay, calls a partition normal if it is a Dirichlet - Voronoi partition, and then it is normal in the sense of Delaunay. A partition of a space into polyhedrons is called a Dirichlet -Voronoi partition, if you can choose a point - center in each separate polyhedron of a partition action, such that this polyhedron of the partition is the Dirichlet - Voronoi region of its center of action with respect to all other centers of action, i.e. each polyhedron is a collection of points of space that are located b others like to the center of action than all other centers of action. " Note that this paragraph has never said that space has a higher dimension, although, judging by the previous one, this is certainly implied. However, the statements and concepts of this paragraph are suitable only in two - dimensional and three - dimensional space. Since the Dirichlet - Voronoi region remains indefinite in space of higher dimension. It will be determined only when the Euler – Poincaré equation is fulfilled for it. But this proof in the theory of Voronoi in the case of space of higher dimension is not.

Further, in the Delaunay theory, groups of motion of partitions of n - dimensional spaces are considered, i.e. symmetry transformations of these partitions are considered, following the Fedorov theory when considering partitions of two - and three - dimensional spaces. However, symmetry transformations of partitions of spaces of higher dimension cannot be considered if symmetry transformations of the constituent figures of higher dimension are unknown. And they, of course, are unknown, since the figures themselves remained unknown. The symmetry transformations of higher - dimensional polytopes were first consecutively considered in the author's monograph (Zhizhin, 2019 a).

Further in the work of Shtogrin follows the definition of a normal partition of an n - dimensional space. "A partition is called normal if for any bodies V_1 and V_2 of this partition in the group G of motions that combine this partition, there is at least one movement that takes the body V_1 to the body V_2. The bodies of the normal partition are called stereohedrons. In particular, if G is a group of parallel translations, then the partition bodies are called parallelohedra. "

In these definitions, it remained unknown what figures of higher dimension should be, that is, what should be their structure in order for these definitions to be fulfilled. Thus, these definitions do not define the essence of these figures. In this regard, for example, despite these definitions and proofs of a number of theorems (Delaunay, 1961), neither Delaunay himself nor his followers were able to give at least one specific example of a stereohedron of higher dimension.

Of great importance in Delaunay geometry is given to the set of points, which is called the "Delaunay system". This concept was introduced in 1924 and published in 1937 (Delaunay, 1937). Following the work of Galiulin (2003), the Delaunay system is the set of points that satisfies the following two axioms:

1) discreteness - the distance from any point of the set to the nearest point of the same set is greater than or equal to some fixed segment of length r,
2) coverage - the distance from any point in space to the point of the system nearest to it is less than or equal to some fixed segment of length R.

In the study of the structure of substances in recent decades, phenomena were discovered that were unknown at the time of David Hilbert. These include spontaneous zooming (scaling processes), for example, during second-order phase transitions (Landau, 1937; Kadanoff, 1966; Wilson, 1971); discovery of crystals with the absence of translational symmetry in three-dimensional space (Shehtman, et al., 1984); detection of the fractal nature of matter (Mandelbrott, 1982); hierarchical filling of space with polytopes of higher dimension (Zhizhin, 2012); the highest dimension of most molecules of chemical compounds (Zhizhin, 2018). All these phenomena cannot be described by Delaunay point systems. Zhizhin (2018) introduced the concept of a discrete system of points in which the distance between points asymptotically tends to zero, without reaching exact equality. However, there is no need to introduce the concept of an "empty ball" introduced in Delaunay systems. The possibility of asymptotically reducing the distance between points of a discrete system of points corresponds to the distribution of points in the diffraction patterns of quasicrystals and to the scaling process, i.e. continuous zooming in on a system discovered in recent decades. In connection with the above, it should be considered that the decisive value attributed to systems of Delaunay for describing a discrete world (Galiulin, 2003) is significantly exaggerated and, moreover, its use may lead to incorrect results. For example, in the work of Ryzhkov and Shushbaev (1981), based on the understanding of Delaunay systems, it is argued that with the help of a 4 – cross - polytope one can obtain a correct partition of four - dimensional space. Let us prove that this is not so. We use the representation of 4 - cross - polytopes in the form proposed by Stringham (Stringham, 1880) Figure 1.

Figure 1. 4 - cross - polytope

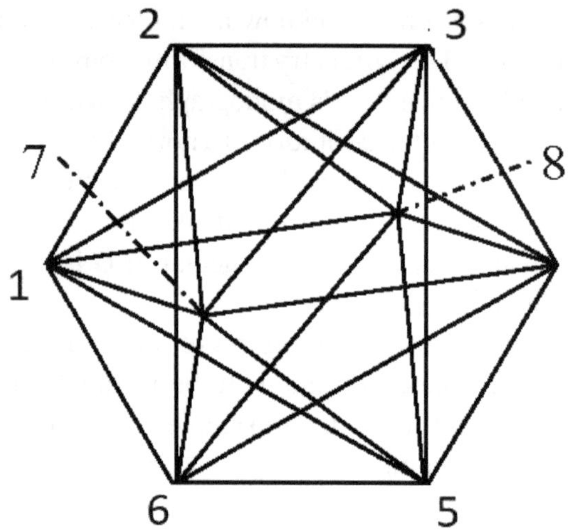

Figure 1 shows that it has 8 vertices, 24 edges, 32 flat triangular faces and 16 tetrahedra as a facet. A feature of cross - polytopes of any dimension is that pairs of oppositely located vertices are not connected by an edge, and there are as many such pairs as the dimension of cross -polytopes. In the case of 4 - cross - polytopes in Figure 7.1, these are pairs of vertices 1 - 4, 2 - 5, 3 - 6, 7 - 8. Let us introduce the origin of the coordinate system of four - dimensional space (x, y, z, t) in the center of some initial 4 – cross - polytopes A_0 (Figure 2).

The coordinates will be directed in the directions between the vertices that are not connected by edges. Let the distance from the origin to any vertex be equal to unity. Then the coordinates of the vertices indicated in Figure 1 are equal 1(-1, 0, 0, 0), 2(0, 1, 0, 0), 3(0, 0, 1, 0), 4(1, 0, 0, 0), 5 (0, - 1, 0, 0), 6(0, 0,- 1, 0), 7(0, 0, 0, -1), 8(0, 0, 0, -1). We will broadcast polytope A_0 in coordinates x, y, z by two units. Then we get three more 4 - cross - polytopes. The coordinate values of the vertices of all four cross - polytopes are determined by the equalities

$$
\begin{aligned}
A_0 &= \{(-1,0,0,0),(0,1,0,0),(0,0,1,0),(1,0,0,0),(0,-1,0,0),(0,0,-1,0),(0,0,0,-1),(0,0,0,1)\},\\
A_1 &= A_0(x+2) = \{(1,0,0,0),(2,1,0,0),(2,0,1,0),(3,0,0,0),(2,-1,0,0),(2,0,-1,0),(2,0,0,-1),(2,0,0,1)\},\\
A_2 &= A_0(y+2) = \{(-1,2,0,0),(0,3,0,0),(0,0,1,0),(1,0,0,0),(0,-1,0,0),(0,0,-1,0),(0,0,0,-1),(0,0,0,1)\},\\
A_3 &= A_0(z+2) = \{(-1,0,0,0),(0,1,0,0),(0,0,1,0),(1,0,0,0),(0,-1,0,0),(0,0,-1,0),(0,0,0,-1),(0,0,0,1)\}.
\end{aligned}
\tag{7.1}
$$

Figure 2. The translation of 4 – cross – polytope at coordinates x, y, z on one step of length 2

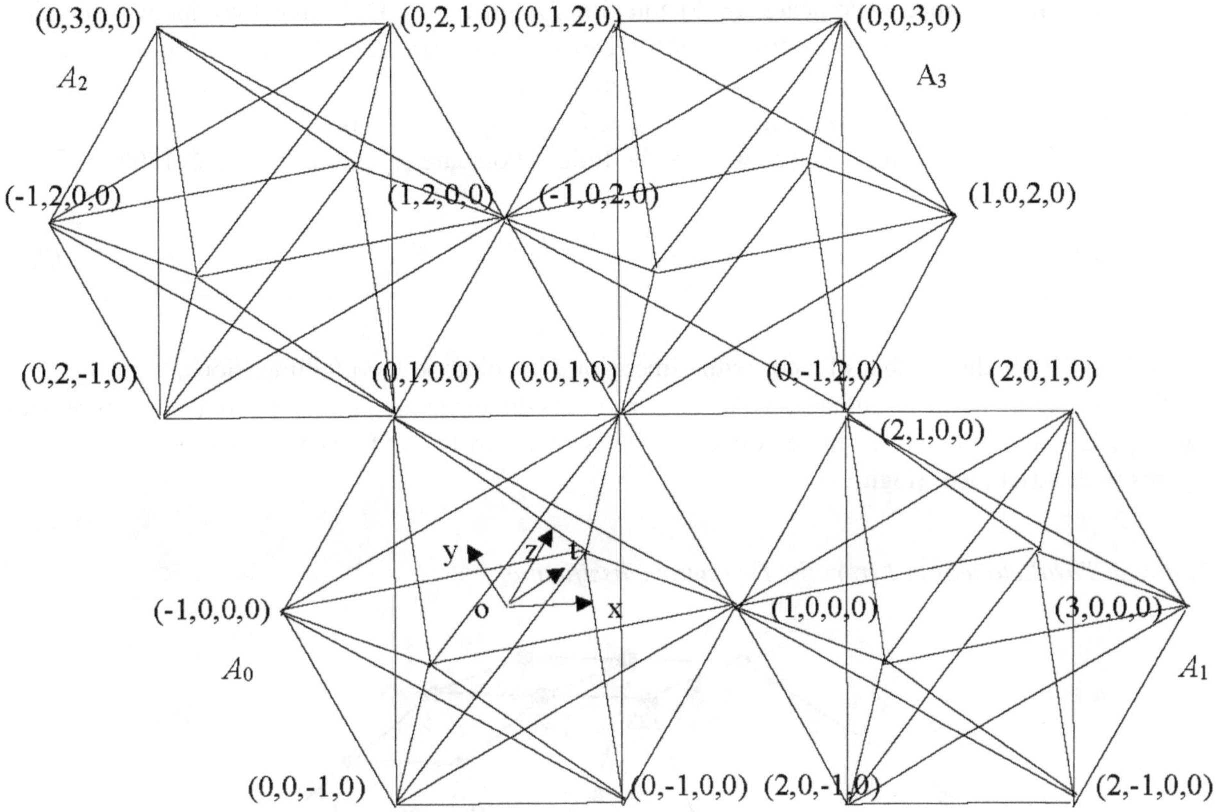

Even without translation along the coordinate *t*, it follows from equalities (7.1) and Figure 7.2 that a 4 – cross - polytope cannot fill a four - dimensional space face – to - face, as polytopes A_1 and A_3, as well as polytopes A_3 and A_2, result from the translation matching vertices with different coordinate values. And besides, between polytopes A_0, A_2, A_3, as well as between polytopes A_0, A_1, A_3, areas arise in which a 4 – cross - polytope cannot fit in any way.

ANALYSIS OF FOUR - DIMENSIONAL DELAUNAY PARALLELOHEDRONS

In 1929, Delaunay gave a classification of parallelohedrons, which he believed had a dimension of 4. This classification includes 51 figures. The principle of constructing four - dimensional parallelohedrons, which Delaunay used, contradicts the existence of polytopes of higher dimension. In particular, in polytopes of higher dimensionality, three - dimensional faces when depicting polytopes in a projection onto a two - dimensional plane, relying on some two -dimensional face, pass through each other (Stringham, 1880; Coxeter, 1963; Grünboum, 1967; Zhizhin, 2019 a). In Delaunay constructions, three - dimensional faces in projections onto the two - dimensional plane of four - dimensional figures are located on opposite sides of the two -dimensional face on which they rely. This applies to all figures in the Delaunay classification. For none of the figures in the Delaunay classification is there evidence that the figure has dimension 4. A necessary condition for the existence of a multidimensional convex figure is the fulfillment of the Euler - Poincaré (Poincaré, 1895) equation for this figure. Delaunay does not even mention this condition. It can be shown that for any of the Delaunay classification figures this condition is not fulfilled; therefore, the figures given by Delaunay are not four - dimensional parallelohedrons. We give this proof for three arbitrarily taken figures from the Delaunay classification.

The dimension of a figure is determined by the Euler - Poincaré equation (Poincaré, 1895)

$$\sum_{i=0}^{d-1}(-1)^i f_i(P) = 1 + (-1)^{d-1} . \qquad (7.2)$$

In (7.2) $f_i(P)$ is the number of faces with dimension i in polytope P with dimension d.

The first figure in the Delaunay classification of four - dimensional parallelohedra is the figure (Delaunay, 1929), composed of 10 three - dimensional figures with 14 flat faces and 20 three -dimensional figures with 8 flat faces (Figure 3).

Figure 3. Parallelohedron 1 from the Delaunay classification.

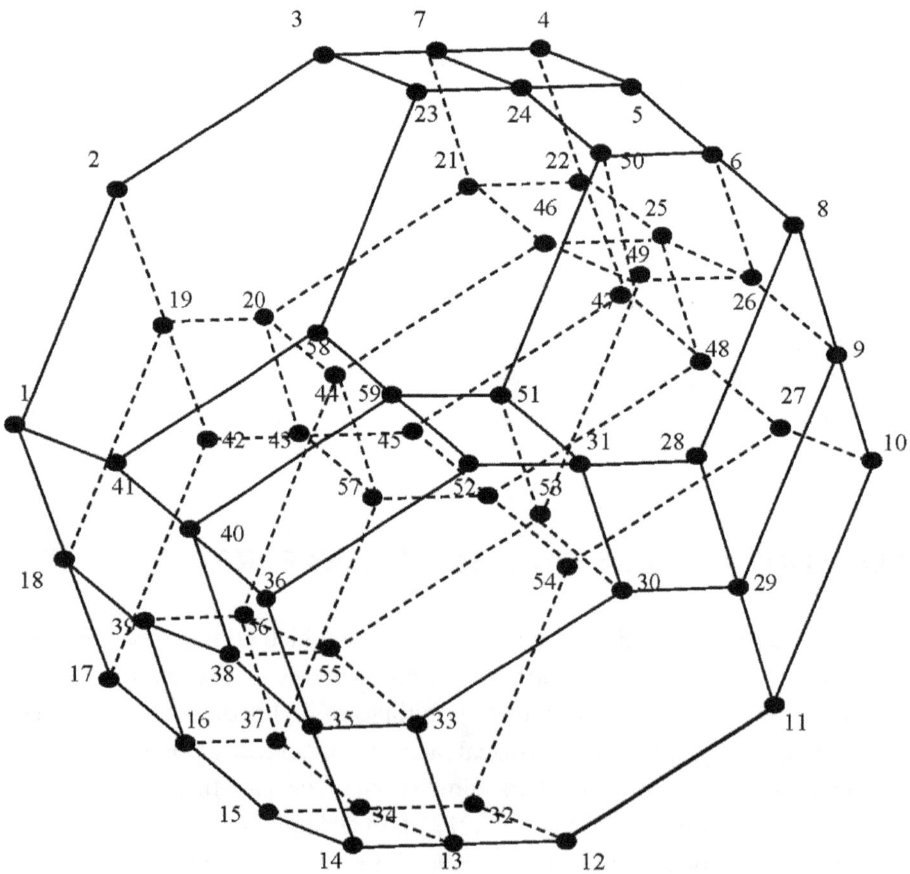

As follows from Figure 7.3, it contains 60 vertices ($f_0(P_1)$=60). The edges in this figure are the following segments:

1 – 2, 1 - 18, 1 - 41, 2 - 3, 2 - 19, 3 - 7, 3 - 23, 4 - 7, 4 - 5, 4 - 22, 5 - 24, 5 -6, 6 - 50, 6 - 26, 6 - 8, 7 - 21, 7 - 24, 8 - 9, 8 - 28, 9 - 26, 9 - 10, 9 - 29, 10 - 27, 10 - 11, 11 - 29, 11 - 12, 12 - 13, 12 - 32, 13 - 33, 13 - 34, 13 - 14, 14 - 35, 14 - 15, 15 - 34, 15 - 16, 16 - 37, 16 -39, 16 - 17, 17 - 42, 17 - 18, 18 - 19, 18 - 39, 19 - 20, 19 - 42, 20 - 43, 20 - 44, 20 - 21, 21 - 46, 21 - 22, 22 - 25, 22 - 47, 23 - 58, 23 - 24, 24 - 50, 25 - 46, 25 - 48, 25 - 26, 26 - 49, 27 - 48, 27 -54, 28 - 31, 28 - 29, 29 - 30, 30 - 33, 30 - 53, 30 - 31, 31 - 51, 31 - 60, 32 - 34, 32 - 54, 33 - 35, 33 - 55, 34 - 37, 35 - 38, 35 - 36, 36 - 40, 36 - 60, 37 - 56, 37 - 57, 38 - 39, 38 - 40, 38 - 55, 39 -56, 40 - 59, 40 - 41, 41 - 58, 42 - 43, 43 - 45, 43 - 57, 44 - 46, 44 - 57, 44 - 56, 45 - 52, 46 - 49, 47 - 45, 47 - 48, 48 - 52, 49 - 53, 49 - 50, 50 - 51, 51 - 59, 51 - 53, 52 - 57, 52 - 54, 53 - 55, 55 -56, 58 - 59, 59 - 60.

Thus, $f_1(P_1)$=107.

The listed edges form the following two - dimensional faces:

1 – 2 – 3 – 23 – 58 - 41, 1 – 41 – 40 – 38 – 39 - 18,1 – 2 – 19 - 18, 2 – 3 – 7 – 21 – 20 - 19, 3 – 7 – 24 - 23, 4 – 7 – 21 - 22, 4 – 7 – 5 -24, 4 – 5 – 6 – 26 – 25 - 22, 5 – 6 – 50 - 24, 6 – 8 – 9 - 26, 7 – 24 – 21 – 50 – 46 - 49, 6 – 8 – 28 -31 – 51 - 50, 8 – 9 – 28 - 29, 9 – 10 – 11 - 29, 9 – 26 – 49 – 53 – 30 - 29, 9 – 10 – 27 – 48 – 25 -26, 10 – 27 – 54 – 32 – 12 - 11, 11 – 29 – 30 – 33 – 13 - 12, 12 – 13 – 34 - 32, 13 – 14 – 15 - 34, 13 – 14 – 33 - 35, 14 – 15 – 35 – 38 – 39 - 16, 15 – 34 – 37 - 16, 16 – 37 – 56 - 39, 16 – 17 – 39 -18, 16 – 37 – 57 – 43 – 42 - 17, 17 – 18 – 19 - 42, 18 – 19 – 20 – 44 – 56 - 39, 19 – 20 – 42 - 43, 20 – 21 – 46 - 44, 20 – 21 – 22 – 47 – 45 - 43, 20 – 43 – 44 - 57, 21 – 22 – 25 - 46, 22 – 25 – 48 -47, 23 – 24 – 50 – 51 – 59 - 58, 25 – 26 – 46 - 49, 25 – 46 – 44 – 57 – 52 - 48, 27 – 48 – 52 - 54, 28 – 29 – 30 - 31, 30 – 31 – 60 – 36 – 35 - 33, 30 – 31 – 51 - 53, 31 – 51 – 59 - 60, 32 – 54 – 52 -57 – 37 - 34, 35 – 37 – 38 - 40, 36 – 40 – 59 - 60, 37 – 56 – 44 - 57, 38 – 39 – 55 - 56, 40 – 42 -58 - 59, 44 – 46 – 49 – 53 – 55 - 56, 45 – 47 – 48 - 52.

Thus, $f_2(P_1)$=50.

It should be noted that when calculating the number of edges, it is assumed that the edge does not contain a vertex inside. Otherwise, we are dealing with two edges, i.e. vertices can only be on border of edges. The same situation applies to flat faces. A two - dimensional region is considered a face only when it is not separated by any edge (Zhizhin, 2019 a).

The number of three - dimensional figures included in P_1 according to Delaunay's order is 30. To them we must add the figure formed by the outer surface of P_1 (Zhizhin, 2019). Consequently, $f_3(P_1)$=31. Substituting the obtained values f_i in the left side of the Euler - Poincaré equation (7.2) we find

$$60 – 107 + 50 – 31 = - 28.$$

The obtained value of the left side of equation (7.2) cannot be equal to the right side of equation (7.2) for any value of dimension d (the right side can be equal to only 2 or 0). Therefore, the figure indicated by the Delaunay parallelohedron 1 is not a four - dimensional parallelohedron. Moreover, since the Euler - Poincaré equation is not satisfied for it, it cannot be considered a convex figure at all.

Consider some simpler figure from the Delaunay classification, for example, figure number 15 (Figure 4).

Figure 4. Parallelohedron 15 from Delaunay classification

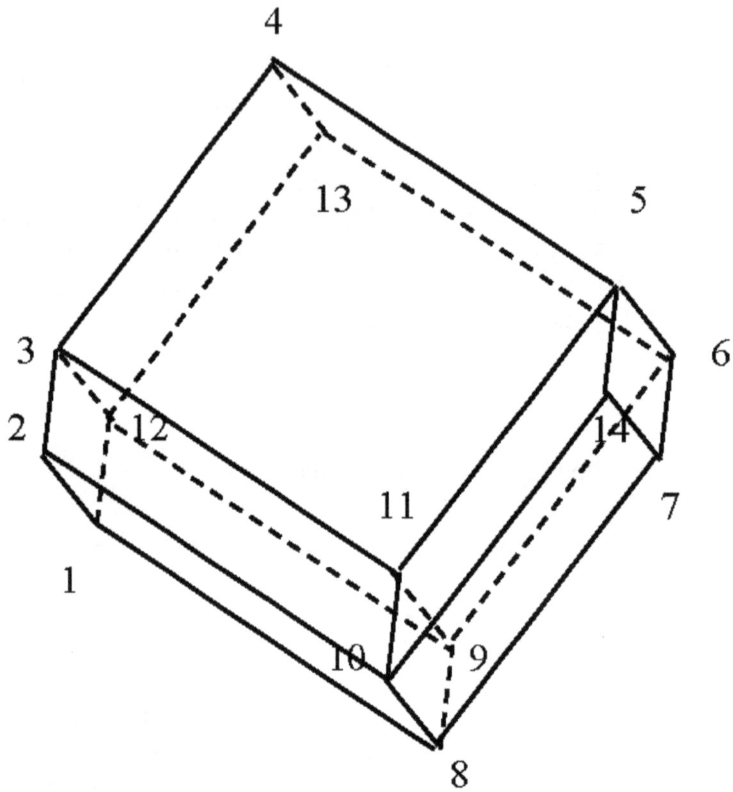

From Figure 7.4 it follows that it has 14 vertices ($f_0(P_{15})$=14). The edges in this figure are the following segments:

1 - 12, 1 - 8, 2 - 3, 2 - 10, 3 - 12, 3 - 4, 3 - 11, 4 - 13, 4 - 5, 5 - 6, 5 - 14, 5 -11, 6 - 13, 6 - 9, 6 - 7, 7 - 14, 7 - 8, 8 - 9, 8 - 10, 9 - 12, 9 - 11, 10 - 14, 10 - 11, 12 - 13.

Thus, $f_1(P_{15})$=25.

The listed edges form the following two - dimensional faces:

1 – 2 – 3 - 12, 1 – 2 – 10 - 8, 1 – 12 – 9 - 8, 2 – 3 – 11 - 10, 3 – 11 – 4 - 5, 3 – 12 – 13 - 4, 4 – 5 – 6 - 13, 5 – 6 – 7 - 14, 5 – 11 – 10 -14, 6 – 9 – 12 - 13, 6 – 7 – 8 - 9, 7 – 8 – 10 - 14, 8 – 9 – 10 - 11.

Thus, $f_2(P_{15})$=13.

From Figure 7.4 it following that it has three - dimensional faces

1 – 2 – 3 – 12 – 9 – 8 – 10 - 11, 8 – 10 – 11 – 9 – 5 – 6 – 7 - 14, 3 – 12 – 9 – 11 – 4 – 13 – 5 - 6.

Besides, the three - dimensional face form outer surface of figure. Thus, $f_3(P_{15})$=4.

Substituting the obtained values f_i in the left side of the Euler - Poincaré equation (7.2) we find

14 – 25 + 13 – 4 = - 2.

The obtained value of the left side of equation (7.2) cannot be equal to the right side of equation (7.2) for any value of dimension d (the right side can be equal to only 2 or 0). Therefore, the figure indicated by the Delaunay parallelohedron 15 is not a four - dimensional parallelohedron. Moreover, since the Euler - Poincaré equation is not satisfied for it, it cannot be considered a convex figure at all. When calculating the number of three - dimensional polyhedrons included in a parallelohedron, Delaunay considered three - dimensional figures formed from the combination of several three - dimensional figures as independent three - dimensional figures, in addition to the connected figures. This, firstly, contradicts the accepted condition for counting the number of figures of various dimensions, and secondly, it only further deviates the value of the left side of equation (7.2) from the value of the right side of this equation.

The simplest figure in the Delaunay classification is figure 13 (Figure 5).

Figure 5. Parallelohedron 13 from Delaunay classification

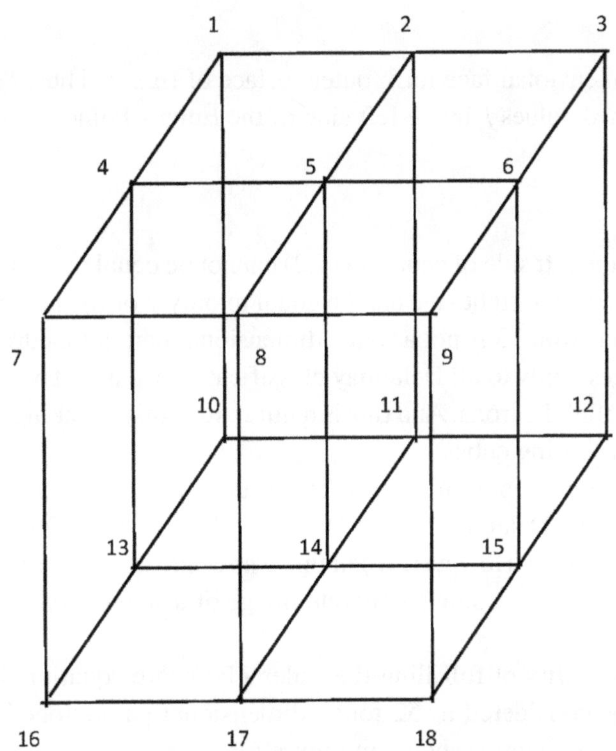

The flaws of this classification are most clearly identified in this figure.

In fact, it is a folded four dice. The number of vertices in this figure is 18 ($f_0(P_{13})$=18). The edges in this figure are the following segments:

1 - 2, 1 - 4, 1 - 10, 2 - 3, 2 - 11, 2 - 5, 3 - 12, 3 - 6, 4 - 13, 4 - 7, 4 - 5, 5 - 6, 5 - 14, 5 - 8, 6 - 15, 6 - 9, 16 - 7, 7 - 8, 8 - 9, 8 - 17, 9 - 18, 10 - 11, 10 - 13, 11 - 12, 14 - 11, 12 - 15, 13 - 14, 13 - 16, 14 - 15, 14 - 17, 15 - 18, 16 - 17, 17 - 18.

Thus, $f_1(P_{13})$=33.

The listed edges form the following two-dimensional faces:

1 – 2 – 4 - 5, 1 – 2 – 10 - 11, 1 – 4 -10 - 13, 2 – 3 – 11 - 12, 2 – 3 – 5 - 6, 2 – 11 – 14 - 5, 3 – 12 – 6 - 15, 4 – 5 – 13 - 14, 4 – 5 – 7 - 8, 4 – 7 – 13 - 16, 5 – 6 – 14 - 15, 5 – 6 – 9 - 8, 5 – 8 – 14 - 17, 6 – 9 – 15 - 18, 7 – 8 – 16 - 17, 8 – 9 – 17 - 18.

Thus, $f_2(P_{13})$=16.

From Figure 7.5 it following that it has four three- dimensional faces

1 – 2 – 4 – 5 – 10 – 11 – 13 -14, 2 – 3 – 5 – 6 – 11 – 12 – 14 - 15, 4 – 5 – 7 – 8 – 13 – 14 – 16 - 17, 5 – 6 – 8 – 9 – 14 – 15 -17 - 18.

Besides, the three - dimensional face form outer surface of figure. Thus, $f_3(P_{13})$=5.

Substituting the obtained values f_i in the left side of the Euler - Poincaré equation (7.2) we find

$$18 - 33 + 16 - 5 = -4.$$

The obtained value of the left side of equation (7.2) cannot be equal to the right side of equation (7.2) for any value of dimension d (the right side can be equal to only 2 or 0). Therefore, the figure indicated by the Delaunay parallelohedron 13 is not a four - dimensional parallelohedron.

The noted disadvantages apply to all Delaunay classification figures. Therefore, all these figures are not four - dimensional parallelohedrons. And this is natural. Roughly speaking, it is impossible to change the dimension of space by folding cubes.

Subsequently, Delaunay attempted to construct the theory of stereohedrons as part of parallelohedrons of higher dimension (Delaunay, 1961; Delaunay, Sandakova, 1961), using the definitions of stereohedrons by Fedorov (1885). However, when this was used, as was later discovered (Zhizhin, 2019 a), erroneous assumptions and not a single concrete image of a stereohedron of dimension higher than three were obtained.

We now verify the possibility of fulfilling the Euler - Poincaré equation for the figure obtained by Shtogrin (1973), which he considered as 52 four - dimensional parallelohedron, i.e. additional to the Delaunay classification. This figure is shown in Figure 6.

As follows from Figure 7.6, it contains 21 vertices ($f_0(P_{52})$=21). The edges in this figure are the following segments:

1 – 10, 1 - 21, 1 - 41, 1 - 2, 2 - 17, 3 - 2, 2 - 19, 2 - 11, 4 - 3, 3 - 16, 3 - 18, 5 - 4, 4 - 16, 4 - 17, 5 - 16, 5 – 6, 5 - 13, 6 - 16, 6 - 15, 6 - 7, 7 - 13, 7 - 19, 7 - 21, 7 - 8, 8 - 14, 8 - 20, 8 - 9, 9 - 12, 9 - 20, 9 - 10, 10 - 20, 10 – 11, 11 - 18, 11 - 12, 11 - 19, 12 - 15, 12 - 19, 13 - 19, 13 - 17, 13 - 14, 14 - 20, 14 - 21, 15 - 16, 15 - 18, 16 - 18, 16 - 19, 17 - 19, 17 - 21, 19 - 20, 20 - 21 .

Figure 6. Four - dimensional Shtogrin parallelohedron

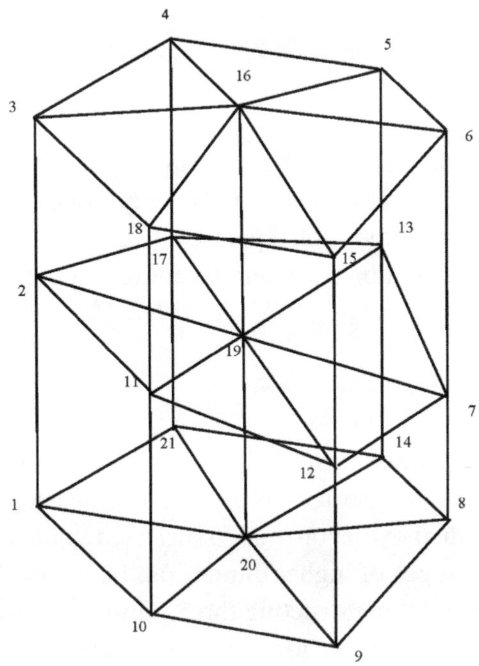

Thus, $f_1(P_{52})$=50.

The listed edges form the following two - dimensional faces:

1 – 2 – 11 – 10, 1 – 21 – 20, 1 – 20 – 10, 2 – 1 – 17 – 21, 3 – 2 – 16 - 19, 2– 3 – 18 - 11, 2 – 17 – 19, 2 – 19 – 11, 3 – 4 – 16, 3 – 16 – 18, 4 – 5 – 13 – 17, 4 – 16 – 17 - 19, 4 – 5 – 16, 5 – 6 – 16, 5 – 6 – 13 – 7, 5 – 16 – 13 – 19, 6 – 7 – 16 – 19, 6 – 15 – 16, 6 – 7 – 12 - 15, 7 – 13 – 19, 7 – 19 – 20 - 8, 7 – 8 – 12 – 9, 7 – 8 – 13 - 14, 8 – 20 – 9, 8 – 14 – 20, 9 – 12 – 19 – 20, 9 – 20 – 10, 9 – 10 – 11 – 12, 10– 11 – 19 - 20, 11 – 18 – 16 - 19, 11 – 19 – 12, 11 – 18 – 15 - 12, 12 – 15 – 16 - 19, 13 – 14 – 19 - 20, 13 – 17 – 19, 14 – 20 – 21, 15 – 16 – 18, 17 – 19 – 20 - 21.

Thus, $f_2(P_{52})$=39.

It should be noted that when calculating the number of edges, it is assumed that the edge does not contain a vertex inside. Otherwise, we are dealing with two edges, i.e. vertices can only be on border of edges. The same situation applies to flat faces. A two - dimensional region is considered a face only when it is not separated by any edge (Zhizhin, 2019 a).

The number of three - dimensional figures included in P_{52} is 13:

1 – 20 – 10 – 2 – 19 – 11, 1 – 21 – 20 – 2 – 17 – 19, 2 – 11 – 19 – 3 – 16 – 18, 2 – 17 – 19 – 3 – 4 – 16, 4 – 5 – 16 – 13 - 17 – 19, 5 – 16 – 6 – 13 – 7 – 19, 6 – 7 – 16 – 15 – 12 – 19, 7 – 13 – 19 – 8 – 14 – 20, 7 – 8 – 19 – 20 – 12 – 9, 9 – 12 – 19 – 20 – 10 – 11, 11 – 12 – 1 9 – 15 – 16 – 18, 13 – 14 – 19 – 20 – 17 – 21, 1 – 2 – 3 – 4 – 5 – 6 – 7 – 8 – 9 – 10 – 11 – 12 – 13 – 14 – 15 – 16 – 17 – 18 – 19 – 20 – 21.

Consequently, $f_3(P_{52})=13$. Substituting the obtained values f_i in the left side of the Euler - Poincaré equation (7.2) we find

$$21-50+39-13=-3.$$

The obtained value of the left side of equation (7.2) cannot be equal to the right side of equation (7.2) for any value of dimension d (the right side can be equal to only 2 or 0). Therefore, the figure indicated by the parallelohedron 52 is not a four - dimensional parallelohedron. Moreover, since the Euler - Poincaré equation is not satisfied for it, it cannot be considered a convex figure at all.

CONCLUSION

A detailed examination of the provisions of the Delaunay theory (Delaunay, 1929, 1937, 1961) showed that in it the geometric properties of constructions of three - dimensional spaces with the help of three - dimensional convex figures studied by Fedorov are transferred to the construction of spaces of higher dimensions with the help of polytopes of higher dimension. However, there are no grounds for such a transfer. When solving the problem of constructing three - dimensional spaces, Fedorov previously investigated and qualified convex three - dimensional figures in detail, studying their symmetry properties and establishing the existence of 230 groups of symmetry transformations. In Delaunay's theory, the properties of polytopes of higher dimension are not considered at all and the conditions of their existence are not taken into account. This also applies to parallelohedrons and to stereohedrons, including for four -dimensional space and four - dimensional figures. It is significant that in the Delaunay theory, when considering the construction of spaces of higher dimension, the Euler – Poincaré equation is not mentioned and is not used, which must be fulfilled without fail for any figure in the space of higher dimension. Discrete Delaunay systems do not take into account important properties of nature, such as scaling (continuous zooming), the existence of so - called quasi - crystalline structures in which there is no translational symmetry when viewed in three - dimensional space, but there is translational symmetry when moving into a space of higher dimension. A direct verification of a number of figures from the classification of 51 Delaunay parallelohedra and one Shtogrin (1973) parallelohedron showed that they all do not satisfy the Euler – Poincaré equation for any dimension. This can be attributed to all the figures of this classification, given the same method of obtaining them. Therefore, the assertion that they are four-dimensional parallelohedra is erroneous. Application of the Delaunay theory leads to incorrect results in other cases. This is demonstrated in this chapter.

This forces us to expand the formulation of the question of constructing n - dimensional spaces, posed by David Hilbert in 1900, and to abandon the methods of considering the construction of spaces by geometric figures that do not take into account the accumulated discoveries in the physics of the structure of substances. Here we should take into account the new paradigm of the discrete world built in the latest works of Zhizhin (Zhizhin, 2018, 2019 b, 2019 a, 2019 c).

The solution of Hilbert's 18th problem of constructing spaces of higher dimension using congruent figures is currently becoming especially relevant in connection with the established fact about the multidimensionality of the nanoworld (Zhizhin, 2013, 2014, 2015, 2018; Shevchenko, Zhizhin, Mackay, 2013; Zhizhin, Dudea, 2016). This, along with the theoretical mathematical value, gives the 18th Hilbert problem technological and, in the future, practical value (Zhizhin, 2019 c).

REFERENCES

Alexandrov, A. D. (1934). The derivation of four - dimensional abnormal parallelohedrons. *Proceedings of the U.S.S.R. Academy of Sciences. Atmospheric and Oceanic Physics, 6*, 803–817.

Alexandrov, A. D. (1954). On filling space with polyhedrons. *Bulletin of the Leningrad University. Mathematics Series., 1*, 33–43.

Alexandrov, P. S. (Ed.). (1969). *Hilbert's problems*. Science.

Coxeter, Y. S. M. (1963). *Regular Polytopes*.

Delaunay, B.N. (1929). Sur la partition ré gulière de l`espace A4 dimensions. *Bulletin de L`Académie des Sciences de L`Urss., 1*, 79 - 110, 147 - 164.

Delaunay, B. N. (1937). Geometry of positive quadratic forms. *Progresses of Mathematical Sciences, 3*, 16–62.

Delaunay, B. N. (1961). Proof of the main theorem of the theory of stereohedrons. *Reports. Academy of Sciences of the USSR, 138*(6), 1270–1272.

Delaunay, B. N., & Sandakova, N. N. (1961). Theory of stereohedrons. *Proceeding Mathematical In. name V.A. Steklov of Academy Sciences of the USSR, 64*, 28–51.

Fedorov, E.S. (1885). Beginning of the teachings on the figures. *Notes of the St. Petersburg Mineralogical Society. 2nd series, 21*, 1 - 277.

Fedorov, E.S. (1889). Symmetry of finite figures. *Notes of the St. Petersburg Mineralogical Society. 2nd series, 25*, 1 - 52.

Fedorov, E.S. (1891). Symmetry of the correct systems of figures. *Notes of the St. Petersburg Mineralogical Society. 2nd series, 28*, 1 - 146.

Galiulin, R. V. (2003). Systems B.N. Delaunay as the mathematical foundation of a discrete world. *Journal of Computational Mathematics and Mathematical Physics, 43*(6), 790–801.

Grünboum, B. (1967). *Convex Polytopes*. Springer.

Hilbert, D. (1901). Gesammelte Abhandlangen. *Archiv f. Mathematics and Physics, III*(1), 44–63, 213–237.

Kadanoff, L. P. (1966). Scaling Laws For Ising Models Near. *Phsics., 2*, 263–272.

Landau, L. V. (1937). Theory of phase transitions 1. *Journal of Experimental and Theoretical Physics, 7*, 19–38.

Mandelbrott, B. B. (1982). *The Fractal Geometry of Nature*. Freeman.

Minkowski, H. (1911). Diskontinuitätshereich für arithemetische Äquivalenz. *Gesammelte Abhandlangen, II*, 70–72.

Poincaré A. (1895). Analysis situs. *J. de é Ecole Polytechnique, 1,* 1 – 121.

Ryzhkov, S.S., & Shushbaev. (1981). Building A - partitions for the second perfect lattice. *Mathematical Collection, 146*(2), 218 - 231.

Schoenflies, A. (1891). Krystallsysteme und Krystallstruktur. Academic Press.

Shafranovsky, I. I. (1963). *Efgraf Stepanovich Fedorov*. Publishing House of the USSR Academy of Sciences.

Shechtman, D., Blech, I., Gratias, D., & Cahn, J. (1984). Metallic phase with longerange orientational order and no translational symmetry. *Physical Review Letters, 53*(20), 1951–1953. doi:10.1103/PhysRevLett.53.1951

Shevchenko, V. Ya., Zhizhin, G. V., & Mackay, A. L. (2013). On the Structure of Quasicrystals in a Higher-Dimensional Space. In M. V. Diudea & C. L. Nagy (Eds.), *Diamond and Related Nanostructures* (pp. 311–320). Springer. doi:10.1007/978-94-007-6371-5_17

Shtogrin, M. I. (1973). Correct Dirichlet - Voronoi partitions for the second triclinic group. *Proceedings of the Mathematical Institute. V.A. Steklova*, 123.

Stott A. B. (1900). On certain series of sections of the regular four - dimensional hyper - solids. *Verhandelingen der Koninklijke Akademie van Wetenschappen te Amsterdam*, 3 - 7.

Stott A. B. (1910). Geometrical deduction of semiregular from regular polytopes and space fillings. *Verhandelingen der Koninklijke Akademie van Wetenschappen te Amsterdam,* 1 – 11.

Stringham, W. I. (1880). Regular figures in *n* - dimensional space. *American Journal of Mathematics, 3*(1), 1–14. doi:10.2307/2369441

Venkov, B. A. (1954). On a class of Euclidean polyhedrons. *Bulletin of the Leningrad University. Mathematics., 2*, 11–31.

Venkov, B.A. (1959). On the design of parallelohedrons. *Mathematical Collection, 49*(2), 207-224.

Voronoi, G. F. (1952). *Studies on primitive parallelohedrons. In Collected Works, 2*. Publishing House of the Academy of Sciences of the Ukrainian SSR.

Wilson, R. G. (1971). Renormalization group and critical phenomena 1. Renormalization group and the Kadanoff scaling picture. *Physics Rev. B., 4*, 3174–3183. doi:10.1103/PhysRevB.4.3174

Zhizhin, G. V. (2012). Hierarchical filling of spaces with polytopes. *Science and human progress.* In *7th St. Petersburg meeting of Nobel Prize winners*. St. Petersburg: Publishing House of the Polytechnic University.

Zhizhin, G. V. (2013). Images of convex regular and semiregular *n* - dimensional polytopes. In *Proceedings of the 9th All-Russian Scientific School "Mathematical Research in the Natural Sciences"*. Apatity: Geological Institute.

Zhizhin, G. V. (2015). *Polytopic prismahedrons – foundamental regions of the* n *- dimension nanostructures.* Report on the International conference "Nanoscience in Chemistry, Physics, Biology and Mathematics", Cluj-Napoca, Romania.

Zhizhin, G. V. (2018). *Chemical Compound Structures and the higher Dimension of Molecules: Emerging research and Opportunities*. IGI Global. doi:10.4018/978-1-5225-4108-0

Zhizhin, G. V. (2019a). *The Geometry of Higher - Dimensional polytopes*. IGI Global. doi:10.4018/978-1-5225-6968-8

Zhizhin, G. V. (2019b). *Attractors and Higher Dimensions in Population and Molecular Biology: Emerging Research and Opportunities*. IGI Global. doi:10.4018/978-1-5225-9651-6

Zhizhin, G. V. (2019c). On the structure of parallelohedrons of higher dimension: Hilbert's 18th problem. *International Journal of Chemoinformatics and Chemical Engineering*, *8*(2), 69–96. doi:10.4018/IJCCE.20190701.oa2

Zhizhin, G. V., & Diudea, M. V. (2016). Spaces of Nanoworld. In M. V. Putz & M. C. Mirica (Eds.), *Sustainable nanosystems, development, properties and applications* (pp. 214–236). IGI Global.

KEY TERMS AND DEFINITIONS

Congruent Polyhedrons: Polyhedrons compatible with each other by movement.

Dimension of the Space: The member of independent parameters needed to describe the change in position of an object in space.

N-Cross-Polytope: The convex polytope of dimension *n* in which opposite related of centrum edges do not have connection of edge.

N-Simplex: The convex polytope of dimension *n* in which each vertex is joined by edges with all remain vertices of polytope.

Parallelogon: A polygon capable of filling a two-dimensional plane without gaps during its translation, with adjacent polygons adjacent to each other on the whole sides of the polygons.

Parallelohedron in Three-Dimensional Space: A polyhedron capable of filling a three-dimensional space without gaps during its translation, with adjacent polyhedrons adjacent to each other along whole flat faces of the polyhedrons.

Polytope: Polyhedron in the space of higher dimension.

Chapter 8
Parallelohedrons of Higher Dimension and Partition of N-Dimensional Spaces

ABSTRACT

For more than 100 years in science, many researchers, when trying to solve Hilbert's 18th problem of constructing n-dimensional space, used the principles of the Delaunay geometric theory. In this book, as a result of a careful analysis of the work in this direction, it is shown that the principles of the Delaunay theory are erroneous. They do not take into account the features of figures of higher dimensionality, do not agree with modern advances in the physics of the structure of matter, and lead to erroneous results. A new approach to solving the 18th Hilbert problem is proposed, based on modern knowledge in the field of the structure of matter and the geometric properties of figures of higher dimension. The basis of the new approach to solving the 18th Hilbert problem is the theory developed by the author on polytopic prismahedrons.

INTRODUCTION

Chapter 6 analytically examined the products of polytopes, which were called polytopic prismahedrons. It was noted that due to the presence of several families of parallel edges in them, they can serve as the basis for the creation of parallelohedra of higher dimension (i.e., dimensions 4 and higher). This is important, since in Chapter 7 it was proved that the figures proposed by Delaunay (Delaunay, 1929) and Shtogrin (Shtogrin, 1973) as parallelohedrons of dimension 4 are not. The fact that polytopic prismahedrons can be the basis for creating parallelohedrons of higher dimension was discovered in the author's previous works (Zhizhin, 2012, 2013, 2014, 2015, 2018, 2019, 2019 a). In this chapter, using the analytical constructions of polytopic prismahedrons for a sufficiently large product of polytopes of various dimensions in Chapter 6, the possibility of constructing spaces of higher dimensions with their help will be investigated by the direct construction of these spaces. Shapes that lead to partitions of higher dimensional spaces by translating them along the coordinates of a higher dimensional space

DOI: 10.4018/978-1-7998-6768-5.ch008

will be considered parallelohedrons of higher dimension (according to the widely used definition of parallelohedrons). Consideration of these features will be carried out graphically, i.e. using the image of the results of the translation of polytopic prismahedrons, and analytically, calculating the coordinates of the vertices of the polytopic prismahedrons when they are translated according to the coordinates of spaces of higher dimension. In this way, we really build parallelohedrons of higher dimension and find out the main features of their structure. This will allow us to avoid the definition of parallelohedra of n - dimensional spaces of the type "butter butted", i.e. a figure is called a parallelohedron if it gives the construction of space during its translation. Under these constructions, naturally, all conditions for the existence of polytopes of higher dimension will be taken into account. Studies will be conducted for a sufficiently large number of polytopic prismahedrons of higher dimension.

N - CUBE AS A POLYTOPIC PRISMAHEDRON

Theorem 8.1. N – cube is n – dimensional parallelohedron in n – dimensional space $(n\ ^3\ 1)$.

Proof. In cases where n is equal to unity, two or three of theorem 8.1 obviously holds. No proof is required. A four - dimensional cube is the simplest version of a polytopic prismahedron, since it is the product of a three - dimensional cube and a one - dimensional segment (Figure 1). Can to introduce one of the vertices of the 4 – cube origin of the four - dimensional space (x, y, z, t). Orient the coordinates, such as indicated in Figure 8.1.

Figure 1. 4 - cube

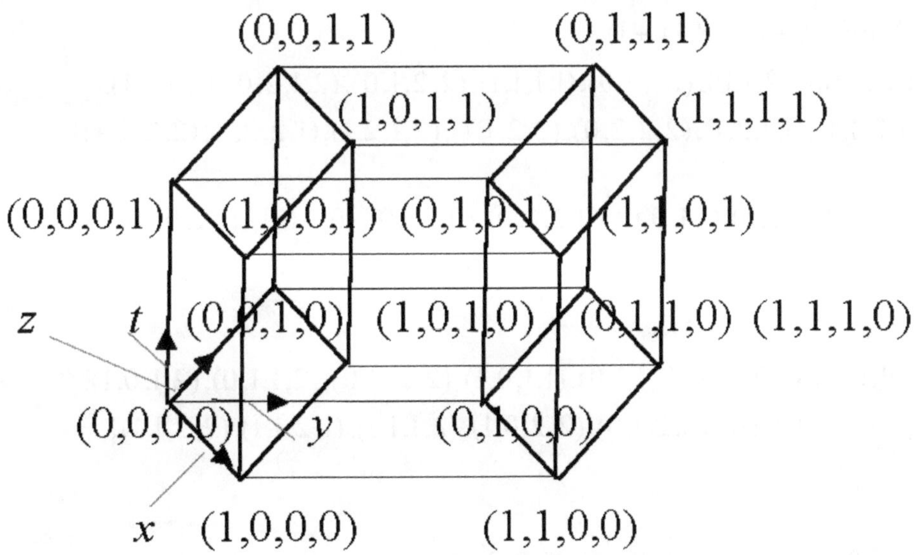

Assume that the length of each edge is equal to 1. Then, each vertex of 4 - cube can be associated with a set of integers (Figure 8.1). Then

$$A_0 = [(0,0,0,0),(1,0,0,0),(0,1,0,0),(0,0,1,0),(0,0,0,1),(1,1,0,0),(0,1,1,0),(0,1,0,1),(0,0,1,1),$$
$$(1,1,1,0),(1,1,0,1),(1,0,1,1),(0,1,1,1),(1,1,1,1)], \tag{8.1}$$

$$A_1 = A_0(x+1) = [(1,0,0,0),(2,0,0,0),(1,1,0,0),(1,0,1,0),(1,0,0,1),(2,1,0,0),(2,0,1,0),$$
$$(2,0,0,1),(1,1,1,0),(1,1,0,1),(1,0,1,1),(2,1,1,0),(2,0,1,1),(1,1,1,1),(2,1,1,1)],$$

$$A_2 = A_0(z+1) = [(0,0,1,0),(1,0,1,0),(0,1,1,0),(0,0,2,0),(0,0,1,1),(1,1,1,0),(1,0,2,0),(1,0,1,1),$$
$$(0,1,2,0),(0,1,1,1),(0,0,2,1),(1,1,2,0),(1,1,1,1),(1,0,2,1),(0,1,2,1),(1,1,2,1)],$$

$$A_3 = A_0(x+1,z+1) = A_1(z+1) =$$
$$[(1,0,1,0),(2,0,1,0),(1,1,1,0),(1,0,2,0),(1,0,1,1,),(2,1,1,0),(2,0,2,0),(2,0,1,1),$$
$$(1,1,2,0),(1,1,1,1),(1,0,2,1),(2,1,2,0),(2,1,1,1),(2,0,2,1),(1,1,2,1),(2,1,2,1)],$$

$$A_4 = A_0(y+1,z+1) = A_2(y+1) =$$
$$[(0,1,1,0),(1,1,1,0),(0,2,1,0),(0,1,2,0),(0,1,1,1),(1,2,1,0),(1,1,2,0),(1,1,1,1),$$
$$(0,2,2,0),(0,2,1,1),(0,1,2,1),(1,2,2,0),(1,2,1,1),(0,2,2,1),(1,2,2,1)],$$

$$A_5 = A_0(x+1,y+1,z+1) = A_3(y+1) =$$
$$[(1,1,1,0),(2,1,1,0),(1,2,1,0),(1,1,2,0),(1,1,1,1),(2,2,1,0),(2,1,2,0),(2,1,1,1),$$
$$(1,2,2,0),((1,2,1,1),(1,1,2,1),(2,2,2,0),(2,2,1,1),(2,1,2,1),(1,2,2,1),(2,2,2,1)],$$

$$A_6 = A_0(y+1) = [(0,1,0,0),(1,1,0,0),(0,2,0,0),(0,1,1,0),(0,1,0,1),(1,2,0,0),(0,2,1,0),(0,2,0,1)],$$

$$A_7 = A_0(y+1,x+1) = A_1(y+1) =$$
$$[(1,1,0,0),(2,1,0,0),(1,2,0,0),(1,1,1,0),(1,1,0,1),(2,2,0,0),(2,1,1,0),(2,1,0,1),$$
$$(1,2,1,0),(1,2,0,1),(1,1,1,1),(2,2,1,0),(2,2,0,1),(2,1,1,1),(1,2,1,1),(2,2,1,1)],$$

$$A_8 = A_0(t+1) =$$
$$[(0,0,0,1),(1,0,0,1),(0,1,0,1),(0,0,1,1),(0,0,0,2),(1,1,0,1),(1,0,1,1),$$
$$(1,0,0,2),(0,1,1,1),(0,1,0,2),(0,0,1,2),(1,1,1,1),(1,1,0,2),(1,0,1,2),(0,1,1,2),(1,1,1,2)],$$

$$A_9 = A_0(y+1,t+1) = A_6(t+1) =$$
$$[(0,1,0,1),(1,1,0,1),(0,2,0,1),(0,1,1,1),(0,1,0,2),(1,2,0,1),(1,1,1,1),(1,1,0,2),$$
$$(0,2,1,1),(0,2,0,2),(0,1,1,2),(1,2,1,1),(1,2,0,2),(1,1,1,2),(0,2,1,2),(1,2,1,2)],$$

$$A_{10} = A_0(x+1,y+1,t+1) = A_7(t+1) =$$
$$[(1,1,0,1),(2,1,0,1),(1,2,0,1),(1,1,1,1),(1,1,0,2),(2,2,0,1),(2,1,1,1),(2,1,0,2),(1,2,1,1),$$
$$(1,2,0,2),(1,1,1,2),(2,2,1,1),(2,2,0,2),(2,1,1,2),(1,2,1,2),(2,2,1,2)],$$

$$A_{11} = A_0(x+1,t+1) = A_1(t+1) =$$
$$[(1,0,0,1),(2,0,0,1),(1,1,0,1),(1,0,1,1),(1,0,0,2),(2,1,0,1),(2,0,1,1),(2,0,0,2),$$
$$(1,1,1,1),(1,1,0,2),(1,0,1,2),(2,1,1,1),(2,1,0,2),(2,0,1,2),(1,1,1,2),(2,1,1,2)],$$

$$A_{12} = A_0(y+1,z+1) = A_2(t+1) =$$
$$[(0,0,1,1),(1,0,1,1),(0,1,1,1),(0,0,2,1),(0,0,1,2),(1,1,1,1),(1,0,2,1),(1,0,1,2),$$
$$(0,1,2,1),(0,1,1,2),(0,0,2,2),(1,1,2,1),(1,1,1,2),(1,0,2,2),(0,1,2,2),(1,1,2,2)],$$

$$A_{13} = A_0(z+1,y+1,t+1) = A_4(t+1) =$$
$$[(0,1,1,1),(1,1,1,1),(0,2,1,1),(0,1,2,1),(0,2,2,1),(0,1,1,2),(1,2,1,1),(0,2,1,2),$$
$$(1,1,2,1),(1,1,1,2),(0,1,2,2),(1,2,2,1),(1,2,1,2),(1,1,2,2),(0,2,2,2),(1,2,2,2)],$$

$$A_{14} = A_0(x+1,y+1,z+1,t+1) = A_5(t+1) =$$
$$[(1,1,1,1),(2,1,1,1),(1,2,1,1),(1,1,2,1),(1,1,1,2),(2,2,1,1),(2,1,2,1),(2,1,1,2),$$
$$(1,2,2,1),(1,2,1,2),(1,1,2,2),(2,2,2,1),(2,2,1,2),(2,1,2,2),(1,2,2,2),(2,2,2,2)],$$

$$A_{15} = A_0(x+1,t+1,z+1) = A_3(t+1) =$$
$$[(1,0,1,1),(2,0,1,1),(1,1,1,1),(1,0,2,1),(1,0,1,2),(2,1,1,1),(2,0,2,1),(2,0,1,2),$$
$$(1,1,2,1),(1,1,1,2),(1,0,2,2),(2,1,2,1),(2,1,1,2),(2,0,2,2),(1,1,2,2),(2,1,2,2)].$$

Representing the 4 - cubes $A_0 \div A_{15}$ dots in three - dimensional space, can get the 4 - cube again (4 - cube A). Moreover, the edges of the 4 - cube A correspond to possible changes in the values of one of the coordinates of the vertices of the 4 - cubes unit (Figure 2).

In addition, each edge of the 4 - cube A in Figure 2 can be considered as an element of the overall two 4 - cubes, connected by an edge. Using the coordinate expression 4 - cubes (8.1) can be analytically determined. In Table 1 geometry elements are common to each pair of the 4 - cubes connected by an edge in Figure 2 are listed.

Figure 2. The 4 - cube A from 16 4 - cubes

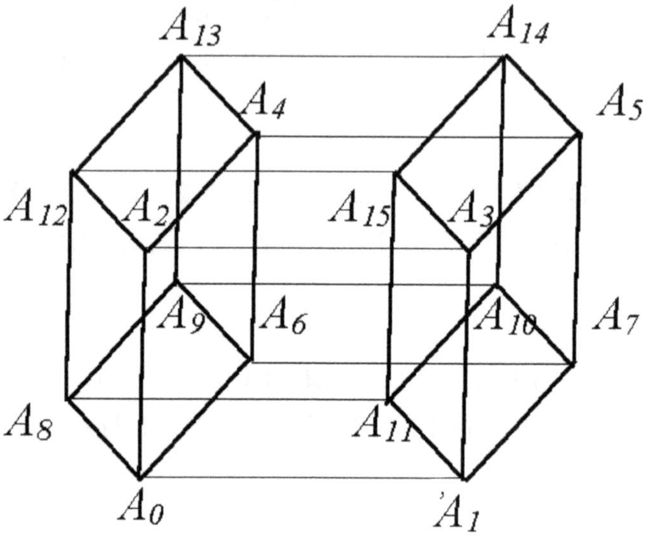

Table 1. Common elements of the 4 - cubes

Edge of the 4-cube A	Common cube of the 4-cubes	Edge of the 4-cube A	Common cube of the 4-cubes
$A_0 A_1$ $A_0 A_2$ $A_0 A_6$ $A_0 A_8$ $A_1 A_3$ $A_7 A_1$ $A_1 A_{11}$ $A_2 A_3$ $A_2 A_4$ $A_2 A_{12}$ $A_3 A_5$ $A_3 A_{15}$ $A_4 A_5$ $A_4 A_6$ $A_4 A_{13}$ $A_7 A_5$	(1,0,0,0)(1,1,0,0)(1,0,1,0)(1,0,0,1) (1,1,1,0)(1,1,0,1)(1,0,1,1)(1,1,1,1) (0,0,1,0)(1,0,1,0)(0,1,1,0)(0,0,1,1) (1,1,1,0)(1,0,1,1)(0,1,1,1)(1,1,1,1) (0,1,0,0)(1,1,0,0)(0,1,1,0)(0,1,0,1) (1,1,1,0)(1,1,0,1)(0,1,1,1)(1,1,1,1) (0,0,0,1)(0,1,0,1)(0,0,1,1)(1,1,0,1) (1,0,1,1)(0,1,1,1)(1,1,1,1)(1,0,0,1) (1,0,1,0)(2,0,1,0)(1,1,1,0)(1,0,1,1) (2,1,1,0)(2,0,1,1)(1,1,1,1)(2,1,1,1) (1,1,0,0,)(2,1,0,0)(1,1,1,0)(1,1,0,1) (2,1,1,0)(2,1,0,1)(1,1,1,1)(2,1,1,1) (1,0,0,1)(2,0,0,1)(1,1,0,1)(1,0,1,1) (2,1,0,1)(2,0,1,1)(1,1,1,1)(2,1,1,1) (1,0,1,0)(1,1,1,0)(1,0,1,1)(0,1,2,0) (1,1,2,0)(1,1,1,1)(1,0,2,1)(1,1,2,0) (0,1,1,0)(1,1,1,0)(0,1,2,0)(0,1,1,1) (1,1,2,0)(1,10,1,1)(0,1,2,1)(1,1,2,1) (0,0,1,1)(0,1,1,1)(0,0,2,1)(1,1,1,1) (1,1,2,1)(0,1,2,1)(1,0,2,1)(1,0,1,1) (1,1,1,0)(2,1,1,0)(1,1,1,1)(2,1,1,1) (1,1,2,1)(2,1,2,1)(2,1,2,0)(2,1,1,1) (2,1,1,1)(1,0,1,1)(2,0,1,1)(1,1,1,1) (1,0,2,1)(2,0,2,1)(1,1,2,1)(2,1,2,1) (1,1,1,0)(1,2,1,0)(1,1,2,0)(1,1,1,1) (1,2,2,0)(1,2,1,1)(1,1,2,1)(1,2,2,0) (1,1,1,0))(0,2,1,0)(0,1,1,1)(1,2,1,0) (1,1,1,1)(0,2,1,1)(1,2,1,1)(0,1,1,0) (0,1,1,1)(1,1,1,1)(0,2,1,1)(0,1,2,1) (1,2.1,1)(1,1,2,1)(0,2,2,1)(1,2,2,1) (1,1,1,0)(2,1,1,0)(1,2,1,0)(1,1,1,1) (2,2,1,0)(2,1,1,1)(1,2,1,1)(2,2,1,1)	$A_5 A_{14}$ $A_6 A_7$ $A_6 A_9$ $A_7 A_{10}$ $A_8 A_{11}$ $A_8 A_9$ $A_8 A_{12}$ $A_9 A_{10}$ $A_9 A_{13}$ $A_{11} A_{10}$ $A_{14} A_{10}$ $A_{11} A_{15}$ $A_{13} A_{12}$ $A_{12} A_{15}$ $A_{13} A_{14}$ $A_{14} A_{15}$	(1,1,1,1)(1,2,1,1)(2,1,1,1)(2,2,1,1) (2,1,2,1)(1,2,2,1)(2,2,2,1)(1,1,2,1) (1,1,1,0)(1,1,0,0)(1,2,0,0)(1,1,0,1) (1,2,1,0)(1,2,0,1)(1,1,1,1)(1,2,1,1) (0,1,0,1)(0,2,0,1)(1,1,0,1)(0,1,1,1) (1,1,1,1)(0,2,1,1)(1,2,1,1)(0,2,0,1) (1,1,0,1)(1,2,0,1)(1,1,1,1)(2,2,0,1) (1,2,1,1)(2,2,1,1)(2,1,0,1)(2,1,1,1) (1,0,0,1)(1,1,0,1)(1,0,1,1)(1,0,0,2) (1,1,1,1)(1,1,0,2)(1,0,1,2)(0,1,1,2) (1,1,1,1)(1,1,0,2)(0,1,1,2)(1,1,1,2) (0,1,0,1)(1,1,0,1)(0,1,1,1)(0,1,0,2) (0,0,1,1)(0,1,1,1)(1,0,1,1)(0,0,1,2) (1,1,1,1)(1,0,1,2)(0,1,1,2)(1,1,1,2) (1,1,0,1)(1,2,0,1)(1,1,1,1)(1,1,0,2) (1,2,1,1)(1,1,1,2)(1,2,0,2)(1,1,1,2) (1,2,1,1)(0,1,1,1)(1,1,1,1)(0,2,1,1) (0,1,1,2)(1,2,1,2)(1,1,1,2)(0,2,1,2) (1,1,1,2)(1,1,0,1)(1,1,1,1)(1,1,0,2) (2,1,1,1)(2,1,0,2)(2,1,1,2)(2,1,0,1) (1,1,1,1)(1,2,1,1)(2,1,1,1)(1,1,1,2) (2,2,1,1)(2,1,1,2)(1,2,1,2)(2,2,1,2) (1,0,1,1)(1,1,1,1)(2,0,1,1)(1,0,1,2) (2,1,1,1)(2,0,1,2)(1,1,1,2)(2,1,1,2) (0,1,1,1)(1,1,1,1)(0,1,2,1)(0,1,1,2) (1,1,2,1)(1,1,1,2)(1,1,2,2)(0,1,2,2) (1,0,1,1)(1,1,1,1)(1,0,1,2)(1,0,2,1) (1,1,1,2)(1,0,2,2)(1,1,2,2)(1,1,2,1) (1,1,1,2)(1,2,1,2)(1,1,2,2)(1,2,2,2) (1,1,1,1)(1,1,2,1)(1,2,1,1)(1,2,2,1) (1,1,1,2)(2,1,2,1)(2,1,1,2)(1,1,2,2) (1,1,1,1)(2,1,1,1)(1,1,2,1)(2,1,2,2)

From Table 8.1 it follows that for two 4 - cubes shifted parallel to each other in the same coordinate by the edge length, a 3 - cube is common, i.e. facet. However, shifting 3 - cubes by the length of an edge over a larger number of coordinates leads to a change in the common element of the cubes. Thus, the shift of 3 - cubes by the length of the edge in two coordinates corresponds to the diagonals of flat faces of the 4 - cube A. Using (8.1), it can be shown that in this case squares are common elements of 4 - cubes. For example, diagonals A_8A_{13}, A_9A_{12} correspond to a common square element (0,1,1,1)(1,1,1,1)(0,1,1,2)(1,1,1,2); diagonals A_0A_4, A_2A_6 correspond to a common square element (0,1,1,0)(1,1,1,0)(0,1,1,1)(1,1,1,1); diagonals A_2A_8, A_0A_{12} correspond to a common square element (0,0,1,1)(1,0,1,1)(0,0,1,1)(1,1,1,1); and go on. Similarly, it can be shown that a shift of 4 - cubes by the length of an edge along three coordinates corresponds to a common element - an edge. When 4 - squares are shifted in four coordinates by the edge length, the common element of 4 - cubes is a vertex. A shift along any coordinate by two edge lengths (or more) leads to the absence of common elements in 4 - cubes.

Since in the vicinity of any 4 - cube at a distance of the length of the edge along each coordinate there is another 4 - cube having a common facet with a central 4 - cube, and on the diagonal of two coordinates there is a 4 - cube having a common flat face with the central 4 - cube, then the 4 - cube during translation fills the entire four - dimensional space without gaps. This means that the 4 - cube is a four - dimensional parallelohedron.

Now let is 5 – cube (Figure 3). It is a product of 4 – cube on one – dimensional segment. A five - dimensional cube is the version of a polytopic prismahedron, since it is the product of a three - dimensional cube and a two - dimensional segment. Can to introduce one of the vertices of the 5 – cube origin of the four - dimensional space (*x, y, z, t, v*). Orient the coordinates, such as indicated in Figure 8.3.

Figure 3. 5 - cube

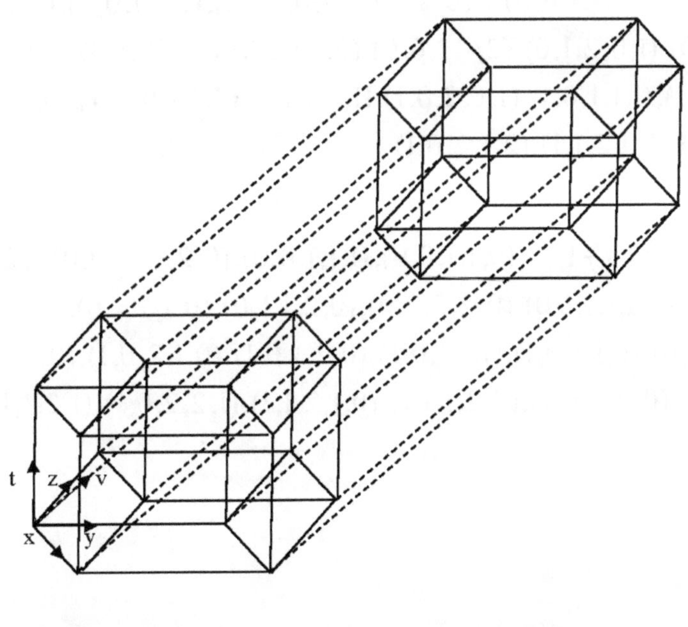

Assume that the length of each edge is equal to 1. Then, each vertex of 5 - cube can be associated with a set of integers (Figure 8.3). Then

$$A_0 = [(0,0,0,0,0),(1,0,0,0,0),(0,1,0,0,0),(0,0,1,0,0),(0,0,0,1,0),(1,1,0,0,0),(0,1,1,0,0),
(0,1,0,1,0),(0,0,1,1,0),(1,1,1,0,0),(1,1,0,1,0),(1,0,1,1,0),(0,1,1,1,0),(1,1,1,1,0),(0,0,0,0,1),
(1,0,0,0,1),(0,1,0,0,1),(0,0,1,0,1),(0,0,0,1,1),(1,1,0,0,1),(0,1,1,0,1),(0,1,0,1,1),(0,0,1,1,1),
(1,1,1,0,0),(1,1,0,1,1),(1,0,1,1,1),(0,1,1,1,1),(1,1,1,1,1)], \tag{8.2}$$

$$A_1 = A_0(x+1) = [(1,0,0,0,0),(2,0,0,0,0),(1,1,0,0,0),(1,0,1,0,0),(1,0,0,1,0),(2,1,0,0,0),
(2,0,1,0,0),(2,0,0,1,0),(1,1,1,0,0),(1,1,0,1,0),(1,0,1,1,0),(2,1,1,0,0),(2,0,1,1,0),(1,1,1,1,0),
(2,1,1,1,0),(1,0,0,0,1),(2,0,0,0,1),(1,1,0,0,1),(1,0,1,0,1),(1,0,0,1,1),(2,1,0,0,1),(2,0,1,0,1),
(2,0,0,1,1),(1,1,1,0,0,1),(1,1,0,1,1),(1,0,1,1,1),(2,1,1,0,1),(2,0,1,1,1),(1,1,1,1,1),(2,1,1,1,1)],$$

$$A_2 = A_0(z+1) = [(0,0,1,0,0),(1,0,1,0,0),(0,1,1,0,0),(0,0,2,0,0),(0,0,1,1,0),(1,1,1,0,0),
(1,0,2,0,0),(1,0,1,1,0),(0,1,2,0,0),(0,1,1,1,0),(0,0,2,1,0),(1,1,2,0,0),(1,1,1,1,0),(1,0,2,1,0),
(0,1,2,1,0),(1,1,2,1,0),(0,0,1,0,1),(1,0,1,0,1),(0,1,1,0,1),(0,0,2,0,1),(0,0,1,1,1),(1,1,1,0,1),
(1,0,2,0,1),(1,0,1,1,1),(0,1,2,0,1),(0,1,1,1,1),(0,0,2,1,1),(1,1,2,0,1),(1,1,1,1,1),(1,0,2,1,1),
(0,1,2,1,1),(1,1,2,1,1)],$$

$$A_3 = A_0(x+1,z+1) = A_1(z+1) =
[(1,0,1,0,0),(2,0,1,0,0),(1,1,1,0,0),(1,0,2,0,0),(1,0,1,1,0),(2,1,1,0,0),(2,0,2,0,0),(2,0,1,1,0),
(1,1,2,0,0),(1,1,1,1,0),(1,0,2,1,0),(2,1,2,0,0),(2,1,1,1,0),(2,0,2,1,0),(1,1,2,1,0),(2,1,2,1,0),
(1,0,1,0,1),(2,0,1,0,1),(1,1,1,0,1),(1,0,2,0,1),(1,0,1,1,1),(2,1,1,0,1),(2,0,2,0,1),(2,0,1,1,1),
(1,1,2,0,1),(1,1,1,1,1),(1,0,2,1,1),(2,1,2,0,1),(2,1,1,1,1),(2,0,2,1,1),(1,1,2,1,1),(2,1,2,1,1)],$$

$$A_4 = A_0(y+1,z+1) = A_2(y+1) = [(0,1,1,0,0),(1,1,1,0,0),(0,2,1,0,0),(0,1,2,0,0),(0,1,1,1,0),
(1,2,1,0,0),(1,1,2,0,0),(1,1,1,1,0),(0,2,2,0,0),(0,2,1,1,0),(0,1,2,1,0),(1,2,2,0,0),(1,2,1,1,0),
(0,2,2,1,0),(1,2,2,1,0),(0,1,1,0,1),(1,1,1,0,1),(0,2,1,0,1),(0,1,2,0,1),(0,1,1,1,1),(1,2,1,0,1),
(1,1,2,0,1),(1,1,1,1,1),(0,2,2,0,1),(0,2,1,1,1),(0,1,2,1,1),(1,2,2,0,1),(1,2,1,1,1),(0,2,2,1,1),
(1,2,2,1,1)],$$

$$A_5 = A_0(x+1, y+1, z+1) = A_3(y+1) =$$
$$[(1,1,1,0,0),(2,1,1,0,0),(1,2,1,0,0),(1,1,2,0,0),(1,1,1,1,0),(2,2,1,0,0),(2,1,2,0,0),(2,1,1,1,0),$$
$$(1,2,2,0,0),(1,2,1,1,0),(1,1,2,1,0),(2,2,2,0,0),(2,2,1,1,0),(2,1,2,1,0),(1,2,2,1,0),(2,2,2,1,0),$$
$$(1,1,1,0,1),(2,1,1,0,1),(1,2,1,0,1),(1,1,2,0,1),(1,1,1,1,1),(2,2,1,0,1),(2,1,2,0,1),(2,1,1,1,1),$$
$$(1,2,2,0,1),(1,2,1,1,1),(1,1,2,1,1),(2,2,2,0,1),(2,2,1,1,1),(2,1,2,1,1),(1,2,2,1,1),(2,2,2,1,1)],$$

$$A_6 = A_0(y+1) = [(0,1,0,0,0),(1,1,0,0,0),(0,2,0,0,0),(0,1,1,0,0),(0,1,0,1,0),(1,2,0,0,0),$$
$$(0,2,1,0,0),(0,2,0,1,0),(0,1,1,1,0),(1,2,1,0,0),(1,1,1,1,0),(0,2,1,1,0),(1,2,1,1,0),(0,1,0,0,1),$$
$$(1,1,0,0,1),(0,2,0,0,1),(0,1,1,0,1),(0,1,0,1,1),(1,2,0,0,1),(0,2,1,0,1),(0,2,0,1,1),(0,1,1,1,1),$$
$$(1,2,1,0,1),(1,1,1,1,1),(0,2,1,1,1),(1,2,1,1,1)],$$

$$A_7 = A_0(y+1, x+1) = A_1(y+1) = [(1,1,0,0,0),(2,1,0,0,0),(1,2,0,0,0),(1,1,1,0,0),(1,1,0,1,0),$$
$$(2,2,0,0,0),(2,1,1,0,0),(2,1,0,1,0),(1,2,1,0,0),(1,2,0,1,0),(1,1,1,1,0),(2,2,1,0,0),(2,2,0,1,0),$$
$$(2,1,1,1,0),(1,2,1,1,0),(2,2,1,1,0),(1,1,0,0,1),(2,1,0,0,1),(1,2,0,0,1),(1,1,1,0,1),(1,1,0,1,1),$$
$$(2,2,0,0,1),(2,1,1,0,1),(2,1,0,1,1),(1,2,1,0,1),(1,2,0,1,1),(1,1,1,1,1),(2,2,1,0,1),(2,2,0,1,1),$$
$$(2,1,1,1,1),(1,2,1,1,1),(2,2,1,1,1)],$$

$$A_8 = A_0(t+1) = [(0,0,0,1,0),(1,0,0,1,0),(0,1,0,1,0),(0,0,1,1,0),(0,0,0,2,0),(1,1,0,1,0),$$
$$(1,0,1,1,0),(1,0,0,2,0),(0,1,1,1,0),(0,1,0,2,0),(0,0,1,2,0),(1,1,1,1,0),(1,1,0,2,0),(1,0,1,2,0),$$
$$(0,1,1,2,0),(1,1,1,2,0),(0,0,0,1,1),(1,0,0,1,1),(0,1,0,1,1),(0,0,1,1,1),(0,0,0,2,1),(1,1,0,1,1),$$
$$(1,0,1,1,1),(1,0,0,2,1),(0,1,1,1,1),(0,1,0,2,1),(0,0,1,2,1),(1,1,1,1,1),(1,1,0,2,1),(1,0,1,2,1),$$
$$(0,1,1,2,1),(1,1,1,2,1)],$$

$$A_9 = A_0(y+1, t+1) = A_6(t+1) = [(0,1,0,1,0),(1,1,0,1,0),(0,2,0,1,0),(0,1,1,1,0),(0,1,0,2,0),$$
$$(1,2,0,1,0),(1,1,1,1,0),(1,1,0,2,0),(0,2,1,1,0),(0,2,0,2,0),(0,1,1,2,0),(1,2,1,1,0),(1,2,0,2,0),$$
$$(1,1,1,2,0),(0,2,1,2,0),(1,2,1,2,0),(0,1,0,1,1),(1,1,0,1,1),(0,2,0,1,1),(0,1,1,1,1),(0,1,0,2,1),$$
$$(1,2,0,1,1),(1,1,1,1,1),(1,1,0,2,1),(0,2,1,1,1),(0,2,0,2,1),(0,1,1,2,1),(1,2,1,1,1),(1,2,0,2,1),$$
$$(1,1,1,2,1),(0,2,1,2,1),(1,2,1,2,1)],$$

$A_{10} = A_0(x+1, y+1, t+1) = A_7(t+1) = [(1,1,0,1,0),(2,1,0,1,0),(1,2,0,1,0),(1,1,1,1,0),(1,1,0,2,0),$
$(2,2,0,1,0),(2,1,1,1,0),(2,1,0,2,0),(1,2,1,1,0),(1,2,0,2,0),(1,1,1,2,0),(2,2,1,1,0),(2,2,0,2,0),$
$(2,1,1,2,0),(1,2,1,2,0),(2,2,1,2,0),(1,1,0,1,1),(2,1,0,1,1),(1,2,0,1,1),(1,1,1,1,1),(1,1,0,2,1),$
$(2,2,0,1,1),(2,1,1,1,1),(2,1,0,2,1),(1,2,1,1,1),(1,2,0,2,1),(1,1,1,2,1),(2,2,1,1,1),(2,2,0,2,1),$
$(2,1,1,2,1),(1,2,1,2,1),(2,2,1,2,1)],$

$A_{11} = A_0(x+1, t+1) = A_1(t+1) = [(1,0,0,1,0),(2,0,0,1,0),(1,1,0,1,0),(1,0,1,1,0),(1,0,0,2,0),$
$(2,1,0,1,0),(2,0,1,1,0),(2,0,0,2,0),(1,1,1,1,0),(1,1,0,2,0),(1,0,1,2,0),(2,1,1,1,0),(2,1,0,2,0),$
$(2,0,1,2,0),(1,1,1,2,0),(2,1,1,2,0),(1,0,0,1,1),(2,0,0,1,1),(1,1,0,1,1),(1,0,1,1,1),(1,0,0,2,1),$
$(2,1,0,1,1),(2,0,1,1,1),(2,0,0,2,1),(1,1,1,1,1),(1,1,0,2,1),(1,0,1,2,1),(2,1,1,1,1),(2,1,0,2,1),$
$(2,0,1,2,1),(1,1,1,2,1),(2,1,1,2,1)],$

$A_{12} = A_0(y+1, z+1) = A_2(t+1) = [(0,0,1,1,0),(1,0,1,1,0),(0,1,1,1,0),(0,0,2,1,0),$
$(0,0,1,2,0),(1,1,1,1,0),(1,0,2,1,0),(1,0,1,2,0),(0,1,2,1,0),(0,1,1,2,0),(0,0,2,2,0),$
$(1,1,2,1,0),(1,1,1,2,0),(1,0,2,2,0),(0,1,2,2,0),(1,1,2,2,0),(0,0,1,1,1),(1,0,1,1,1),$
$(0,1,1,1,1),(0,0,2,1,1),(0,0,1,2,1),(1,1,1,1,1),(1,0,2,1,1),(1,0,1,2,1),(0,1,2,1,1),$
$(0,1,1,2,1),(0,0,2,2,1),(1,1,2,1,1),(1,1,1,2,1),(1,0,2,2,1),(0,1,2,2,1),(1,1,2,2,1)],$

$A_{13} = A_0(z+1, y+1, t+1) = A_4(t+1) = [(0,1,1,1,0),(1,1,1,1,0),(0,2,1,1,0),(0,1,2,1,0),(0,2,2,1,0),$
$(0,1,1,2,0),(1,2,1,1,0),(0,2,1,2,0),(1,1,2,1,0),(1,1,1,2,0),(0,1,2,2,0),(1,2,2,1,0),(1,2,1,2,0),$
$(1,1,2,2,0),(0,2,2,2,0),(1,2,2,2,0),(0,1,1,1,1),(1,1,1,1,1),(0,2,1,1,1),(0,1,2,1,1),(0,2,2,1,1),$
$(0,1,1,2,1),(1,2,1,1,1),(0,2,1,2,1),(1,1,2,1,1),(1,1,1,2,1),(0,1,2,2,1),(1,2,2,1,1),(1,2,1,2,1),$
$(1,1,2,2,1),(0,2,2,2,1),(1,2,2,2,1)],$

$A_{14} = A_0(x+1, y+1, z+1, t+1) = A_5(t+1) = [(1,1,1,1,0),(2,1,1,1,0),(1,2,1,1,0),(1,1,2,1,0),$
$(1,1,1,2,0),(2,2,1,1,0),(2,1,2,1,0),(2,1,1,2,0),(1,2,2,1,0),(1,2,1,2,0),(1,1,2,2,0),(2,2,2,1,0),$
$(2,2,1,2,0),(2,1,2,2,0),(1,2,2,2,0),(2,2,2,2,0),(1,1,1,1,1),(2,1,1,1,1),(1,2,1,1,1),(1,1,2,1,1),$
$(1,1,1,2,1),(2,2,1,1,1),(2,1,2,1,1),(2,1,1,2,1),(1,2,2,1,1),(1,2,1,2,1),(1,1,2,2,1),(2,2,2,1,1),$
$(2,2,1,2,1),(2,1,2,2,1),(1,2,2,2,1),(2,2,2,2,1)],$

$$A_{15} = A_0(x+1,t+1,z+1) = A_3(t+1) = [(1,0,1,1,0),(2,0,1,1,0),(1,1,1,1,0),(1,0,2,1,0),$$
$$(1,0,1,2,0),(2,1,1,1,0),(2,0,2,1,0),(2,0,1,2,0),(1,1,2,1,0),(1,1,1,2,0),(1,0,2,2,0),$$
$$(2,1,2,1,0),(2,1,1,2,0),(2,0,2,2,0),(1,1,2,2,0),(2,1,2,2,0),(1,0,1,1,1),(2,0,1,1,1),$$
$$(1,1,1,1,1),(1,0,2,1,1),(1,0,1,2,1),(2,1,1,1,1),(2,0,2,1,1),(2,0,1,2,1),(1,1,2,1,1),$$
$$(1,1,1,2,1),(1,0,2,2,1),(2,1,2,1,1),(2,1,1,2,1),(2,0,2,2,1),(1,1,2,2,1),(2,1,2,2,1)],$$

$$A_{16} = A_0(v+1) = [(0,0,0,0,1),(1,0,0,0,1),(0,1,0,0,1),(0,0,1,0,1),(0,0,0,1,1),(1,1,0,0,1),(0,1,1,0,1),$$
$$(0,1,0,1,1),(0,0,1,1,1),(1,1,1,0,1),(1,1,0,1,1),(1,0,1,1,1),(0,1,1,1,1),(1,1,1,1,1),(0,0,0,0,2),$$
$$(1,0,0,0,2),(0,1,0,0,2),(0,0,1,0,2),(0,0,0,1,2),(1,1,0,0,2),(0,1,1,0,2),(0,1,0,1,2),(0,0,1,1,2),$$
$$(1,1,1,0,2),(1,1,0,1,2),(1,0,1,1,2),(0,1,1,1,2),(1,1,1,1,2)],$$

$$A_{17} = A_1(v+1) = [(1,0,0,0,1),(2,0,0,0,1),(1,1,0,0,1),(1,0,1,0,1),(1,0,0,1,1),(2,1,0,0,1),$$
$$(2,0,1,0,1),(2,0,0,1,1),(1,1,1,0,1),(1,1,0,1,1),(1,0,1,1,1),(2,1,1,0,1),(2,0,1,1,1),(1,1,1,1,1),$$
$$(2,1,1,1,1),(1,0,0,0,2),(2,0,0,0,2),(1,1,0,0,2),(1,0,1,0,2),(1,0,0,1,2),(2,1,0,0,2),(2,0,1,0,2),$$
$$(2,0,0,1,2),(1,1,1,0,0,2),(1,1,0,1,2),(1,0,1,1,2),(2,1,1,0,2),(2,0,1,1,2),(1,1,1,1,2),(2,1,1,1,2)],$$

$$A_{18} = A_2(v+1) = [(0,0,1,0,1),(1,0,1,0,1),(0,1,1,0,1),(0,0,2,0,1),(0,0,1,1,1),(1,1,1,0,1),$$
$$(1,0,2,0,1),(1,0,1,1,1),(0,1,2,0,1),(0,1,1,1,1),(0,0,2,1,1),(1,1,2,0,1),(1,1,1,1,1),(1,0,2,1,1),$$
$$(0,1,2,1,1),(1,1,2,1,1),(0,0,1,0,2),(1,0,1,0,2),(0,1,1,0,2),(0,0,2,0,2),(0,0,1,1,2),(1,1,1,0,2),$$
$$(1,0,2,0,2),(1,0,1,1,2),(0,1,2,0,2),(0,1,1,1,2),(0,0,2,1,2),(1,1,2,0,2),(1,1,1,1,2),(1,0,2,1,2),$$
$$(0,1,2,1,2),(1,1,2,1,2)],$$

$$A_{19} = A_3(v+1) = [(1,0,1,0,1),(2,0,1,0,1),(1,1,1,0,1),(1,0,2,0,1),(1,0,1,1,1),(2,1,1,0,1),(2,0,2,0,1),$$
$$(2,0,1,1,1),(1,1,2,0,1),(1,1,1,1,1),(1,0,2,1,1),(2,1,2,0,1),(2,1,1,1,1),(2,0,2,1,1),(1,1,2,1,1),(2,1,2,1,1),$$
$$(1,0,1,0,2),(2,0,1,0,2),(1,1,1,0,2),(1,0,2,0,2),(1,0,1,1,2),(2,1,1,0,2),(2,0,2,0,2),(2,0,1,1,2),$$
$$(1,1,2,0,2),(1,1,1,1,2),(1,0,2,1,2),(2,1,2,0,2),(2,1,1,1,2),(2,0,2,1,2),(1,1,2,1,2),(2,1,2,1,2)],$$

$$A_{20} = A_4(v+1) = [(0,1,1,0,1),(1,1,1,0,1),(0,2,1,0,1),(0,1,2,0,1),(0,1,1,1,1),(1,2,1,0,1),$$
$$(1,1,2,0,1),(1,1,1,1,1),(0,2,2,0,1),(0,2,1,1,1),(0,1,2,1,1),(1,2,2,0,1),(1,2,1,1,1),(0,2,2,1,1),$$
$$(1,2,2,1,1),(0,1,1,0,2),(1,1,1,0,2),(0,2,1,0,2),(0,1,2,0,2),(0,1,1,1,2),(1,2,1,0,2),(1,1,2,0,2),$$
$$(1,1,1,1,2),(0,2,2,0,2),(0,2,1,1,2),(0,1,2,1,2),(1,2,2,0,2),(1,2,1,1,2),(0,2,2,1,2),(1,2,2,1,2)],$$

$$A_{21} = A_5(v+1) = [(1,1,1,0,1),(2,1,1,0,1),(1,2,1,0,1),(1,1,2,0,1),(1,1,1,1,1),(2,2,1,0,1),(2,1,2,0,1)$$
$$(2,1,1,1,1),(1,2,2,0,1),(1,2,1,1,1),(1,1,2,1,1),(2,2,2,0,1),(2,2,1,1,1),(2,1,2,1,1),(1,2,2,1,1),(2,2,2,1,1),$$
$$(1,1,1,0,2),(2,1,1,0,2),(1,2,1,0,2),(1,1,2,0,2),(1,1,1,1,2),(2,2,1,0,2),(2,1,2,0,2),(2,1,1,1,2),$$
$$(1,2,2,0,2),(1,2,1,1,2),(1,1,2,1,2),(2,2,2,0,2),(2,2,1,1,2),(2,1,2,1,2),(1,2,2,1,2),(2,2,2,1,2)],$$

$$A_{22} = A_6(v+1) = [(0,1,0,0,1),(1,1,0,0,1),(0,2,0,0,1),(0,1,1,0,1),(0,1,0,1,1),(1,2,0,0,1),$$
$$(0,2,1,0,1),(0,2,0,1,1),(0,1,1,1,1),(1,2,1,0,1),(1,1,1,1,1),(0,2,1,1,1),(1,2,1,1,1),(0,1,0,0,2),$$
$$(1,1,0,0,2),(0,2,0,0,2),(0,1,1,0,2),(0,1,0,1,2),(1,2,0,0,2),(0,2,1,0,2),(0,2,0,1,2),(0,1,1,1,2),$$
$$(1,2,1,0,2),(1,1,1,1,2),(0,2,1,1,2),(1,2,1,1,2)],$$

$$A_{23} = A_7(v+1) = [(1,1,0,0,1),(2,1,0,0,1),(1,2,0,0,1),(1,1,1,0,1),(1,1,0,1,1),$$
$$(2,2,0,0,1),(2,1,1,0,1),(2,1,0,1,1),(1,2,1,0,1),(1,2,0,1,1),(1,1,1,1,1),(2,2,1,0,1),(2,2,0,1,1),$$
$$(2,1,1,1,1),(1,2,1,1,1),(2,2,1,1,1),(1,1,0,0,2),(2,1,0,0,2),(1,2,0,0,2),(1,1,1,0,2),(1,1,0,1,2),$$
$$(2,2,0,0,2),(2,1,1,0,2),(2,1,0,1,2),(1,2,1,0,2),(1,2,0,1,2),(1,1,1,1,2),(2,2,1,0,2),(2,2,0,1,2),$$
$$(2,1,1,1,2),(1,2,1,1,2),(2,2,1,1,2)],$$

$$A_{24} = A_8(v+1) = [(0,0,0,1,1),(1,0,0,1,1),(0,1,0,1,1),(0,0,1,1,1),(0,0,0,2,1),(1,1,0,1,1),$$
$$(1,0,1,1,1),(1,0,0,2,1),(0,1,1,1,1),(0,1,0,2,1),(0,0,1,2,1),(1,1,1,1,1),(1,1,0,2,1),(1,0,1,2,1),$$
$$(0,1,1,2,1),(1,1,1,2,1),(0,0,0,1,2),(1,0,0,1,2),(0,1,0,1,2),(0,0,1,1,2),(0,0,0,2,2),(1,1,0,1,2),$$
$$(1,0,1,1,2),(1,0,0,2,2),(0,1,1,1,2),(0,1,0,2,2),(0,0,1,2,2),(1,1,1,1,2),(1,1,0,2,2),(1,0,1,2,2),$$
$$(0,1,1,2,2),(1,1,1,2,2)],$$

$$A_{25} = A_9(v+1) = [(0,1,0,1,1),(1,1,0,1,1),(0,2,0,1,1),(0,1,1,1,1),(0,1,0,2,1),$$
$$(1,2,0,1,1),(1,1,1,1,1),(1,1,0,2,1),(0,2,1,1,1),(0,2,0,2,1),(0,1,1,2,1),(1,2,1,1,1),(1,2,0,2,1),$$
$$(1,1,1,2,1),(0,2,1,2,1),(1,2,1,2,1),(0,1,0,1,2),(1,1,0,1,2),(0,2,0,1,2),(0,1,1,1,2),(0,1,0,2,2),$$
$$(1,2,0,1,2),(1,1,1,1,2),(1,1,0,2,2),(0,2,1,1,2),(0,2,0,2,2),(0,1,1,2,2),(1,2,1,1,2),(1,2,0,2,2),$$
$$(1,1,1,2,2),(0,2,1,2,2),(1,2,1,2,2)],$$

$$A_{26} = A_{10}(v+1) = [(1,1,0,1,1),(2,1,0,1,1),(1,2,0,1,1),(1,1,1,1,1),(1,1,0,2,1),$$
$$(2,2,0,1,1),(2,1,1,1,1),(2,1,0,2,1),(1,2,1,1,1),(1,2,0,2,1),(1,1,1,2,1),(2,2,1,1,1),(2,2,0,2,1),$$
$$(2,1,1,2,1),(1,2,1,2,1),(2,2,1,2,1),(1,1,0,1,2),(2,1,0,1,2),(1,2,0,1,2),(1,1,1,1,2),(1,1,0,2,2),$$
$$(2,2,0,1,2),(2,1,1,1,2),(2,1,0,2,2),(1,2,1,1,2),(1,2,0,2,2),(1,1,1,2,2),(2,2,1,1,2),(2,2,0,2,2),$$
$$(2,1,1,2,2),(1,2,1,2,2),(2,2,1,2,2)],$$

$$A_{27} = A_{11}(v+1) = [(1,0,0,1,1),(2,0,0,1,1),(1,1,0,1,1),(1,0,1,1,1),(1,0,0,2,1),$$
$$(2,1,0,1,1),(2,0,1,1,1),(2,0,0,2,1),(1,1,1,1,1),(1,1,0,2,1),(1,0,1,2,1),(2,1,1,1,1),(2,1,0,2,1),$$
$$(2,0,1,2,1),(1,1,1,2,1),(2,1,1,2,1),(1,0,0,1,2),(2,0,0,1,2),(1,1,0,1,2),(1,0,1,1,2),(1,0,0,2,2),$$
$$(2,1,0,1,2),(2,0,1,1,2),(2,0,0,2,2),(1,1,1,1,2),(1,1,0,2,2),(1,0,1,2,2),(2,1,1,1,2),(2,1,0,2,2),$$
$$(2,0,1,2,2),(1,1,1,2,2),(2,1,1,2,2)],$$

$$A_{28} = A_{12}(v+1) = [(0,0,1,1,1),(1,0,1,1,1),(0,1,1,1,1),(0,0,2,1,1),$$
$$(0,0,1,2,1),(1,1,1,1,1),(1,0,2,1,1),(1,0,1,2,1),(0,1,2,1,1),(0,1,1,2,1),(0,0,2,2,1),$$
$$(1,1,2,1,1),(1,1,1,2,1),(1,0,2,2,1),(0,1,2,2,1),(1,1,2,2,1),(0,0,1,1,2),(1,0,1,1,2),$$
$$(0,1,1,1,2),(0,0,2,1,2),(0,0,1,2,2),(1,1,1,1,2),(1,0,2,1,2),(1,0,1,2,2),(0,1,2,1,2),$$
$$(0,1,1,2,2),(0,0,2,2,2),(1,1,2,1,2),(1,1,1,2,2),(1,0,2,2,2),(0,1,2,2,2),(1,1,2,2,2)],$$

$$A_{29} = A_{13}(v+1) = [(0,1,1,1,1),(1,1,1,1,1),(0,2,1,1,1),(0,1,2,1,1),(0,2,2,1,1),$$
$$(0,1,1,2,1),(1,2,1,1,1),(0,2,1,2,1),(1,1,2,1,1),(1,1,1,2,1),(0,1,2,2,1),(1,2,2,1,1),(1,2,1,2,1),$$
$$(1,1,2,2,1),(0,2,2,2,1),(1,2,2,2,1),(0,1,1,1,2),(1,1,1,1,2),(0,2,1,1,2),(0,1,2,1,2),(0,2,2,1,2),$$
$$(0,1,1,2,2),(1,2,1,1,2),(0,2,1,2,2),(1,1,2,1,2),(1,1,1,2,2),(0,1,2,2,2),(1,2,2,1,2),(1,2,1,2,2),$$
$$(1,1,2,2,2),(0,2,2,2,2),(1,2,2,2,2)],$$

$$A_{30} = A_{14}(v+1) = [(1,1,1,1,1),(2,1,1,1,1),(1,2,1,1,1),(1,1,2,1,1),$$
$$(1,1,1,2,1),(2,2,1,1,1),(2,1,2,1,1),(2,1,1,2,1),(1,2,2,1,1),(1,2,1,2,1),(1,1,2,2,1),(2,2,2,1,1),$$
$$(2,2,1,2,1),(2,1,2,2,1),(1,2,2,2,1),(2,2,2,2,1),(1,1,1,1,2),(2,1,1,1,2),(1,2,1,1,2),(1,1,2,1,2),$$
$$(1,1,1,2,2),(2,2,1,1,2),(2,1,2,1,2),(2,1,1,2,2),(1,2,2,1,2),(1,2,1,2,2),(1,1,2,2,2),(2,2,2,1,2),$$
$$(2,2,1,2,2),(2,1,2,2,2),(1,2,2,2,2),(2,2,2,2,2)],$$

$$A_{31} = A_{15}(v+1) = [(1,0,1,1,1),(2,0,1,1,1),(1,1,1,1,1),(1,0,2,1,1),$$
$$(1,0,1,2,1),(2,1,1,1,1),(2,0,2,1,1),(2,0,1,2,1),(1,1,2,1,1),(1,1,1,2,1),(1,0,2,2,1),$$
$$(2,1,2,1,1),(2,1,1,2,1),(2,0,2,2,1),(1,1,2,2,1),(2,1,2,2,1),(1,0,1,1,2),(2,0,1,1,2),$$
$$(1,1,1,1,2),(1,0,2,1,2),(1,0,1,2,2),(2,1,1,1,2),(2,0,2,1,2),(2,0,1,2,2),(1,1,2,1,2),$$
$$(1,1,1,2,2),(1,0,2,2,2),(2,1,2,1,2),(2,1,1,2,2),(2,0,2,2,2),(1,1,2,2,2),(2,1,2,2,2)].$$

When passing from four - dimensional space to five - dimensional space, as follows from a comparison of equalities (8.1) and (8.2), in the values of the coordinates of the vertices in each of the polytopes 4 - the cube $A_0 \div A_{15}$ additionally appears the values of the fifth coordinate v equal to 0 and 1. The number of vertices in each of the polytopes this doubles due to the addition of two values of this coordinate 0 and 1. In addition, the number of vertices of the polytopes 4 - cube due to a shift along the fifth coordinate by the length of the edge doubles, i.e. four - dimensional cubes $A_{16} \div A_{31}$ appear. In these four - dimensional cubes, the coordinates of the vertices also have five values. The values of the fifth coordinate equal to 1 and 2 were added to the four coordinates in four - dimensional space. Representing the 5 - cubes $A_0 \div A_{31}$ dots in five - dimensional space, can get the 5 - cube again (5 - cube A). Moreover, the edges of the 5 - cube A correspond to possible changes in the values of one of the coordinates of the vertices of the 5 - cubes unit (Figure 4.

Figure 4. The 5 - cube A from 32 5 - cubes

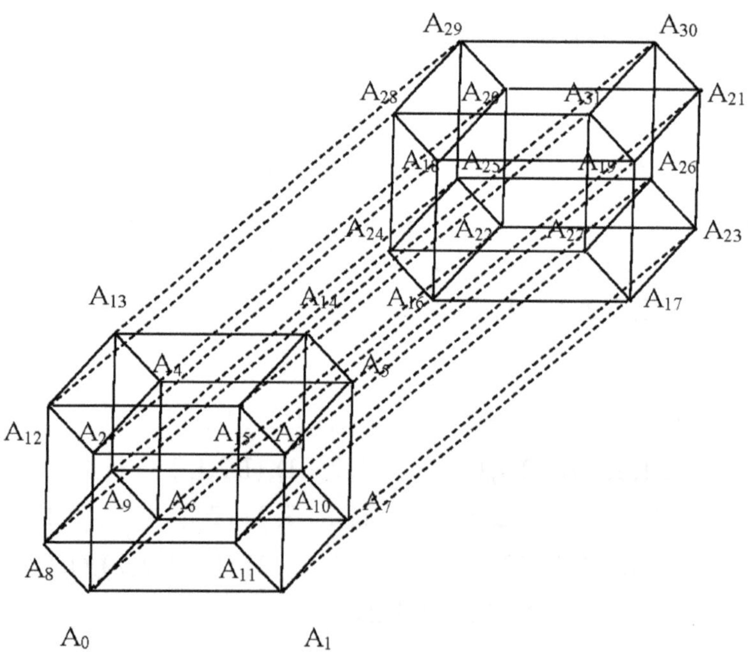

Compare two polytopes in four - dimensional space and two polytopes in five - dimensional space. In four - dimensional space they are defined by the equalities

$$A_0 = [(0,0,0,0),(1,0,0,0),(0,1,0,0),(0,0,1,0),(0,0,0,1),(1,1,0,0),(0,1,1,0),(0,1,0,1),(0,0,1,1),$$
$$(1,1,1,0),(1,1,0,1),(1,0,1,1),(0,1,1,1),(1,1,1,1)],$$

$$A_1 = [(1,0,0,0),(2,0,0,0),(1,1,0,0),(1,0,1,0),(1,0,0,1),(2,1,0,0),(2,0,1,0),$$
$$(2,0,0,1),(1,1,1,0),(1,1,0,1),(1,0,1,1),(2,1,1,0),(2,0,1,1),(1,1,1,1),(2,1,1,1)].$$

In five - dimensional space they are defined by the equalities

$$A_0 = [(0,0,0,0,0),(1,0,0,0,0),(0,1,0,0,0),(0,0,1,0,0),(0,0,0,1,0),(1,1,0,0,0),(0,1,1,0,0),$$
$$(0,1,0,1,0),(0,0,1,1,0),(1,1,1,0,0),(1,1,0,1,0),(1,0,1,1,0),(0,1,1,1,0),(1,1,1,1,0),(0,0,0,0,1),$$
$$(1,0,0,0,1),(0,1,0,0,1),(0,0,1,0,1),(0,0,0,1,1),(1,1,0,0,1),(0,1,1,0,1),(0,1,0,1,1),(0,0,1,1,1),$$
$$(1,1,1,0,0),(1,1,0,1,1),(1,0,1,1,1),(0,1,1,1,1),(1,1,1,1,1)],$$

$$A_1 = [(1,0,0,0,0),(2,0,0,0,0),(1,1,0,0,0),(1,0,1,0,0),(1,0,0,1,0),(2,1,0,0,0),$$
$$(2,0,1,0,0),(2,0,0,1,0),(1,1,1,0,0),(1,1,0,1,0),(1,0,1,1,0),(2,1,1,0,0),(2,0,1,1,0),(1,1,1,1,0),$$
$$(2,1,1,1,0),(1,0,0,0,1),(2,0,0,0,1),(1,1,0,0,1),(1,0,1,0,1),(1,0,0,1,1),(2,1,0,0,1),(2,0,1,0,1),$$
$$(2,0,0,1,1),(1,1,1,0,0,1),(1,1,0,1,1),(1,0,1,1,1),(2,1,1,0,1),(2,0,1,1,1),(1,1,1,1,1),(2,1,1,1,1)].$$

Between A_0 and A_1 in four - dimensional space there is a common element

$$(1, 0, 0, 0), (1, 1, 0, 0), (1, 0, 1, 0), (1, 0, 0, 1), (1, 1, 1, 0), (1, 1, 0, 1), (1, 0, 1, 1), (1, 1, 1, 1). \tag{8.3}$$

There is a common element between A_0 and A_1 in five - dimensional space

$$(1, 0, 0, 0, 0), (1, 1, 0, 0, 0), (1, 0, 1, 0, 0), (1, 0, 0, 1, 0), (1, 1, 1, 0, 0), (1, 1, 0, 1, 0), (1, 0, 1, 1, 0), (1, 1, 1, 1, 0), (1, 0, 0, 0, 1), (1, 1, 0, 0, 1), (1, 0, 1, 0, 1), (1, 0, 0, 1, 1), (1, 1, 1, 0, 1), (1, 1, 0, 1, 1), (1, 0, 1, 1, 1), (1, 1, 1, 1, 1). \tag{8.4}$$

From a comparison of (8.3) and (8.4) it follows that the existence of a common element between polytopes A_0 and A_1 in a five - dimensional space is determined by the existence of a common element between polytopes A_0 and A_1 in a four - dimensional space. Moreover, if the common element between the polytopes A_0 and A_1 in four - dimensional space is a three - dimensional cube, then the common element between the polytopes A_0 and A_1 in five - dimensional space is a four - dimensional cube.

It is easy to verify that a similar situation is repeated along other edges of the polytope 5 - cube A, since they are all edges connecting polytopes of dimension one less. Thus, a common element between polytopes of the 5 - cube when shifting by the length of the edge is 4 - cube. This is true for all edges of the 5 - cube A (Figure 8.3). When a polytope is shifted, a 5 - cube is diagonal of a polytope 5 - cube A; the common elements change their dimension, just as in a polytope 4 - cube A. But in this case, in a polytope 5 - cube A, the dimension of the common elements is one more than in the polytope 4 - cube A. The dimension of the common elements depends on the number of variables along which the shift takes place: when shifting along one variable, the dimension of the common element is 4 (4 - cube), when shifting along two variables the dimension of the common element is 3 (3 - cube), with shift in three variables, the dimension of the common element is 2 (square), when shifting in four variables, the dimension of the common element is 1 (edge), when shifting in five variables, the dimension of the common element is 0 (vertex). When a polytope 5 - cube is shifted by two (or more) edge lengths of at least one of the variables, polytope 5 - cube common elements are absent.

Since in the vicinity of any 5 - cube at a distance of the length of the edge along each coordinate there is another 5 - cube having a common facet (4 – cube) with a central 5 - cube, and on the diagonal of two coordinates there is a 5 - cube having a common three - dimension face with the central 5 - cube, and on the diagonal of three coordinates there is a 5 – cube having a common two – dimension face with the central 5 – cube, and on diagonal of four coordinates there is a 5 - cube having a common edge with the central 5 – cube, and on diagonal of five coordinates there is a 5 – cube having a common vertices with the central 5 - cube. Then the 5 - cube during translation fills the entire five - dimensional space without gaps. This means that the 5 - cube is a five - dimensional parallelohedron.

Obviously, with an increase in the dimension of the cube, the marked regularities remain. The presence of common elements between polytopes n - cube when a polytope is shifted along one of the coordinates of space is determined by the presence of a common element between polytopes (n - 1) - cube when a polytope is shifted along the same coordinate of space and the dimension of the common element is (n - 1), i.e. this is (n - 1) - cube. When shifting along several coordinates by the edge length, the dimension of the common elements decreases by the number of coordinates along which the shift occurs. Neighboring polytopes surround the original polytope without forming gaps, in contact with the initial polytope in elements of various dimensions. Moreover, during shifts, all polytopes are naturally in parallel positions. Thus, an n - cube polytope is an n - dimensional parallelohedron for any value of n of equal or greater unity. Q.E.D.

Notes. The carried proof is based on the algebraic record of changes in the coordinates of a polytope when shifting by the length of an edge. It is precisely proved that cubes of higher dimension with such a shift completely fill the space and the contact between n - cubes over integer elements of lower dimension. The dimension values of these elements (cubes) are determined by the number of coordinates along which the shift is carried out. When shifting along one of the coordinates, the bosonic elements between adjacent n - cubes is a cube of dimension (n - 1), while shifting along a larger number of coordinates, the dimension of the common elements decreases accordingly. The separation of the arguments of previous researchers (Minkowski, 1911; Delaunay, 1929, 1937, 1961; Delaunay, & Sandakova,1961; Voronoi, 1952; Alexandrov, 1934, 1954; Venkov, 1954, 1959) from specific forms of polytopes of higher dimension did not allow them to obtain results that determine the patterns of partitioning of n - dimensional spaces by polytopes of higher dimension, in particular, to define the n - cube as a parallelohedron of n - dimensional space. In connection with the theorem proved by Theorem 8.1, the definition used earlier on the normal partition of an n - dimensional space should be corrected. It is necessary to exclude from this definition an indispensable condition for contact of the contact of n - dimensional figures in this partition only for polytopes of dimension n - 1.

A normal partition of an n - dimensional space by higher - dimensional figures will be a partition of space in which the component of the partition of polytopes of higher dimension are in contact with each other over integer elements of dimension (n -1), (n - 2), (n - 3), …, 1, 0.

POLYTOPIC PRISMAHEDRON ON BASES OF SIMPLEXES

Partition of Space by the 4 - Quadrangular Prism

The simplest element of a polytope of simplex type that does not coincide with elements of a polytope of type cube is a triangle. If we multiply the triangle by a one - dimensional segment, then we get a triangular prism. Is it possible to fill the space using a triangular prism, applying them to each other along whole flat faces? Can. If the triangle is equipotential, which can be assumed, then as a result of applying triangular prisms to each other along the side faces, we obtain a quadrangular prism, each of the bases of which are composed of two triangles (Figure 5). Denote this quadrangular prism 4 - quadrangular prism

Figure 5. 4 - quadrangular prism

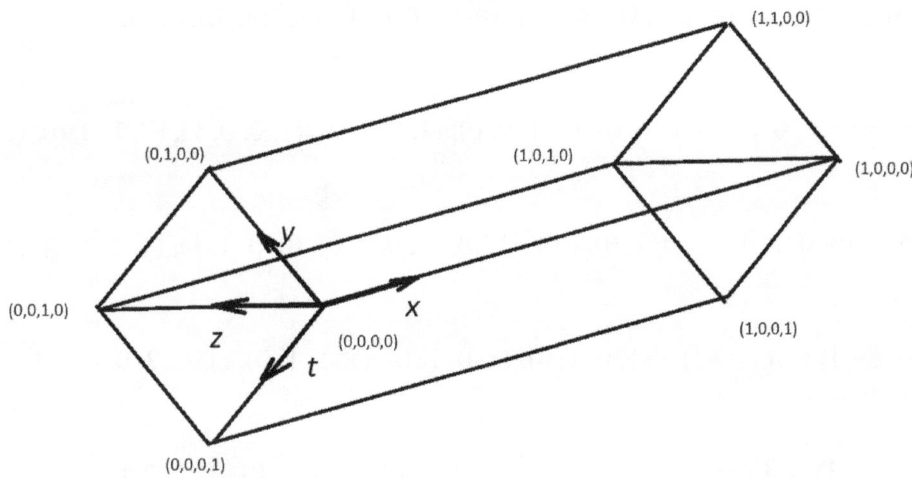

It is easy to verify that the dimension of such a construction is 4. Indeed, The number of vertices in this figure 5 is 8, i.e. f_0=5; the number of edges is 14, i.e. f_1= 14 ; the number of planar faces is 9, i.e. f_2= 9 ; the number of three - dimensional figures is 3 (two triangular prisms and one quadrangular prism), i.e. f_3= 3 . To prove the statement, we use the Euler - Poincaré (Poincaré, 1895) equation

$$\sum_{i=0}^{n-1}(-1)^i f_i(P) = 1 + (-1)^{n-1} . \tag{8.5}$$

In (8.5) $f_i(P)$ is the number of faces with dimension i in polytope P with dimension n.
Substituting the obtained values into the Euler - Poincaré equation, we obtain

$$8 - 14 + 9 - 3 = 0.$$

We see that equation (8.5) is satisfied for $n = 4$. This proves that two triangular prisms folded along the lateral face form a figure with dimension 4. Obviously, this figure is a four - dimensional parallelohedron, since its translation will fill the whole space without gaps. Interestingly, in the case of translation, four - dimensional parallelohedra in this case adjoin each other along two -dimensional faces, either quadrangular or triangular. The neighboring four - dimensional parallelohedra have common two - dimensional faces. This once again proves that the fallacy of the assertion of the Delaunay theory that neighboring four - dimensional parallelohedra when dividing the space must necessarily have common elements of dimension one less than the dimension of parallelohedra, i.e. be three - dimensional. Neighboring four - dimensional parallelohedra can have common triangular faces not during translation, but when they are rotated by 30 degrees relative to each other.

Can to introduce one of the vertices of the 4 - quadrangular prism origin of the four - dimensional space (*x, y, z, t*). Orient the coordinates, such as indicated in Figure 8.5.

Assume that the length of each edge is equal to 1. Then, each vertex of the 4 - quadrangular prism can be associated with a set of integers (Figure 5). Then

A_0= (0, 0, 0, 0), (1, 0, 0, 0), (0, 1, 0, 0), (0, 0, 1, 0), (0, 0, 0, 1), (1, 1, 0, 0), (1, 0, 0, 1) (8.6)

$A_1 = A_0(x+1) =$ (1, 0, 0, 0), (2, 0, 0, 0), (1, 1, 0, 0), (1, 0, 1, 0), (1, 0, 0, 1), (2, 1, 0, 0), (2, 0, 1, 0), (2, 0, 0, 1),

$A_2 = A_0(z+1) =$ (0, 0, 1, 0), (1, 0, 1, 0), (0, 1, 1, 0), (0, 0, 2, 0), (0, 0, 1, 1), (1, 1, 1, 0), (1, 0, 2, 0), (1, 0, 1, 1),

$A_3 = A_0(x+1, z+1) = A_1(z+1) =$ (1, 0, 1, 0), (2, 0, 1, 0), (1, 1, 1, 0), (1, 0, 2, 0), (1, 0, 1, 1), (2, 1, 1, 0), (2, 0, 2, 0), (2, 0, 1, 1),

$A_4 = A_0(y+1, z+1) = A_2(y+1) =$ (0, 1, 1, 0), (1, 1, 1, 0), (0, 2, 1, 0), (0, 1, 2, 0), (0, 1, 1, 1), (1, 2, 1, 0), (1, 1, 2, 0), (1, 1, 1, 1),

$A_5 = A_0(x+1, y+1, z+1) = A_3(y+1) =$ (1, 1, 1, 0), (2, 1, 1, 0), (1, 2, 1, 0), (1, 1, 2, 0), (1, 1, 1, 1), (2, 2, 1, 0), (2, 1, 2, 0), (2, 1, 1, 1),

$A_6 = A_0(y+1) =$ (0, 1, 0, 0), (1, 1, 0, 0), (0, 2, 0, 0), (0, 1, 1, 0), (0, 1, 0, 1), (1, 2, 0, 0), (1, 1, 1, 0), (1, 1, 0, 1),

$A_7 = A_0(y+1, x+1) = A_1(y+1) =$ (1, 1, 0, 0), (2, 1, 0, 0), (1, 2, 0, 0), (1, 1, 1, 0), (1, 1, 0, 1), (2, 2, 0, 0), (2, 1, 1, 0), (2, 1, 0, 1),

$A_8 = A_0(t+1) =$ (0, 0, 0, 1), (1, 0, 0, 1), (0, 1, 0, 1), (0, 0, 1, 1), (0, 0, 0, 2), (1, 1, 0, 1), (1, 0, 1, 1), (1, 0, 0, 2),

$A_9 = A_0(y+1, t+1) = A_6(t+1) =$ (0, 1, 0, 1), (1, 1, 0, 1), (0, 2, 0, 1), (0, 1, 1, 1), (0, 1, 0, 2), (1, 2, 0, 1), (1, 1, 1, 1), (1, 1, 0, 2),

$A_{10} = A_0(x+1, y+1, t+1) = A_7(t+1) =$ (1, 1, 0, 1), (2, 1, 0, 1), (1, 2, 0, 1), (1, 1, 1, 1), (1, 1, 0, 2), (2, 2, 0, 1), (2, 1, 1, 1), (2, 1, 0, 2),

$A_{11} = A_0(x+1, t+1) = A_1(t+1) =$ (1, 0, 0, 1), (2, 0, 0, 1), (1, 1, 0, 1), (1, 0, 1, 1), (1, 0, 0, 2), (2, 1, 0, 1), (2, 0, 1, 1), (2, 0, 0, 2),

$A_{12} = A_0(y+1, z+1) = A_2(t+1) =$ (0, 0, 1, 1), (1, 0, 1, 1), (0, 1, 1, 1), (0, 0, 2, 1), (0, 0, 1, 2), (1, 1, 1, 1), (1, 0, 2, 1), (1, 0, 1, 2),

$A_{13} = A_0(z+1, y+1, t+1) = A_4(t+1) =$ (0, 1, 1, 1), (1, 1, 1, 1), (0, 2, 1, 1), (0, 1, 2, 1), (0, 2, 2, 1), (0, 1, 1, 2), (1, 2, 1, 1), (0, 2, 1, 2),

$A_{14} = A_0(x+1, y+1, z+1, t+1) = A_5(t+1) =$ (1, 1, 1, 1), (2, 1, 1, 1), (1, 2, 1, 1), (1, 1, 2, 1), (1, 1, 1, 2), (2, 2, 1, 1), (2, 1, 2, 1), (2, 1, 1, 2),

$A_{15} = A_0(x+1, t+1, z+1) = A_3(t+1) =$ (1, 0, 1, 1), (2, 0, 1, 1), (1, 1, 1, 1), (1, 0, 2, 1), (1, 0, 1, 2), (2, 1, 1, 1), (2, 0, 2, 1), (2, 0, 1, 2).

Representing of 4 - quadrangular prisms $A_0 \div A_{15}$ dots in four - dimensional space, can get the 4 - cube again (4 - cube A1). Moreover, the edges of the 4 - cube A1 correspond to possible changes in the values of one of the coordinates of the vertices of the 4 - quadrangular prisms unit (4 - cube A1, Figure 3, in which the polytopes at vertices correspondent the equality (8.6)).

Table 2. Common elements of the 4 - quadrangular prisms

Edge of the 4-cube A1	Common elements of the 4 - quadrangular prisms	Edge of the 4-cube A1	Common cube of the 4 - quadrangular prisms
$A_0 A_1$	(1,0,0,0)(1,1,0,0)(1,0,1,0)(1,0,0,1)	$A_5 A_{14}$	(1,1,1,1)(2,1,1,1)
$A_0 A_2$	(0,0,1,0)(1,0,1,0)	$A_6 A_7$	(1,1,1,0)(1,1,0,0)(1,2,0,0)(1,1,0,1)
$A_0 A_6$	(0,1,0,0)(1,1,0,0)	$A_6 A_9$	(0,1,0,1)(1,1,0,1)
$A_0 A_8$	(0,0,0,1)(1,0,1,0)(1,0,0,1)	$A_7 A_{10}$	(1,1,0,1)(2,1,0,1)
$A_1 A_3$	(1,0,1,0)(2,0,1,0)	$A_8 A_{11}$	(1,0,0,1)(1,1,0,1)(1,0,1,1)(1,0,0,2)
$A_7 A_1$	(1,1,0,0,)(2,1,0,0)	$A_8 A_9$	(0,1,0,1)(1,1,0,1)
$A_1 A_{11}$	(1,0,0,1)(2,0,0,1)	$A_8 A_{12}$	(0,0,1,1)(1,0,1,1)
$A_2 A_3$	(1,0,1,0)(1,0,2,0)(1,0,1,1)	$A_9 A_{10}$	(1,1,0,1)(1,2,0,1)(1,1,1,1)(1,1,0,2)
$A_2 A_4$	(0,1,1,0)(1,1,1,0)	$A_9 A_{13}$	(0,1,1,1)(1,1,1,1)
$A_2 A_{12}$	(0,0,1,1)(1,0,1,1)	$A_{11} A_{10}$	(1,1,0,1)(2,1,0,1)
$A_3 A_5$	(1,1,1,0)(2,1,1,0)	$A_{14} A_{10}$	(1,1,1,1)(2,1,1,1)
$A_3 A_{15}$	(1,0,1,1)(2,0,1,1)	$A_{11} A_{15}$	(1,0,1,1)(2,0,1,1)
$A_4 A_5$	(1,1,1,0)(1,2,1,0)(1,1,2,0)(1,1,1,1)	$A_{13} A_{12}$	(0,1,1,1)(1,1,1,1)
$A_4 A_6$	(1,1,1,0))(0,1,1,0)	$A_{12} A_{15}$	(1,0,1,1)(1,1,1,1)(1,0,1,2)(1,0,2,1)
$A_4 A_{13}$	(0,1,1,1)(1,1,1,1)	$A_{13} A_{14}$	(1,1,1,1)(1,2,1,1)
$A_7 A_5$	(1,1,1,0)(2,1,1,0)	$A_{14} A_{15}$	(1,1,1,1)(2,1,1,1)

In addition, each edge of the 4 - cube A1 in Figure 6 can be considered as an element of the overall two the 4 - quadrangular prisms, connected by an edge. Using the coordinate expression of the 4 - quadrangular prisms (8.6) can be analytically determined. In Table 2 geometry elements are common to each pair of the 4 - quadrangular prisms connected by an edge in Figure 6 are listed.

From Table 2 it follows that the really neighboring 4 - quadrangular prisms, when translating them to the length of the edge along one of the coordinates, share quadrangular and triangular flat faces and edges with common elements. When translating over the length of an edge along several coordinates, as follows from analytical expressions (8.6), the dimension of common elements is even smaller. And when broadcasting on two or more edge lengths, common elements of 4 - quadrangular prisms are absent.

Since in the vicinity of any 4 - quadrangular prisms at a distance of the length of the edge along each coordinate there is another 4 - quadrangular prisms having a common faces with a central 4 - quadrangular prisms, and on the diagonal of two coordinates there is a 4 - quadrangular prisms having a common face with the central 4 - quadrangular prisms, then the 4 - quadrangular prisms during translation fills the entire four - dimensional space without gaps. This means that the 4 - quadrangular prisms is a four - dimensional parallelohedron.

Partition of Space by the Polytopic Prism

One of the simplest types of the polytopic prismahedrons is the tetrahedral prism, i.e. a product tetrahedron by the segment (Chapter 6). Its structural formula has form

$$P_4^3(4F_3^2)\times P_2^1 = P_8^4[4P_6^3(3F_4^2,2F_3^2),2F_4^3(4F_3^2)]. \tag{8.7}$$

Here, the subscript in the polytope P and his face F indicates the number of vertices, and the superscript indicates the dimension of the corresponding polytope or faces. The right side of (8.6) describing the structural formula of the product, the facet indicated by the symbol of the polytope to specify which polytopes of dimension n - 1 is composed work polytopes. Thus, P_4^3 - the tetrahedron, P_2^1 - the segment, P_6^3 - the triangular prism, P_3^2 - the triangle, P_4^2 - the quadrilateral, P_8^4 - the tetrahedral prism. The dimension of the tetrahedral prism is equal to 4, it has 8 vertices, 16 edges, 14 faces two - dimensional, 6 three - dimensional faces (2 tetrahedrons, 4 triangular prisms). Image tetrahedral prism is shown in Figure 6.

Figure 6. The tetrahedral prism

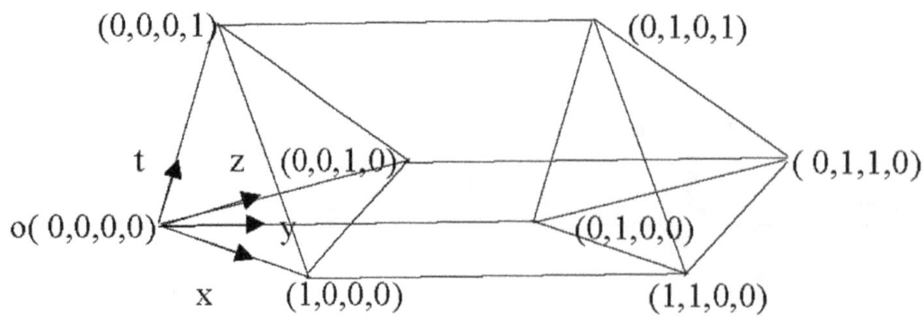

Can to introduce one of the vertices of the tetrahedral prism origin of the four - dimensional space (x, y, z, t). Orient the coordinates, such as indicated in Figure 6.

Assume that the length of each edge is equal to 1. Then, each node tetrahedral prism can be associated with a set of integers (Figure 6). Translating tetrahedral prism along the coordinates x, y, z, t, we obtain the lattice vertices. Let A_0 tetrahedral prism with the values of vertex coordinates in Figure 8.6. Then

A_0= (0, 0, 0, 0), (1, 0, 0, 0), (0, 1, 0, 0), (0, 0, 1, 0), (0, 0, 0, 1), (1, 1, 0, 0), (0, 1, 1, 0), (0, 1, 0, 1), (8.8)

$A_1 = A_0(x+1) =$ (1, 0, 0, 0), (2, 0, 0, 0), (1, 1, 0, 0), (1, 0, 1, 0), (1, 0, 0, 1), (2, 1, 0, 0), (1, 1, 1, 0), (1, 1, 0, 1),

$A_2 = A_0(z+1) =$ (0, 0, 1, 0), (1, 0, 1, 0), (0, 1, 1, 0), (0, 0, 2, 0), (0, 0, 1, 1), (1, 1, 1, 0), (0, 1, 2, 0), (0, 1, 1, 1),

$A_3 = A_0(x+1, z+1) = A_1(z+1) =$ (1, 0, 1, 0), (2, 0, 1, 0), (1, 1, 1, 0), (1, 0, 2, 0), (1, 0, 1, 1), (2, 1, 1, 0), (1, 1, 2, 0), (1, 1, 1, 1),

$A_4 = A_0(y+1, z+1) = A_2(y+1) =$ (0, 1, 1, 0), (1, 1, 1, 0), (0, 2, 1, 0), (0, 1, 2, 0), (0, 1, 1, 1), (1, 2, 1, 0), (0, 2, 2, 0), (0, 2, 1, 1),

$A_5 = A_0(x+1, y+1, z+1) = A_3(y+1) =$ (1, 1, 1, 0), (2, 1, 1, 0), (1, 2, 1, 0), (1, 1, 2, 0), (1, 1, 1, 1), (2, 2, 1, 0), (1, 2, 2, 0), (1, 2, 1, 1),

$A_6 = A_0(y+1) =$ (0, 1, 0, 0), (1, 1, 0, 0), (0, 2, 0, 0), (0, 1, 1, 0), (0, 1, 0, 1), (1, 2, 0, 0), (0, 2, 1, 0), (0, 2, 0, 1),

$A_7 = A_0(y+1, x+1) = A_1(y+1) =$ (1, 1, 0, 0), (2, 1, 0, 0), (1, 2, 0, 0), (1, 1, 1, 0), (1, 1, 0, 1), (2, 2, 0, 0), (1, 2, 1, 0), (1, 2, 0, 1),

$A_8 = A_0(t+1) =$ (0, 0, 0, 1), (1, 0, 0, 1), (0, 1, 0, 1), (0, 0, 1, 1), (0, 0, 0, 2), (1, 1, 0, 1), (0, 1, 1, 1), (0, 1, 0, 2),

$A_9 = A_0(y+1, t+1) = A_6(t+1) =$ (0, 1, 0, 1), (1, 1, 0, 1), (0, 2, 0, 1), (0, 1, 1, 1), (0, 1, 0, 2), (1, 2, 0, 1), (0, 2, 1, 1), (0, 2, 0, 2),

$A_{10} = A_0(x+1, y+1, t+1) = A_7(t+1) =$ (1, 1, 0, 1), (2, 1, 0, 1), (1, 2, 0, 1), (1, 1, 1, 1), (1, 1, 0, 2), (2, 2, 0, 1), (1, 2, 1, 1), (1, 2, 0, 2),

$A_{11} = A_0(x+1, t+1) = A_1(t+1) =$ (1, 0, 0, 1), (2, 0, 0, 1), (1, 1, 0, 1), (1, 0, 1, 1), (1, 0, 0, 2), (2, 1, 0, 1), (1, 1, 1, 1), (1, 1, 0, 2),

$A_{12} = A_0(y+1, z+1) = A_2(t+1) =$ (0, 0, 1, 1), (1, 0, 1, 1), (0, 1, 1, 1), (0, 0, 2, 1), (0, 0, 1, 2), (1, 1, 1, 1), (0, 1, 2, 1), (0, 1, 1, 2),

$A_{13} = A_0(z+1, y+1, t+1) = A_4(t+1) =$ (0, 1, 1, 1), (1, 1, 1, 1), (0, 2, 1, 1), (0, 1, 2, 1), (0, 2, 2, 1), (0, 1, 1, 2), (1, 2, 1, 1), (0, 2, 1, 2),

$A_{14} = A_0(x+1, y+1, z+1, t+1) = A_5(t+1) =$ (1, 1, 1, 1), (2, 1, 1, 1), (1, 2, 1, 1), (1, 1, 2, 1), (1, 1, 1, 2), (2, 2, 1, 1), (1, 2, 2, 1), (1, 2, 1, 2),

$A_{15} = A_0(x+1, t+1, z+1) = A_3(t+1) =$ (1, 0, 1, 1), (2, 0, 1, 1), (1, 1, 1, 1), (1, 0, 2, 1), (1, 0, 1, 2), (2, 1, 1, 1), (1, 1, 2, 1), (1, 1, 1, 2).

Representing the tetrahedral prisms $A_0 \div A_{15}$ dots in four - dimensional space, can get the hypercube. Moreover, the edges of the hypercube correspond to possible changes in the values of one of the coordinates of the vertices of the tetrahedral prism unit (4 - cube A2, Figure 8.3, in which the polytopes at vertices correspondent the equality (8.8)).

In addition, each edge of the hypercube in Figure 8.3 in this case can be considered as an element of the overall two tetrahedral prisms, connected by an edge. Using the coordinate expression tetrahedral prisms (8.8) can be analytically determined. In Table 3 geometry elements are common to each pair of the tetrahedral prisms connected by an edge in Figure 3 are listed.

Diagonals flat faces in the hypercube correspond to a simultaneous change in the unit values of the two coordinates of the vertices of the tetrahedral prisms (Table 3). Diagonals 8 cubes in the hypercube in Figure 8.3 in this case correspond to a simultaneous change in the unit of some three coordinates. Common elements of the tetrahedral prisms with such change vertex coordinates is either a vertex or the empty set. When you change the same unit coordinates all four common elements in the tetrahedral prisms not.

Between A_i type tetrahedral prisms arranged tetrahedral prism type B_i (see Figure 7) having a closest to them tetrahedral prisms A_i general quadrangular two - dimensional face.

The tetrahedral prisms A_i and B_i have common edges, which are parties to the general quadrilateral faces. The tetrahedral prisms A_i and B_i are connected symmetry transformation - turning the 180^0 around a common edge of the tetrahedron. Forms A and B are two enantiomorphic forms of a tetrahedral prism. Let us prove that the complex of two pairs of enantiomorphic forms of tetrahedral prisms (Figure 7) has dimension 5. As follows from Figure 7, this complex is not a convex polytope. However, for its analysis, the Euler – Poincaré equation can be used, since Poincare showed that this equation is applicable to all connected forms (Poincaré, 1895).

The polytope in Figure 8.7 has 18 vertices, f_0=18. The polytope in Figure 8.7 has 47 edges:

0 – 1, 0 – 8, 0 – 7, 0 – 10, 1 – 7, 1 – 8, 1 – 3, 1 – 2, 1 – 11, 2 – 3, 2 – 8, 2 – 12, 3 – 8, 3 – 5, 3 – 4, 3 – 13, 4 – 8, 4 – 5, 4 – 14, 5 – 8, 5 – 6, 5 – 15, 6 – 8, 6 – 7, 6 – 16, 7 – 8, 7 – 17, 8 – 9, 9 – 10, 9 – 11, 9 – 12, 9 – 13, 9 – 14, 9 – 15, 9 – 16, 9 – 17, 10 – 11, 10 – 17, 11 – 12, 11 – 13, 12 – 13, 13 – 14, 13 – 15, 14 – 15, 15 – 16, 15 – 17, 16 – 17, i.e. f_1=47

Figure 7. The complex of the tetrahedral prisms

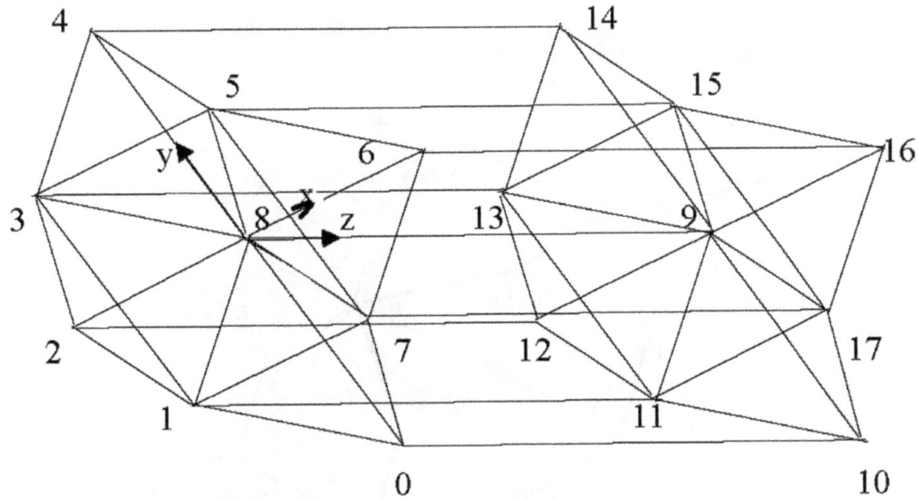

The polytope in Figure 8.7 has 50 two – dimensional elements:

0 – 8 – 9 – 10, 0 – 7 – 17 – 10, 0 – 1 – 11 – 10, 0 – 8 – 7, 0 – 1 – 8, 0 – 1 – 7, 1 – 8 – 7, 1 – 2 – 8, 1 – 2 – 3, 1 – 3 – 8, 1 – 2 – 12 – 11, 1 – 8 – 9 – 11, 1 – 17 – 7 – 11, 1 – 3 – 11 – 13, 2 – 3 – 13 – 12, 2 – 3 – 9 – 12, 2 – 3 – 8, 3 – 5 – 15 – 13, 4 – 3 – 14 – 13, 3 – 4 – 5, 3 – 5 – 8, 3 – 4 – 8, 4 – 5 – 8, 4 – 14 – 8 – 9, 4 – 5 – 15 – 14, 5 – 6 – 7, 5 – 8 – 7, 5 – 6 – 8, 5 – 15 – 8 – 9, 5 – 17 – 7 – 15, 5 – 6 – 15 – 16, 6 – 7 – 8, 6 – 16 – 7 – 17, 6 – 16 – 8 – 9, 7 – 8 – 9 – 17, 9 – 13 – 14, 9 – 13 – 15, 9 – 14 – 15, 9 – 15 – 16, 9 – 15 – 17, 9 – 16 – 17, 9 – 17 – 10, 9 – 10 – 11, 9 – 11- 17, 10 – 17 – 11, 11 – 12 – 13, 13 – 14 – 15, 15 – 10 – 17, 3 – 5 – 7 – 1,11 – 13 – 15 – 17, i.e. f_2=50

The polytope in Figure 7 has 25 three – dimensional elements:

3 – 4 – 5 – 8, 14 – 15 – 13 – 9, 5 – 6 – 7 – 8,15 – 16 – 17 – 9, 1 – 2 – 3 – 8, 11 – 12 – 13 – 9, 8 – 7 – 0 – 1, 10 – 11 – 17 – 9, 0 – 8 – 7 – 10 – 17 – 9, 0 – 1 – 7 – 10 – 11 – 17, 0 – 1 – 8 – 10 – 11 – 9, 1 – 8 – 7 – 11 – 9 – 17, 1 – 2 8 – 11 – 12 – 9, 1 – 2 – 3 – 13 – 12 – 11, 2 – 3 – 8 – 12 – 13 – 9, 3 – 5 – 8 – 13 – 15 – 9, 3 – 4 – 5 – 13 – 14 – 15, 4 – 5 – 8 – 14 – 15 – 9, 5 – 8 – 6 – 15 – 16 – 9, 5 – 6 – 7 – 15 – 16 – 17, 5 – 8 – 7 – 15 – 9 – 17, 3 – 5 – 1 – 7 – 8, 13 – 15 – 17 – 11 – 9, 3 – 5 – 7 – 1 – 13 – 15 – 17 – 11, 1 – 2 – 3 – 4 – 5 – 6 – 7 – 8 – 9 – 10 – 11 – 12 – 13 – 14 – 15 – 16 – 17 – 18 (three - dimensional body bounded by the outer surface of the complex), i.e. f_3=25

The polytope in Figure 8.7 has 6 four – dimensional elements: four of the tetrahedral prisms and two bases of the complex, each of which consists of four tetrahedrons. Consider one of these bases (Figure 8). It has 9 vertices, i.e. f_0=9, 20 edges:

0 – 1, 1 – 2, 2 – 3, 3 – 4, 4 – 5, 5 – 6, 6 – 7, 0 – 7, 8 – 3, 8 – 4, 8 – 5, 8 – 6, 8 – 7, 8 – 0, 8 – 2, 8 – 1, 3 – 5, 1 – 7, 3 – 1, 5 – 7, f_1=20

Figure 8. The base of the tetrahedral prisms complex

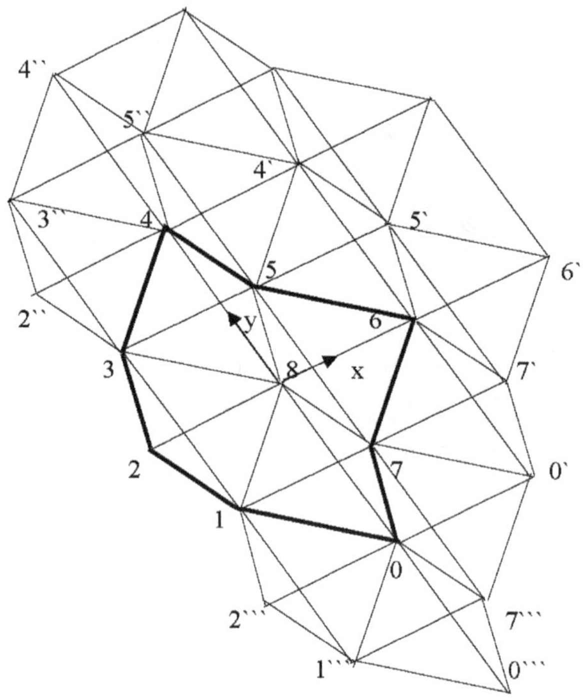

The complex from four tetrahedrons has 17 two – dimensional faces:

3 – 4 – 5, 3 – 5 – 8, 3 – 4 – 8, 3 – 8 – 2, 3 – 8 – 1, 3 – 1 – 2, 4 – 5 – 8, 5 – 6 – 8, 5 – 6 – 7, 5 – 8 – 7, 6 – 7 – 8, 1 – 7 – 8, 1 – 7 – 0, 1 – 8 – 0, 1 – 8 – 2, 1 – 3 – 5 – 7, i.e. $f_2=17$

The complex from four tetrahedrons has 6 three – dimensional faces: four tetrahedrons, one pyramid 3 – 5 – 7 – 1 – 8 and three - dimensional body bounded by the outer surface of the complex, i.e. $f_3=6$. Substituting the obtained values into the Euler - Poincaré equation (8.5), we obtain 9 – 20 + 17 – 6 = 0. The Euler - Poincaré equation for the base of the tetrahedral prism complex is satisfied for a dimension equal to 4, therefore this base has dimension 4. Then for the entire complex of tetrahedral prisms in Figure 8.7, the number of elements with dimension 4 is 6, i.e. $f_4=6$. Substituting the obtained values for the complex of tetrahedral prisms in the Euler - Poincaré equation (8.5), we obtain

$18 - 47 - 50 - 25 + 6 = 2.$

This means that in this case the Euler - Poincaré equation is satisfied for a dimension equal to 5, i.e. the tetrahedral prism complex in Figure 8.7 has dimension 5. It should be noted that not all edges in the complex of tetrahedral prisms form continuous lines, taking into account their possible continuation in neighboring tetrahedral prisms. Therefore, in this case, when choosing a coordinate system, it is necessary to select as the directions of the coordinate variables those edges that form continuous straight lines. There are three such directions, as can be seen from Figure 8.7. This is the direction along the generators of the complex of tetrahedral prisms (z) and two directions perpendicular to these generators (*x, y*) (Figure 8.7). Assuming that the length of the edges of the complex of tetrahedral prisms is equal to unity, we assign coordinates (*x, y, z*) to its vertices

0(0, -1, 0), 1(-1/2, -1/2, 0), 2(-1, 0, 0), 3(-1/2, 1/2, 0), 4(0, 1, 0), 5(1/2, 1/2, 0), 6(1, 0, 0), 7(1/2, -1/2, 0), 8(0, 0, 0), 9(0, 0, 1), 10(0, -1, 1), 11(-1/2, -1/2, 1),12(-1, 0, 1), 13(-1/2, 1/2, 1), 14(0, 1, 1), 15(1/2, 1/2, 1), 16(1, 0, 1), 17(1/2, -1/2, 1). (8.9)

The distribution of tetrahedral prisms B_i can be determined in the same way as the distribution of tetrahedral prisms A_i (see Zhizhin, 2019). However, in this case, of great interest is the question of the spatial motion of the entire complex of tetrahedral prisms, consisting of two pairs of enantiomorphic tetrahedral prisms.

Theorem 8.2: The complex, consisting of two pairs of enantiomorphic tetrahedral prisms adjacent to each other over whole quadrangular plane sections, is a parallelohedron of dimension 5.

Proof. The fact that the complex consisting of two pairs of enantiomorphic tetrahedral prisms has dimension 5 has already been proved. It remains to prove that it is a parallelohedron. For this, we assume that the initial state of the complex is that which is determined by the values of its vertices (8.9). Let us depict one of the bases (for example, the left) of the tetrahedral prism complex in a separate figure in bold line segments (Figure 8.8). We will not depict the generators of the tetrahedral prism complex so as not to clutter up the pattern. We remember that the generators come from each vertex of the base and remain parallel to themselves and the axis *z*. We will not depict the right base either, since it is parallel to the left base. We increase the values of the coordinate *x* by one step (one rib length) at all the vertices of the left base of the tetrahedral prism complex. This will correspond to the translation of the complex of tetrahedral prisms by one step along the *x* axis. Then the following mapping of the vertices of the left base of the complex

0(0, -1, 0) → 0`(1, -1, 0), 1(-1/2, -1/2, 0) → 7(1/2, -1/2, 0), 2(-1, 0, 0) → 8(0, 0, 0), 3(-1/2, 1/2, 0) → 5(1/2, 1/2, 0), 4(0, 1, 0) → 4`(1, 1, 0), 5(1/2, 1/2, 0) → 5`(3/2, 1/2, 0), 6(1, 0, 0) → 6`(2, 0, 0), 7(1/2, -1/2, 0) → 7`(3/2,-1/2, 0), 8(0, 0, 0) → 6(1, 0, 0).

As a result of the mapping, a new base of the complex of tetrahedral prisms was formed 0` - 7 – 8 – 5 – 4` - 5` - 6` - 7` - 6. This base has dimension 4 and common tetrahedron 7 – 8 – 5 – 6 with initial left base.

Now we increase the values of the coordinate *y* by one step (one rib length) at all the vertices of the left base of the tetrahedral prism complex. This will correspond to the translation of the complex of tetrahedral prisms by one step along the *y* axis. Then the following mapping of the vertices of the left base of the complex

0(0, -1, 0) → 8(0, 0, 0), 1(-1/2, -1/2, 0) → 3(-1/2, 1/2, 0), 2(-1, 0, 0) → 2``(-1, 1, 0), 3(-1/2, 1/2, 0) → 3``(-1/2, 3/2, 0), 4(0, 1, 0) → 4``(0, 2, 0), 5(1/2, 1/2, 0) → 5``(1/2, 3/2, 0), 6(1, 0, 0) → 4`(1, 1, 0), 7(1/2, -1/2, 0) → 5(1/2, 1/2, 0), 8(0, 0, 0) → 4(0, 1, 0).

As a result of the mapping, a new base of the complex of tetrahedral prisms was formed 8 – 3 – 2`` – 3`` – 4`` - 5`` - 4` - 5 - 4. This base has dimension 4 and common tetrahedron 3 – 8 – 5 – 4 with initial left base.

Now we decrease the values of the coordinate y by one step (one rib length) at all the vertices of the initial left base of the tetrahedral prism complex. This will correspond to the translation of the complex of tetrahedral prisms by one step opposite the y axis. Then the following mapping of the vertices of the left base of the complex

0(0, -1, 0) → 0```(0, -2, 0), 1(-1/2, -1/2, 0) → 1```(-1/2,-3/2, 0), 2(-1, 0, 0) → 2```(-1, -1, 0), 3(-1/2, 1/2, 0) → 1(-1/2, -1/2, 0), 4(0, 1, 0) → 8(0, 0, 0), 5(1/2, 1/2, 0) → 7(1/2, -1/2, 0), 6(1, 0, 0) → 0`(1, -1, 0), 7(1/2, -1/2, 0) → 7```(1/2,-3/2, 0), 8(0, 0, 0) → 0(0, -1, 0).

As a result of the mapping, a new base of the complex of tetrahedral prisms was formed 0```– 1``` – 2``` – 1 – 8 - 7 - 0` - 7``` - 0. This base has dimension 4 and common tetrahedron 1 – 8 – 7 – 0 with initial left base.

Similar broadcasts in the coordinate x and coordinate y can be continued. In all cases, we will receive new instances of the left bases of the tetrahedral prism complex. Moreover, each of these bases will have a common tetrahedron with any of the neighboring bases. Now recall that from each vertex of the base of the tetrahedral prism complex, the generators of the tetrahedral prism complex emanate. Therefore, the translation of the left base of the complex along the coordinates x and y corresponds to the translation of the entire complex of tetrahedral prisms along these coordinates. In these translations of the tetrahedral prism complex, the common between neighboring complexes will no longer be tetrahedrons, but tetrahedral prisms of dimension 4. In addition, there is still the possibility of translating the complex of tetrahedral prisms along the coordinate z. With this translation (one step), the left base of the complex takes the place of the right base of the initial position of the complex. Each of them has dimension 4. From the analysis it follows that the translation of the five - dimensional complex of tetrahedral prisms leads to a complete filling of the space without gaps with these complexes, which are in a parallel position relative to each other. This proves that the five - dimensional complex of tetrahedral prisms is a parallelohedron.

Partition of Space of the Triangular Prismahedron

The triangular prismahedron is the product of a triangle by a triangle (see Chapter 5). The dimension of this polytope equal 4, it has 9 vertices and 6 triangle prisms (Chapter 5).

Figure 9. The triangular prismahedron

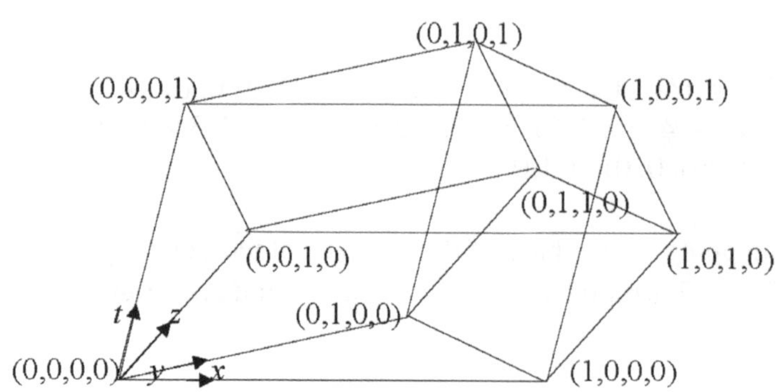

One direct the coordinates of the four - dimensional space along the edges of the triangular prismahedron (see the Figure 9).

Assume that the length of each edge is equal to 1. Then, each node triangular prismahedron can be associated with a set of integers (Figure 8.9). Translating the triangular prismahedron along the coordinates *x, y, z, t*, can to obtain the lattice vertices. Let E_0 triangular prismahedron with the values of vertex coordinates in Figure 8.9. Then

E_0= (0, 0, 0, 0), (1, 0, 0, 0), (0, 1, 0, 0), (0, 0, 1, 0), (0, 0, 0, 1), (1, 0, 1, 0), (0, 1, 1, 0), (1, 0, 0, 1), (0, 1, 0, 1), (8.10)

$E_1 = E_0(x+1) =$ (1, 0, 0, 0), (2, 0, 0, 0), (1, 1, 0, 0), (1, 0, 1, 0), (1, 0, 0, 1), (2, 0, 1, 0), (1, 1, 1, 0), (2, 0, 0, 1), (1, 1, 0, 1),

$E_2 = E_0(z+1) =$ (0, 0, 1, 0), (1, 0, 1, 0), (0, 1, 1, 0), (0, 0, 2, 0), (0, 0, 1, 1), (1, 0, 2, 0), (0, 1, 2, 0), (1, 0, 1, 1), (0, 1, 1, 1),

$E_3 = E_0(x+1, z+1) = E_1(z+1) =$ (1, 0, 1, 0), (2, 0, 1, 0), (1, 1, 1, 0), (1, 0, 2, 0), (1, 0, 1, 1), (2, 0, 2, 0), (1, 1, 2, 0), (2, 0, 1, 1), (1, 1, 1, 1),

$E_4 = E_0(y+1, z+1) = E_2(y+1) =$ (0, 1, 1, 0), (1, 1, 1, 0), (0, 2, 1, 0), (0, 1, 2, 0), (0, 1, 1, 1), (1, 1, 2, 0), (0, 2, 2, 0), (1, 1, 1, 1), (0, 2, 1, 1),

$E_5 = E_0(x+1, y+1, z+1) = E_3(y+1) =$ (1, 1, 1, 0), (2, 1, 1, 0), (1, 2, 1, 0), (1, 1, 2, 0), (1, 1, 1, 1), (2, 1, 2, 0), (1, 2, 2, 0), (2, 1, 1, 1), (1, 2, 1, 1),

$E_6 = E_0(y+1) =$ (0, 1, 0, 0), (1, 1, 0, 0), (0, 2, 0, 0), (0, 1, 1, 0), (0, 1, 0, 1), (1, 1, 1, 0), (0, 2, 1, 0), (1, 1, 0, 1),

$E_7 = E_0(y+1, \mathrm{x}+1) = E_1(y+1) =$ (1, 1, 0, 0), (2, 1, 0, 0), (1, 2, 0, 0), (1, 1, 1, 0), (1, 1, 0, 1), (2, 1, 1, 0), (1, 2, 1, 0), (2, 1, 0, 1), (1, 2, 0, 1),

$E_8 = E_0(t+1) =$ (0, 0, 0, 1), (1, 0, 0, 1), (0, 1, 0, 1), (0, 0, 1, 1), (0, 0, 0, 2), (1, 0, 1, 1), (0, 1, 1, 1), (1, 0, 0, 2), (0, 1, 0, 2),

$E_9 = E_0(y+1, t+1) = E_6(t+1) =$ (0, 1, 0, 1), (1, 1, 0, 1), (0, 2, 0, 1), (0, 1, 1, 1), (0, 1, 0, 2), (1, 1, 1, 1), (0, 2, 1, 1), (1, 1, 0, 2), (0, 2, 0, 2),

$E_{10} = E_0(x+1, y+1, t+1) = E_7(t+1) =$ (1, 1, 0, 1), (2, 1, 0, 1), (1, 2, 0, 1), (1, 1, 1, 1), (1, 1, 0, 2), (2, 1, 1, 1), (1, 2, 1, 1), (2, 1, 0, 2), (1, 2, 0, 2),

$E_{11} = E_0(x+1, t+1) = E_1(t+1) =$ (1, 0, 0, 1), (2, 0, 0, 1), (1, 1, 0, 1), (1, 0, 1, 1), (1, 0, 0, 2), (2, 0, 1, 1), (1, 1, 1, 1), (2, 0, 0, 2), (1, 1, 0, 2),

$E_{12} = E_0(y+1, z+1) = E_2(t+1) =$ (0, 0, 1, 1), (1, 0, 1, 1), (0, 1, 1, 1), (0, 0, 2, 1), (0, 0, 1, 2), (1, 0, 2, 1), (0, 1, 2, 1), (1, 0, 1, 2), (0, 1, 1, 2),

$E_{13} = E_0(z+1, y+1, t+1) = E_4(t+1) =$ (0, 1, 1, 1), (1, 1, 1, 1), (0, 2, 1, 1), (0, 1, 2, 1), (0, 1, 1, 2), (1, 1, 2, 1), (1, 1, 1, 2), (0, 2, 1, 2),

$E_{14} = E_0(x+1, y+1, z+1, t+1) = E_5(t+1) =$ (1, 1, 1, 1), (2, 1, 1, 1), (1, 2, 1, 1), (1, 1, 2, 1), (1, 1, 1, 2), (2, 1, 2, 1), (1, 2, 2, 1), (2, 1, 1, 2), (1, 2, 1, 2),

$E_{15} = E_0(x+1, t+1, z+1) = E_3(t+1) =$ (1, 0, 1, 1), (2, 0, 1, 1), (1, 1, 1, 1), (1, 0, 2, 1), (1, 0, 1, 2), (2, 0, 2, 1), (1, 1, 2, 1), (2, 0, 1, 2), (1, 1, 1, 2),

Representing the triangular prismahedrons $E_0 \div E_{15}$ dots in four - dimensional space, can get the hypercube. Moreover, the edges of the hypercube correspond to possible changes in the values of one of the coordinates of the vertices of the triangular prismahedron unit (Figure 10).

Figure 10. The hypercube from 16 triangular prismahedrons

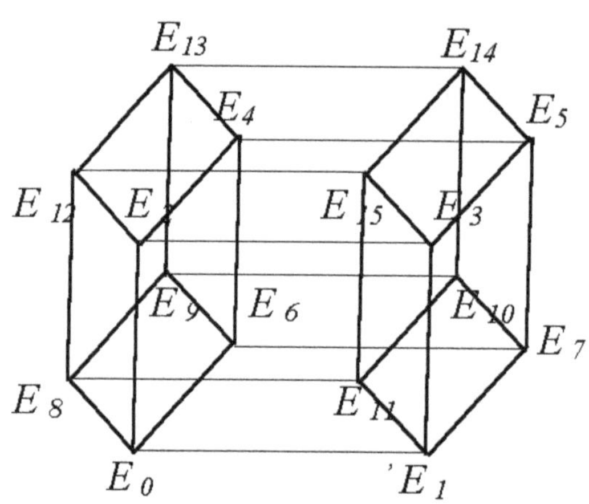

In addition, each edge of the hypercube in Figure 8.10 can be considered as an element of the overall two triangular prismahedrons, connected by an edge. Using the coordinate expression triangular prismahedrons (8.10) can be analytically determined. In Table 8.3 geometry elements are common to each pair of triangular prismahedrons connected by an edge in Figure 8.10 are listed. From Table 3 it follows that triangular prismahedrons of type *E* when translating along the coordinates of a four-dimensional space form a network, adjoining each other along entire triangles. The direct construction of this network shows (see Figure 11) that between these prismhedrons formed closed areas of three more types, which are also triangular prismahedrons *E*, but differing in the different arrangement of the triangles, as the factors of the product.

The types of these triangular prismahedrons (denoted by *L, C, D*) are presented separately in the Figure 12, Figure 83, Figure 14.

Table 3. Common elements of the triangular prismahedrons $E_0 \div E_{15}$

Edge of the Hypercube	Common element of the triangular prismahedrons	Edge of the hypercube	Common element of the triangular prismahedrons
$E_0 E_1$	triangle (1,0,0,0)(1,0,1,0)(1,0,0,0)	$E_5 E_{14}$	triangle (1,1,1,1)(1,2,1,1)(2,1,1,1)
$E_0 E_2$	triangle (0,1,1,0)(0,0,1,0)(1,0,1,0)	$E_6 E_7$	triangle (1,1,1,0)(1,1,0,0)(1,1,0,1)
$E_0 E_6$	triangle (0,1,0,0)(0,1,1,0)(0,1,0,1)	$E_6 E_9$	triangle (0,1,0,1)(0,2,0,1)(1,1,0,1)
$E_0 E_8$	triangle (0,1,0,1)(0,0,0,1)(1,0,0,1)	E_7E_{10}	triangle (1,1,0,1)(2,1,0,1)(1,2,0,1)
E_1E_3	triangle (1,0,1,0)(1,1,1,0)(2,0,1,0)	$E_8 E_{11}$	triangle (1,0,0,1)(1,0,1,1)(1,0,0,2)
E_7E_1	triangle (1,1,0,0,)(1,1,1,0)(1,1,0,1)	E_8E_9	triangle (0,1,0,1)(0,1,1,1)(1,0,0,2)
E_1E_{11}	triangle (1,0,0,1)(1,1,0,1)(2,0,0,1)	E_8E_{12}	triangle (0,0,1,1)(1,0,1,1)(1,0,0,2)
E_2E_3	triangle (1,0,1,0)(1,0,2,0)(1,0,1,1)	E_9E_{10}	triangle (1,1,0,1)(1,1,0,2)(1,1,1,1)
E_2E_4	triangle (0,1,1,0)(0,1,1,1)(0,1,2,0)	E_9E_{13}	triangle (0,2,1,1)(0,1,1,1)(1,1,1,1)
E_2E_{12}	triangle (0,1,1,1)(0,0,1,1)(1,0,1,1)	$E_{11}E_{10}$	triangle (1,1,0,1)(1,1,0,2)(1,1,1,1)
E_3E_5	triangle (1,1,1,0)(1,1,2,0)(0,1,1,1)	$E_{14}E_{10}$	triangle (1,1,1,1)(1,2,1,1)(2,1,1,1)
E_3E_{15}	triangle (1,0,1,1)(1,1,1,1)(2,0,1,1)	$E_{11}E_{15}$	triangle (1,0,1,1)(1,1,1,1)(2,0,1,1)
E_4E_5	triangle (1,1,1,0)(1,1,2,0)(1,1,1,1)	$E_{13}E_{12}$	triangle (0,1,1,1)(0,1,2,1)(0,1,1,2)
E_4E_6	triangle (0,1,1,0)(0,2,1,0)(1,1,1,0)	$E_{12}E_{15}$	triangle (1,0,1,1)(1,0,2,1)(1,1,1,2)
E_4E_{13}	triangle (0,1,1,1)(1,1,1,1)(0,2,1,1)	$E_{13}E_{14}$	triangle (1,1,1,1)(1,1,2,1)(1,1,1,2)
E_7E_5	triangle (1,1,1,0)(1,2,1,0)(2,1,1,0)	$E_{14}E_{15}$	triangle (1,1,1,1)(1,1,2,1)(1,1,1,2)

Figure 11. The complex of enantiomorph forms of the triangular prismahedrons

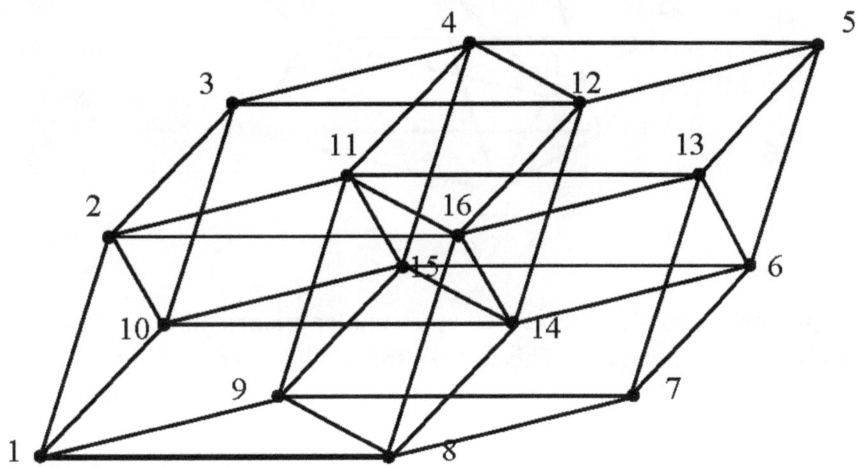

Figure 12. The triangular prismahedron L

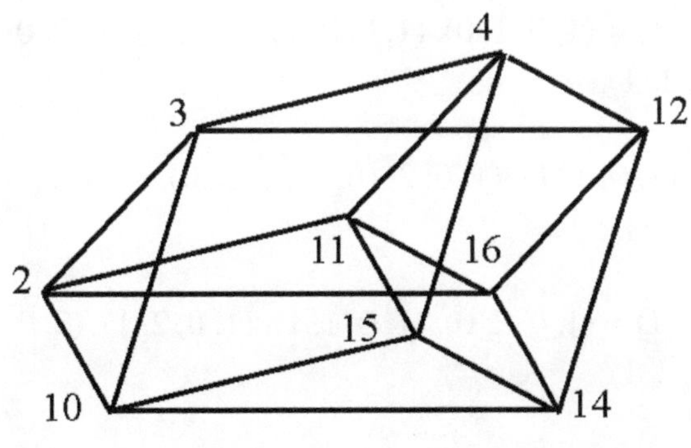

Figure 13. The triangular prismahedron C

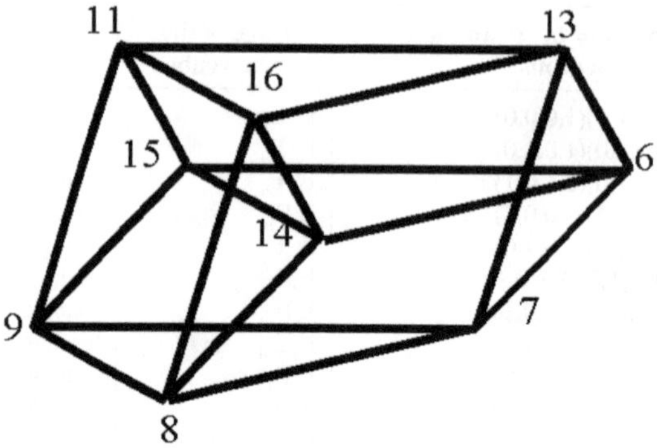

Figure 14. The triangular prismahedron D

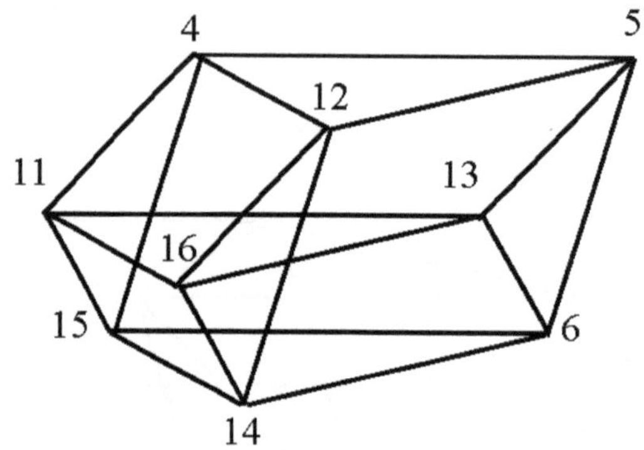

The triangular prismahedrons *L, C, D*, as well as triangular prismahedrons *E* form networks in which they adjoin each other along entire faces. This follows from indicated the triangular prismahedrons *L, C, D* in coordinates of vertices of network and tables 8.4, 8.5, 8.6 of common elements in each of network.

The indicated of the triangular prismahedrons type *L*

L_0= (0, 0, 1, 0), (0, 0, 0, 1), (0, 0, 1, 1), (1, 0, 1, 0), (0, 1, 1, 0), (0, 1, 0, 1), (1, 0, 0, 1), (0, 1, 1, 1), (1, 0, 1, 1),

$L_1 = L_0(x+1) = L_1 = L_0(x+1) =$ (1, 0, 1, 0), (1, 0, 0, 1), (1, 0, 1, 1), (2, 0, 1, 0), (1, 1, 1, 0), (1, 1, 0, 1), (2, 0, 0, 1), (1, 1, 1, 1), (2, 0, 1, 1),

$L_2 = L_0(z+1) =$(1, 0, 2, 0), (1, 0, 1, 1), (1, 0, 2, 0), (2, 0, 2, 0), (1, 1, 2, 0), (1, 1, 1, 1), (2, 0, 1, 1), (1, 1, 2, 1), (2, 0, 2, 1),

$L_3 = L_0(x+1, z+1) = L_1(z+1) =$ (1, 0, 2, 0), (1, 0, 1, 1), (1, 0, 2, 1), (2, 0, 2, 0), (1, 1, 2, 0), (1, 1, 1, 1), (2, 0, 1, 1), (1, 1, 2, 1), (2, 0, 2, 1),

$L_4 = L_0(y+1, z+1) = L_2(y+1) =$ (1, 1, 2, 0), (1, 1, 1, 1), (1, 1, 2, 1), (2, 1, 2, 0), (1, 2, 2, 0), (1, 2, 1, 1), (2, 1, 1, 1), (1, 2, 2, 1), (2, 1, 2, 1),

$L_5 = L_0(x+1, y+1, z+1) = L_3(y+1) =$ (1, 1, 2, 0), (1, 1, 1, 1), (1, 1, 2, 1), (2, 1, 2, 0), (1, 2, 2, 0), (1, 2, 1, 1), (2, 1, 1, 1), (1, 2, 2, 1), (2, 1, 2, 1),

$L_6 = L_0(y+1) =$ (0, 1, 1, 0), (0, 1, 0, 1), (0, 1, 1, 1), (1, 1, 1, 0), (0, 2, 1, 0), (0, 2, 0, 1), (1, 1, 0, 1), (0, 2, 1, 1), (1, 1, 1, 1),

$L_7 = L_0(y+1, x+1) = L_1(y+1) =$ (1, 1, 1, 0), (1, 1, 0, 1), (1, 1, 1, 1), (2, 1, 1, 0), (1, 2, 1, 0), (1, 2, 0, 1), (2, 1, 0, 1), (1, 2, 1, 1), (2, 1, 1, 1),

$L_8 = L_0(t+1) =$ (0, 0, 1, 1), (0, 0, 0, 2), (0, 0, 1, 2), (1, 0, 1, 2), (0, 1, 1, 1), (0, 1, 0, 2), (1, 0, 0, 2), (0, 1, 1, 2), (1, 0, 1, 2),

$L_9 = L_0(y+1, t+1) = L_6(t+1) =$ (0, 1, 1, 1), (0, 1, 0, 2), (0, 1, 1, 2), (1, 1, 1, 1), (0, 2, 1, 1), (0, 2, 0, 2), (1, 1, 0, 2), (0, 2, 1, 2), (1, 1, 1, 2),

$L_{10} = L_0(x+1, y+1, t+1) = L_7(t+1) =$ (1, 1, 1, 1), (1, 1, 0, 2), (1, 1, 1, 2), (2, 1, 1, 1), (1, 2, 1, 1), (1, 2, 0, 2), (2, 1, 0, 2), (1, 2, 1, 2), (2, 1, 1, 2),

$L_{11} = L_0(x+1, t+1) = L_1(t+1) =$ (1, 0, 1, 1), (1, 0, 0, 2), (1, 0, 1, 2), (2, 0, 1, 1), (1, 1, 1, 1), (1, 1, 0, 2), (2, 0, 0, 2), (1, 1, 1, 2), (2, 0, 1, 2),

$L_{12} = L_0(y+1, z+1) = L_2(t+1) =$ (1, 0, 2, 1), (1, 0, 1, 2), (1, 0, 2, 2), (2, 0, 2, 1), (1, 1, 2, 1), (1, 1, 1, 2), (2, 0, 1, 2), (1, 1, 2, 2), (2, 0, 2, 2),

$L_{13} = L_0(z+1, y+1, t+1) = L_4(t+1) =$ (1, 1, 2, 1), (1, 1, 1, 2), (1, 1, 2, 2),(2, 1,2, 1), (1, 2, 2, 1), (1, 2, 1, 2),(2, 1, 1, 2), (1, 2, 2, 2), (2, 1, 2, 2),

$L_{14} = L_0(x+1, y+1, z+1, t+1) = L_5(t+1) =$ (1, 1, 2, 1), (1, 1, 1, 2), (1, 1, 2, 2), (2, 1, 2, 1), (1, 2, 2, 1), (1, 2, 1, 2), (2, 1, 1, 2), (1, 2, 2, 2), (2, 1, 2, 2),

$L_{15} = L_0(x+1, t+1, z+1) = L_3(t+1) =$ (1, 0, 2, 1), (1, 0, 1, 2), (1, 0, 2, 2), (2, 0, 2, 1), (1, 1, 2, 1), (1, 1, 1, 2), (2, 0, 1, 2), (1, 1, 2, 2), (2, 0, 2, 2),

The indicated of the triangular prismahedrons type C

C_0= (0, 1, 0, 0), (1, 0, 0, 0), (0, 1, 0, 1), (0, 1, 1, 0), (1, 0, 1, 0), (1, 1, 0, 0), (1, 0, 0, 1), (1, 1, 1, 0), (1, 1, 0, 1),

$C_1 = C_0(x+1) =$ (1, 1, 0, 0), (2, 0, 0, 0), (1, 1, 0, 1), (1, 1, 1, 0), (2, 0, 1, 0), (2, 1, 0, 0), (2, 0, 0, 1), (2, 1, 1, 0), (2, 1, 0, 1),

$C_2 = C_0(z+1) =$ (0, 1, 1, 0), (1, 0, 1, 0), (0, 1, 1, 1), (0, 1, 2, 0), (1, 0, 2, 0), (1, 1, 1, 0), (1, 0, 1, 1), (1, 1, 2, 0), (1, 1, 1, 1),

$C_3 = C_0(x+1, z+1) = C_1(z+1) =$ (1, 1, 1, 0), (2, 0, 1, 0), (1, 1, 1, 1), (1, 1, 2, 0), (2, 0, 2, 0), (2, 1, 1, 0), (2, 0, 1, 1), (2, 1, 2, 0), (2, 1, 1, 1),

$C_4 = C_0(y+1, z+1) = C_2(y+1) =$ (0, 2, 1, 0), (1, 1, 1, 0), (0, 2, 1, 1), (0, 2, 2, 0), (1, 1, 2, 0), (1, 2, 1, 0), (1, 1, 1, 1), (1, 2, 2, 0), (1, 2, 1, 1),

$C_5 = C_0(x+1, y+1, z+1) = C_3(y+1) =$ (1, 2, 1, 0), (2, 1, 1, 0), (1, 2, 1, 1), (1, 2, 2, 0), (2, 1, 2, 0), (2, 2, 1, 0), (2, 1, 1, 1), (2, 2, 2, 0), (2, 2, 1, 1),

$C_6 = C_0(y+1) =$ (2, 2, 0, 0), (1, 1, 0, 0), (0, 2, 0, 1), (0, 2, 1, 0), (1, 1, 1, 0), (1, 2, 0, 0), (1, 1, 0, 1), (1, 2, 1, 0), (1, 2, 0, 1),

$C_7 = C_0(y+1, x+1) = C_1(y+1) =$ (1, 2, 0, 0), (2, 1, 0, 0), (1, 2, 0, 1), (1, 2, 1, 0), (2, 1, 1, 0), (2, 2, 0, 0), (2, 1, 0, 1), (2, 2, 1, 0), (2, 2, 0, 1),

$C_8 = C_0(t+1) =$ (0, 1, 0, 1), (1, 0, 0, 1, (0, 1, 0, 2), (0, 1, 1, 1), (1, 0, 1, 1), (1, 1, 0, 1), (1, 0, 0, 2), (1, 1, 1, 1), (1, 1, 0, 2),

$C_9 = C_0(y+1, t+1) = C_6(t+1) =$ (0, 2, 0, 1), (1, 1, 0, 1), (0, 2, 0, 2), (0, 2, 1, 1), (1, 1, 1, 1), (1, 2, 0, 1), (1, 1, 0, 2), (1, 2, 1, 1), (1, 2, 0, 2),

$C_{10} = C_0(x+1, y+1, t+1) = C_7(t+1) =$ (1, 2, 0, 1), (2, 1, 0, 1), (1, 2, 0, 2), (1, 2, 1, 1), (2, 1, 1, 1), (2, 2, 0, 1), (2, 1, 0, 2), (2, 2, 1, 1), (2, 2, 0, 2),

$C_{11} = C_0(x+1, t+1) = C_1(t+1) =$ (1, 1, 0, 1), (2, 0, 0, 1), (1, 1, 0, 2), (1, 1, 1, 1), (2, 0, 1, 1), (2, 1, 0, 1), (2, 0, 0, 2), (2, 1, 1, 1), (2, 1, 0, 2),

$C_{12} = C_0(y+1, z+1) = C_2(t+1) =$ (0, 1, 1, 1), (1, 0, 1, 1), (0, 1, 1, 2), (0, 1, 2, 1), (1, 0, 2, 1), (1, 1, 1, 1), (1, 0, 1, 2), (1, 1, 2, 1), (1, 1, 1, 2),

$C_{13} = C_0(z+1, y+1, t+1) = C_4(t+1) =$ (0, 2, 1, 1), (1, 1, 1, 1), (0, 2, 1, 2),(0, 2, 2, 1), (1, 1, 2, 1), (1, 2, 1, 1),(1, 1, 1, 2), (1, 2, 2, 1), (1, 2, 1, 2),

$C_{14} = C_0(x+1, y+1, z+1, t+1) = C_5(t+1) =$ (1, 2, 1, 1), (2, 1, 1, 1), (1, 2, 1, 2), (1, 2, 2, 1), (2, 1, 2, 1), (2, 2, 1, 1), (2, 1, 1, 2), (2, 2, 2, 1), (2, 2, 1, 2),

$C_{15} = C_0(x+1, t+1, z+1) = C_3(t+1) =$ (1, 1, 1, 1), (2, 0, 1, 1), (1, 1, 1, 2), (1, 1, 2, 1), (2, 0, 2, 1), (2, 1, 1, 1), (2, 0, 1, 2), (2, 1, 2, 1), (2, 1, 1, 2),

The indicated of the triangular prismahedrons type D

$D_0 =$ (0, 1, 1, 0), (1, 0, 1, 0), (1, 0, 0, 1), (0, 1, 0, 1), (0, 1, 1, 1), (1, 0, 1, 1), (1, 1, 0, 1), (1, 1, 1, 0), (1, 1, 1, 1),

$D_1 = D_0(x+1) =$ (1, 1, 1, 0), (2, 0, 1, 0), (2, 0, 0, 1), (1, 1, 0, 1), (1, 1, 1, 1), (2, 0, 1, 1), (2, 1, 0, 1), (2, 1, 1, 0), (2, 1, 1, 1),

$D_2 = D_0(z+1) =$ (0, 1, 2, 0), (1, 0, 2, 0), (1, 0, 1, 1), (0, 1, 1, 1), (0, 1, 2, 1), (1, 1, 1, 1), (1, 1, 2, 0), (1, 1, 2, 1), (1, 0, 2, 1),

$D_3 = D_0(x+1, z+1) = D_1(z+1) =$ (1, 1, 2, 0), (2, 0, 2, 0), (2, 0, 1, 1), (1, 1, 1, 1), (1, 1, 2, 1), (2, 0, 2, 1), (2, 1, 1, 1), (2, 1, 2, 0), (2, 1, 2, 1),

$D_4 = D_0(y+1, z+1) = D_2(y+1) =$ (0, 2, 2, 0), (1, 1, 2, 0), (1, 1, 1, 1), (0, 2, 1, 1), (0, 2, 2, 1), (1, 1, 2, 1), (1, 2, 1, 1), (1, 2, 2, 0), (1, 2, 2, 1),

$D_5 = D_0(x+1, y+1, z+1) = D_3(y+1) =$ (1, 2, 2, 0), (2, 1, 2, 0), (2, 1, 1, 1), (1, 2, 1, 1), (1, 2, 2, 1), (2, 1, 2, 1), (2, 2, 1, 1), (2, 2, 2, 0), (2, 2, 2, 1),

$D_6 = D_0(y+1) =$ (0, 2, 1, 0), (1, 1, 1, 0), (1, 1, 0, 1), (0, 2, 0, 1), (0, 2, 1, 1), (1, 1, 1, 1), (1, 2, 0, 1), (1, 2, 1, 0), (1, 2, 1, 1),

$D_7 = D_0(y+1, x+1) = D_1(y+1) =$ (1, 2, 1, 0), (2, 1, 1, 0), (2, 1, 0, 1), (1, 2, 0, 1), (1, 2, 1, 1), (2, 1, 1, 1), (2, 2, 0, 1), (2, 2, 1, 0), (2, 2, 1, 1),

$D_8 = D_0(t+1) =$ (0, 1, 1, 1), (1, 0, 1, 1), (1, 0, 0, 2), (0, 1, 0, 2), (0, 1, 1, 2), (1, 0, 1, 2), (1, 1, 0, 2), (1, 1, 1, 1), (1, 1, 1, 2),

$D_9 = D_0(y+1, t+1) = D_6(t+1) =$ (0, 2, 1, 1), (1, 1, 1, 1), (1, 1, 0, 2), (0, 2, 0, 2), (0, 2, 1, 2), (1, 1, 1, 2), (1, 2, 0, 2), (1, 2, 1, 1), (1, 2, 1, 2),

$D_{10} = D_0(x+1, y+1, t+1) = D_7(t+1) =$ (1, 2, 1, 1), (2, 1, 1, 1), (2, 1, 0, 2), (1, 2, 0, 2), (1, 2, 1, 2), (2, 1, 1, 2), (2, 2, 1, 1), (2, 2, 1, 2), (2, 1, 1, 2),

$D_{11} = D_0(x+1, t+1) = D_1(t+1) =$ (1, 1, 1, 1), (2, 0, 1, 1), (2, 0, 0, 2), (1, 1, 0, 2), (1, 1, 1, 2), (2, 0, 1, 2), (2, 1, 0, 2), (2, 1, 1, 1), (2, 1, 1, 2),

$D_{12} = D_0(y+1, z+1) = D_2(t+1) =$ (0, 1, 2, 1), (1, 0, 2, 1), (1, 0, 1, 2), (0, 1, 1, 2), (0, 1, 2, 2), (1, 1, 1, 2), (1, 1, 2, 1), (1, 1, 2, 2), (1, 0, 2, 2),

$D_{13} = D_0(z+1, y+1, t+1) = D_4(t+1) =$ (1, 2, 2, 1), (1, 1, 2, 1), (1, 1, 1, 2),(0, 2, 1, 2), (0, 2, 2, 2), (1, 1, 2, 2),(1, 2, 1, 2), (1, 2, 2, 1), (1, 2, 2, 2),

$D_{14} = D_0(x+1, y+1, z+1, t+1) = D_5(t+1) =$ (1, 2, 2, 1), (2, 1, 2, 1), (2, 1, 1, 2), (1, 2, 1, 2), (1, 2, 2, 2), (2, 1, 2, 2), (2, 2, 1, 2), (2, 2, 2, 1), (2, 2, 2, 2),

$D_{15} = D_0(x+1, t+1, z+1) = D_3(t+1) =$ (1, 1, 2, 1), (2, 0, 2, 1), (2, 0, 1, 2), (1, 1, 1, 2), (1, 1, 2, 2), (2, 0, 2, 2), (2, 1, 1, 2), (2, 1, 2, 1), (2, 1, 2, 2),

Defining the intersections of the triangular prismahedrons in each of the sets of prismahedrons *E, L, C* or *D* from their coordinate expressions it is easy to determine that the common part of any pair of triangular prismahedrons belonging to a given set of prismahedrons (*E, L, C* or *D*) is some triangular face of the chosen pair of prismahedrons. Representing the triangular prismahedrons $L_0 \div L_{15}$, $C_0 \div C_{15}$, $D_0 \div D_{15}$ dots in the space, for each type triangular prismahedrons can get the hypercube. Moreover, the edges of the hypercube correspond to possible changes in the values of one of the coordinates of the vertices of the triangular prismahedrons of given type unit (see Figure 8.6 for the triangular prismahedrons of type *E*).

Using the coordinate expressions of the triangular prismahedrons, can establish that triangular prismahedrons belonging to different types can have common geometric elements of different dimensions that enter into triangular prismahedrons. Their intersections can be empty (that is, they do not have common elements), for example, E_2 Ç D_1= Ø. Their intersection can be single edge, for example, E_3 Ç $L_0 = (1,0,1,0)(1,0,1,1)$,

E_3 Ç $C_1 = (1,0,1,0)(1,1,1,0)$, E_2 Ç $D_0 = (0,1,1,0)(1,0,1,1)$.

Their intersection can be square face, for example,

E_7 Ç D_1= (1,1,1,0)(1,1,0,1)(2,1,1,0)(2,1,0,1), E_3 Ç D_2= (1,0,2,0)(1,0,1,1)(1,1,2,0)(1,1,1,1).

Their intersection can be triangular prism also, for example,

E_3Ç $L_1 = (1,0,1,0)(2,0,1,0)(1,1,1,0)(1,0,1,1)(2,0,1,1)(1,1,1,1)$,

L_1 Ç $D_0 = (1,0,1,0)(1,0,0,1)(1,0,1,1)(1,1,1,0)(1,1,0,1)(1,1,1,1)$.

Let us consider the set of the triangular prismahedrons of four types *E,L,C,D* of least - cubed numbers close to the origin from the considered set of the triangular prismahedrons. This set consists of the triangular prismahedrons E_0, L_0, C_0, D_0. By their coordinate expressions we establish that the pairs of the triangular prismahedrons E_0 and C_0, E_0 and L_0, C_0 and D_0, L_0 and D_0 have common trigonal prisms, and the pairs of the triangular prismahedrons E_0 and D_0, C_0 and L_0 have common planar faces of the four coal form

$$E_0 \c{C} C_0 = (0,1,0,0,)(1,0,0,0)(0,1,0,1)(1,0,0,1)(0,1,1,0)(1,0,1,0),$$

$$E_0 \c{C} L_0 = (0,0,1,0,)(0,1,1,0)(1,0,1,0)(0,0,0,1)(0,1,0,1)(1,0,0,1),$$

$$C_0 \c{C} D_0 = (0,1,0,1,)(1,1,0,1)(1,0,0,1)(0,1,1,0)(1,1,1,0)(1,0,1,0),$$

$$L_0 \c{C} D_0 = (0,1,1,1,)(1,0,1,1)(0,1,0,1)(1,0,0,1)(0,1,1,0)(1,0,1,0),$$

$$E_0 \c{C} D_0 = (0,1,0,1)(1,0,0,1)(0,1,1,0)(1,0,1,0),$$

$$L_0 \c{C} C_0 = (0,1,0,1)(1,0,0,1)(0,1,1,0)(1,0,1,0).$$

Thus, the triangular prismahedrons form closed complex of the triangular prismahedrons (see Figure 8.11). Figure 8.11 shows that the complex of tetrahedral prisms is a union of four enantiomorphic trigonal forms (in the general case, not related to the origin). Moreover, the common elements between enantiomorphic forms are determined by the equalities

$$E \bigcap L = 2 - 11 - 16 - 10 - 15 - 14, \text{ prism},$$

$$E \bigcap C = 15 - 14 - 8 - 9 - 16 - 11, \text{ prism},$$

$$E \bigcap D = 15 - 14 - 16 - 11, \text{ square},$$

$$L \bigcap C = 15 - 14 - 16 - 11, \text{ square},$$

$$L \bigcap D = 15 - 14 - 11 - 16 - 4 - 12, \text{ prism},$$

$$D \bigcap C = 11 - 15 - 14 - 16 - 13 - 6, \text{ prism}.$$

Theorem 8.3. The complex of the enantiomorph trigonal prismahedrons *E, L, C, D* (Figure 8.11) is a parallelohedron with dimension 6.

Proof. The complex of the enantiomorph trigonal prismahedrons *E, L, C, D* (Figure 8.11) has 16 vertices, i.e. f_0=16, 40 edges:

1 – 2, 1 – 10, 1 – 9, 1 – 8, 2 – 3, 2 – 11, 2 – 16, 2 – 10, 3 – 4, 3 – 12, 3 – 10, 4 – 5, 4 – 12, 4 – 11, 4 – 15, 5 – 12, 5 – 6, 5 – 13, 6 – 13, 6 – 15, 6 – 14, 6 – 7, 7 – 8, 7 – 9, 7 – 13, 8 – 9, 8 – 14, 8 – 16, 9 – 11, 9 – 15, 10 – 14, 10 – 15, 11 – 15, 11 – 16, 11 – 13, 12 – 14, 12 – 16, 13 – 16, 14 – 15, 14 – 16, i.e. f_1=40

The complex of the enantiomorph trigonal prismahedrons *E, L, C, D* (Figure 8.11) has 34 two – dimension faces:

1 – 2 – 10, 1 – 2 – 11 – 9, 1 – 10 – 14 – 8, 1 – 9 – 8, 1 – 10 – 15 – 9, 2 – 3 – 10, 2 – 3 – 12 – 16, 2 – 3 – 4 – 11, 2 – 11 – 16, 2 – 11 – 10 – 15, 2 – 10 – 16 – 14, 3 – 4 – 12, 3 – 12 – 10 – 14, 3 – 4 – 10 – 15, 4 – 5 – 12, 4 – 12 – 11 – 16, 4 – 12 – 14 – 15, 4 – 5 – 6 – 15, 5 – 6 – 13, 5 – 12 – 6 – 14, 6 – 7 – 13, 6 – 13 – 11 – 15, 6 – 13 – 14 – 16, 6 – 7 – 15 – 9, 6 – 15 – 14, 7 – 8 – 9, 7 – 9 – 11 – 13, 7 – 8 – 16 – *13*, 8 – 14 – 16, 8 – 9 – 11 – 16, 8 – 9 – 14 – 15, 9 – 11 – 15, 10 – 14 – 15, 11- 15 – 14 – 16, i.e. f_2=34

The complex of the enantiomorph trigonal prismahedrons *E, L, C, D* (Figure 8.11) has 17 three – dimensional faces:

1 – 2 – 10 – 8 – 14 – 16, 1 – 2 – 10 – 11 – 15 – 9, 1 – 10 – 15 – 14 – 8 – 9, 2 – 3 – 4 – 15 – 11 – 10, 2 – 10 – 11 – 15 – 16 – 14, 2 – 3 – 10 – 12 – 16 – 14, 3 – 4 – 12 – 10 – 15 – 14, 4 – 12 – 14 – 16 – 15 – 11, 4 – 12 – 5 – 6 – 14 – 15, 4 – 11 – 15 – 5 – 6 – 13, 5 – 6 – 13 – 12 – 16 – 14, 6 – 13 – 14 – 16 – 11 – 15, 6 – 7 – 13 – 9 – 15 – 11, 6 – 7 – 13 – 8 – 14 – 16, 7 – 8 – 9 13 – 11 – 16, 8 – 9 – 11 – 15 – 14 – 16, 1 – 10 – 2 – 3 – 4 – 11 – 5 – 13 – 6 – 7 8 – 14 – 9 -15 (external surface), i.e. f_3=17

The complex of the enantiomorph trigonal prismahedrons *E, L, C, D* (Figure 8.11) has 11 four – dimensional faces. For this four are trigonal prismahedrons *E, L, C, D* and seven 4 - quadrangular prisms (Figure 8.5):

1 – 2 – 3 – 10 – 11 – 4 – 15 – 9, 1 – 2 – 3 – 10 – 12 – 16 – 14 – 8, 9 – 11 – 15 – 4 – 8 – 16 – 14 – 12, 3 – 4 – 12 – 10 – 15 – 14 – 5 – 6, 5 – 6 – 13 – 7 – 12 – 16 – 14 – 8, 5 – 6 – 13 – 7 – 4 – 11 – 15 – 9, 1 – 2 – 11 – 9 – 8 – 16 – 13 – 7, i.e. f_4=11

The complex of the enantiomorph trigonal prismahedrons *E, L, C, D* (Figure 8.11) has 4 five – dimensional faces. Fist of them (P_1) presented on Figure 15.

Let us prove that Figure 8.15 has dimension 5. It has 12 vertices (f_0=12) and 27 edges:

1 – 2, 1 – 9, 1 – 8, 1 – 10, 2 – 10, 2 – 16, 6 – 15, 6 – 14, 6 – 13, 6 – 7, 7 – 8, 7 – 9, 8 – 9, 8 – 14, 8 – 16, 9 – 15, 9 – 11, 10 – 14, 10 – 15, 11 – 15, 11 – 16, 14 – 15, 14 – 16, 11 – 13, 13 – 16, 7 – 13, i.e. f_1=27.

The polytope P_1 has 25 two – dimensional faces:

1 – 2 – 10, 1 – 10 – 14 - 8, 1 – 9 – 8, 1 – 10 – 15 – 9, 1 – 2 – 11 – 9, 1 – 2 – 16 – 8, 2 – 11 – 16, 2 – 11 – 10 – 15, 2 – 10 – 16 – 14, 6 – 14 – 16 – 13, 6 – 14 – 15, 6 – 13 – 11 – 15, 6 – 7 – 9 – 15, 6 – 7 – 8 – 14, 6 – 13 – 7, 7 – 8 – 9, 7 – 9 – 11 – 13, 7 – 8 – 16 – 13, 8 – 14 – 16, 8 – 9 – 11 – 16, 8 – 9 – 15 – 14, 9 – 15 – 11, 10 – 14 – 15, 11 – 15 – 14 – 16, 11 – 13 – 16, i.e. f_2=25

Figure 15. The polytope P_1

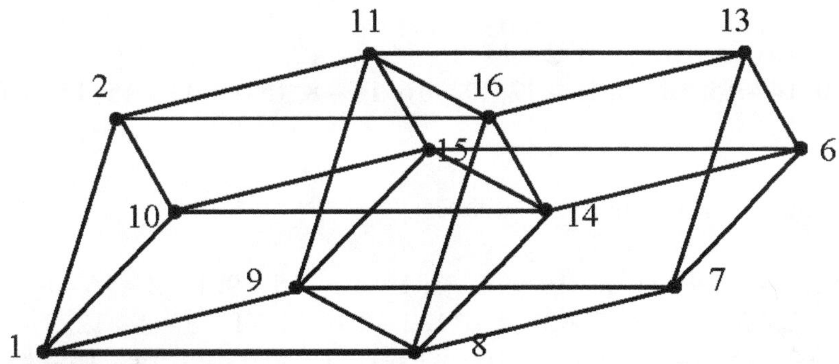

The polytope P_1 has 10 three – dimensional faces:

1 – 2 – 11 – 9 – 10 – 15, 1 – 9 – 8 – 10 – 15 – 14, 2 – 10 – 14 – 16 – 11 – 15, 11 – 13 – 6 – 15 – 16 – 14, 7 – 8 – 9 – 15 – 14 – 6, 8 – 16 – 13 – 7 – 14 – 6, 9 – 8 – 11- 16 – 15 – 14, 1 – 9 – 8 – 2 - 11 – 16, 7 – 8 – 9 – 11 – 13 – 16, i.e. f_3=10

The polytope P_1 has 2 four – dimensional faces: two trigonal prismahedrons, i.e. f_4=2. Substituting the obtained values f_i for the polytope P_1 in the Euler-Poincaré equation (8.5), we obtain

12 – 27 – 25 – 10 + 2 = 2.

This means that in this case the Euler - Poincaré equation is satisfied for a dimension equal to 5, i.e. the polytope P_1in Figure 8.15 has dimension 5.

Second of them (P_2) presented on Figure 16.

Figure 16. The polytope P_2

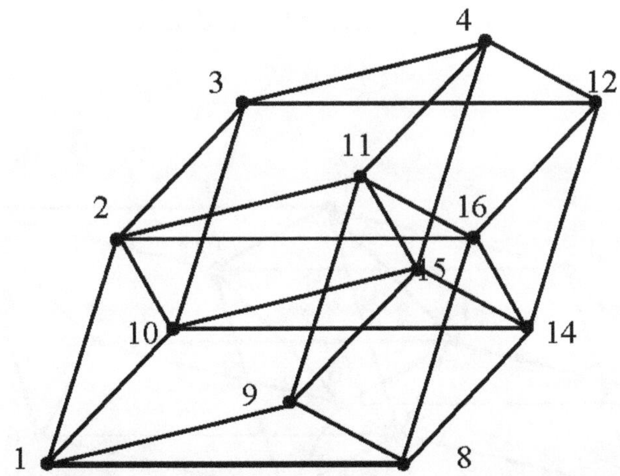

Let us prove that Figure 8.16 has dimension 5. It has 12 vertices (f_0=12) and 27 edges:

1 – 2, 1 – 9, 1 – 8, 1 – 10, 2 – 10, 2 – 3, 2 – 11, 2 – 16, 3 – 12, 3 – 4, 3 – 10, 4 – 12, 4 – 11, 4 – 7, 14 – 16, 14 – 12, 14 – 10, 14 – 15, 14 – 8, 16 – 12, 11 – 16, 16 – 8, 15 – 9, 11 – 15, 15– 10, 8 – 9, 9 – 11, i.e. f_1=27

The polytope P_2 has 24 two – dimensional faces:

1 – 9 – 8, 1 – 2 – 10, 1 – 2 – 9 - 11, 1 – 10 – 14 – 8, 1 – 10 – 15 – 9, 1 – 2 – 16 – 8, 2 – 11 – 16, 2 – 3 – 10, 2 – 10 – 16 – 14, 2 – 3 - 16 – 12, 2 – 3 – 4 – 11, 2 – 10 – 11 – 15, 3 – 4 – 12, 3 – 12 – 10 – 14, 3 – 4 – 15 – 10, 4 – 12 – 14 – 15, 4 – 15 – 11, 4 – 16 – 11 – 12, 14 – 16 – 12, 14 – 16 – 15 – 11, 8 – 14 – 16, 8 – 9 – 14 – 15, 8 – 9 – 11 – 16, 9 – 15 – 11, i.e. f_2=24

The polytope P_2 has 11 three – dimensional faces:

3 – 4 – 12 – 14 – 10 – 15, 4 – 12 – 16 – 14 – 15 – 11, 1 – 2 – 10 – 14 – 16 – 8, 1 – 10 – 15 – 9 – 14 – 8, 1 – 10 – 2 – 9 – 15 – 11, 8 – 16 – 11 – 15 – 14 – 9, 2 – 10 – 11- 16 – 15 – 14, 2 – 3 – 10 – 4 - 11 – 15, 2 – 3 – 10 – 16 – 14 – 12, 2 – 3 – 4 – 11 – 12 – 16, exterior surface, i.e. f_3=11

The polytope P_2 has 4 four – dimensional faces:

1 – 2 – 11 – 16 – 14 – 10 – 15 – 9 – 8, 1 – 2 – 3 – 10 – 9 – 15 – 11 – 4, 9 – 15 – 11 – 4 – 8 – 14 – 16 – 12, 1 – 2 – 3 – 10 – 16 – 14 – 8 - 12, i.e. f_4=4

Substituting the obtained values f_i for the polytope P_2 in the Euler-Poincaré equation (8.5), we obtain

12 – 27 + 24 – 11 + 4 = 2.

This means that in this case the Euler - Poincaré equation is satisfied for a dimension equal to 5, i.e. the polytope P_2 in Figure 8.16 has dimension 5.

The third polytope with dimension 5 (P_3) presented on Figure 8.17.

Figure 17. The polytope P_3

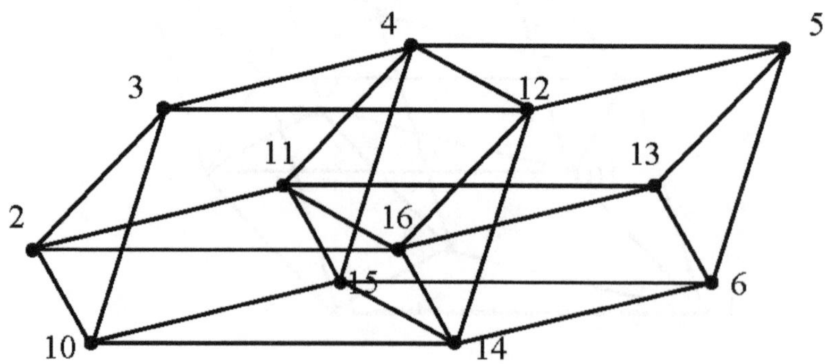

Let us prove that Figure 8.17 has dimension 5. It has 12 vertices (f_0=12) and 27 edges:

1 – 2, 1 – 3, 1 – 158, 1 – 14, 2 – 10, 2 – 3, 2 – 11, 2 – 16, 3 – 12, 3 – 4, 4 – 5, 4 – 12, 4 – 11, 4 – 15, 5 – 6, 5 – 12, 5 – 13, 6 – 13, 14 – 6, 16 – 11, 11 – 15, 16 – 12, 12 – 14, 16– 13, 15– 14, 14 – 16, 6 – 15, i.e. f_1=27

The polytope P_3 has 24 two – dimensional faces:

1 – 2 – 16 – 14, 1 – 2 – 3, 1 – 3 – 12 - 14, 1 – 15 – 14, 1 – 2 – 15 – 11, 1 – 3 – 15 – 4, 2 – 3 – 4 – 11, 3 – 4 – 12, 4 – 12 – 15 – 14, 4 – 11 – 12 – 16, 4 – 5 – 11 – 13, 4 – 11 – 15, 4 – 5 – 12, 4 – 5 – 6 – 15, 5 – 6 – 13, 5 – 12 – 13 – 16, 6 – 13 – 14 – 16, 6 – 13 – 11 – 15, 6 – 15 – 14, 11 – 15 – 16 – 14, 12 – 16 – 14, 11 – 13 – 16, i.e. f_2=24

The polytope P_3 has 9 three – dimensional faces:

1 – 2 – 3 – 14 – 16 – 12, 1 – 2 – 3 – 4 – 11 – 15, 4 – 11 – 15 – 12 – 16 – 14, 14 – 16 – 12 – 5 – 6 – 13, 1 – 15 – 14 – 2 – 11 – 16, 4 – 15 – 16 – 14 – 13 – 6, 4 – 11 – 15 – 5 – 13 – 6, 4 – 15 – 12 – 14 – 5 – 6, exterior surface, i.e. f_3=9

The polytope P_3 has 2 four – dimensional faces: two triangle polyhedrons, i.e. f_4=2. Substituting the obtained values f_i for the polytope P_3 in the Euler - Poincaré equation (8.5), we obtain

$$12 - 27 + 24 - 9 + 2 = 2.$$

This means that in this case the Euler - Poincaré equation is satisfied for a dimension equal to 5, i.e. the polytope P_3 in Figure 8.17 has dimension 5.

The fourth polytope with dimension 5 (P_4) presented on Figure 8.18.

Figure 18. The polytope P_4

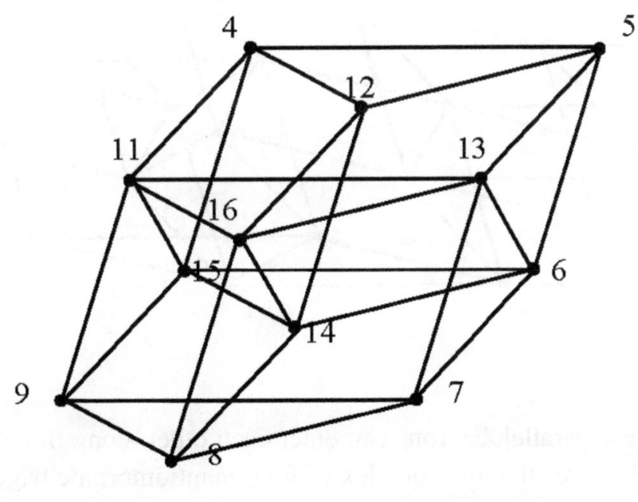

It can be verified that this polytope has as many elements of different dimensions as the polytope in figure 16. Therefore, its dimension is also 5.

Substituting the obtained values f_i for the polytope on Figure 8.11 in the Euler - Poincaré equation (8.5), we obtain

$$16 - 40 + 34 - 17 + 11 - 4 = 0.$$

This means that in this case the Euler - Poincaré equation is satisfied for a dimension equal to 6, i.e. the polytope in Figure 8.11 has dimension 6.

Let us prove now that this convex polytope is a parallelohedron. And from Figure 8.11 it can be seen that each of the flat faces of this polytope, with the exception of one central face 11 – 14 – 15 – 16, has one or more parallel faces. This means that during the translation of this polytope, faces of different dimensions can coincide with the corresponding faces of the original polytope. Moreover, common to these two polytopes can be both flat faces, and three - dimensional and four -dimensional faces. There are a lot of options for such translations. All of them can give a partition of six - dimensional space into separate blocks without gaps. Figure 19 shows, as an example, the partition of space into six - dimensional parallelohedra that adhere to each other along flat faces.

Figure 19. Partition of the space of complex from four trigonal prismahedrons.

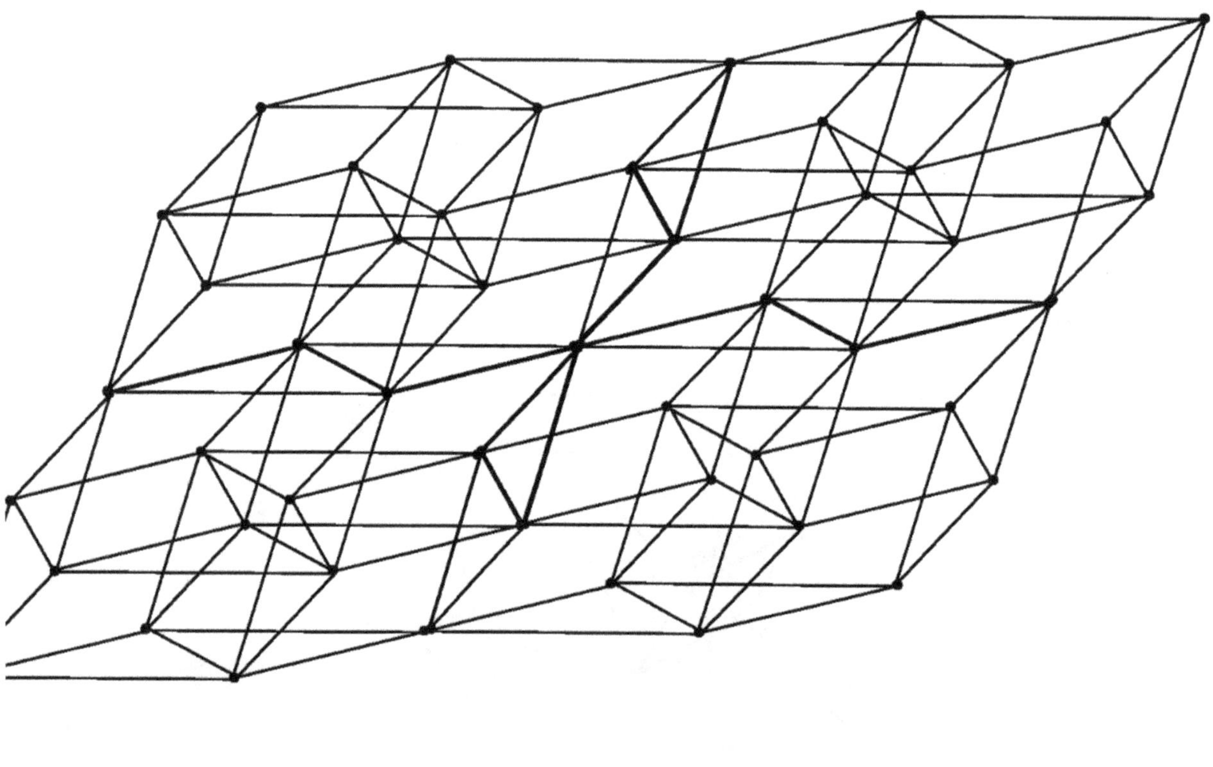

It can be seen that these parallelohedrons can enter each other along three - dimensional and four - dimensional faces. This proves that the complex of four enantiomorphic trigonal prismahedrons is a six - dimensional parallelohedron.

POLYTOPIC PRISMAHEDRON ON BASES OF CROSS – POLYTOPES

The smallest dimension of a cross - polytope is 4. A polytopic prismahedron of 4 – cross -polytope can be obtained by multiplying it by a one - dimensional segment. However, for the further task of constructing a space with the participation of a cross - polytope, multiply 4 – cross - polytope by a parallelogram. The result of such a multiplication is shown in Figure 20.

Figure 20. Product 4 - cross - polytope on a parallelogram.

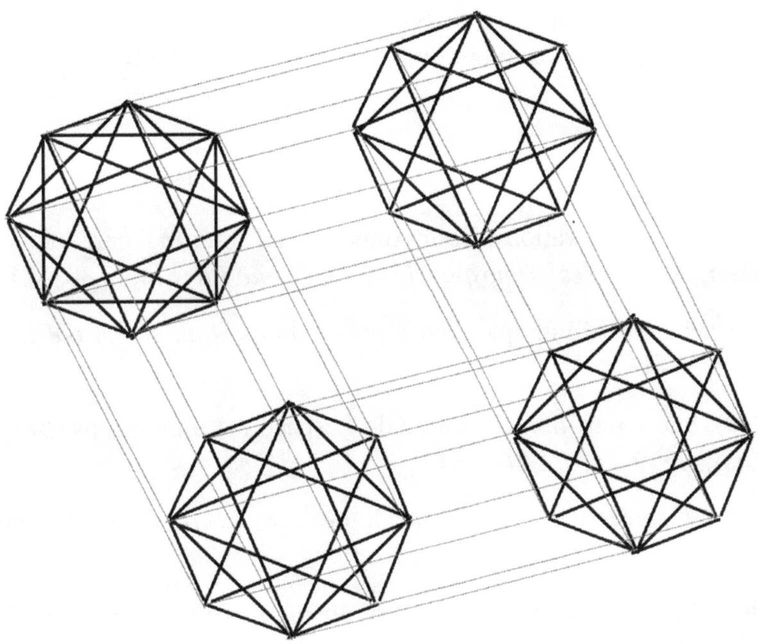

The product of a 4 – cross - polytope on a parallelogram is a polytope of dimension 6. As can be seen from Figure 8.20, this polytopic prismahedron can construct all spaces without gaps, combining every two polytopic prismahedrons along a polytope of dimension 5, which is a product of 4 - cross - polytope into a one - dimensional segment. In this way. the product of 4 - cross - polytopes on a parallelogram is a parallelohedron of dimension 6. it is clear that it is possible to construct parallelohedra based on cross - polytopes of a larger dimension.

CONCLUSION

The study showed that polytopic prismahedrons, due to their geometric structure, provide the ability to construct parallelohedrons of higher dimension. The basis of these parallelohedrons of higher dimension are the well - known figures of higher dimension: *n* - cube, *n* - simplex, *n* - cross - polytope. Moreover, neighboring *n* - parallelohedrons can have common elements of different dimensions. In the case of the existence of enantiomorphic forms of polytopic prismahedrons, parallelohedrons are formed from several polytopic prismahedrons - enantiomorphic forms. Thus, taking into account the developed method for constructing parallelohedrons of the highest dimension (in principle, of any higher dimension), the obtained concrete examples of their construction, we can assume that Hilbert's 18th problem of constructing n - dimensional spaces using congruent figures is solved. Although the construction of parallelohedrons with their large dimensions can present some technical difficulties. It should be noted that the solution obtained corresponds to the ideas of Fedorov (Fedorov, 1885, 1889, 1891) about the need to take into account the properties of the figures of space when solving the problem of constructing space using congruent figures. All the work of Delaunay and his school (Delaunay, 1929, 1937, 1961) did not satisfy this condition when considering the construction of spaces of higher dimension, which caused the lack of concrete results of these works.

Currently, this solution can be used to obtain new nanomaterials with useful properties.

REFERENCES

Alexandrov, A. D. (1934). The derivation of four - dimensional abnormal parallelohedrons. *Proceedings of the U.S.S.R. Academy of Sciences. Atmospheric and Oceanic Physics*, *6*, 803–817.

Alexandrov, A. D. (1954). On filling space with polyhedrons. *Bulletin of the Leningrad University. Mathematics Series.*, *1*, 33–43.

Delaunay, B.N. (1929). Sur la partition ré gulière de l`espace A4 dimensions. *Bulletin de L`Académie des Sciences de L`Urss.*, *1*, 79 - 110, 147 - 164.

Delaunay, B. N. (1937). Geometry of positive quadratic forms. *Progresses of Mathematical Sciences*, *3*, 16–62.

Delaunay, B. N. (1961). Proof of the main theorem of the theory of stereohedrons. *Reports. Academy of Sciences of the USSR*, *138*(6), 1270–1272.

Delaunay, B. N., & Sandakova, N. N. (1961). Theory of stereohedrons. *Proceeding Mathematical In. name V.A. Steklov of Academy Sciences of the USSR*, *64*, 28–51.

Fedorov, E.S. (1885). Beginning of the teachings on the figures. *Notes of the St. Petersburg Mineralogical Society. 2nd series*, *21*, 1 - 277.

Fedorov, E.S. (1889). Symmetry of finite figures. *Notes of the St. Petersburg Mineralogical Society. 2nd series*, *25*, 1 - 52.

Fedorov, E.S. (1891). Symmetry of the correct systems of figures. *Notes of the St. Petersburg Mineralogical Society. 2nd series*, *28*, 1 - 146.

Hilbert, D. (1901). Gesammelte Abhandlangen. *Archiv f. Mathematics and Physics*, *III*(1), 44–63, 213–237.

Poincare A. (1895). Analysis situs. *J. de é Ecole Polytechnique*, *1*, 1 – 121.

Shtogrin, M. I. (1973). Correct Dirichlet - Voronoi partitions for the second triclinic group. *Proceedings of the Mathematical Institute. V.A. Steklova*, 123.

Zhizhin, G. V. (2012). Hierarchical filling of spaces with polytopes. In *Science and human progress. 7th St. Petersburg meeting of Nobel Prize winners*. St. Petersburg: Publishing House of the Polytechnic University.

Zhizhin, G. V. (2013). Images of convex regular and semiregular *n* - dimensional polytopes. In *Proceedings of the 9th All-Russian Scientific School "Mathematical Research in the Natural Sciences"*. Apatity: Geological Institute.

Zhizhin, G. V. (2015). *Polytopic prismahedrons – foundamental regions of the* n *- dimension nanostructures*. Report on the International conference "Nanoscience in Chemistry, Physics, Biology and Mathematics", Cluj-Napoca, Romania.

Zhizhin, G. V. (2018). *Chemical Compound Structures and the higher Dimension of Molecules: Emerging research and Opportunities*. IGI Global. doi:10.4018/978-1-5225-4108-0

Zhizhin, G. V. (2019). *The Geometry of Higher - Dimensional polytopes*. IGI Global. doi:10.4018/978-1-5225-6968-8

Zhizhin, G. V. (2019a). On the Structure of Parallelohedrons of Higher Dimension. Hilbert's 18th problem. *International Journal of Chemoinformatics and Chemical Engineering*, *8*(2), 69–96. doi:10.4018/IJCCE.20190701.oa2

KEY TERMS AND DEFINITIONS

Congruent Polyhedrons: Polyhedrons compatible with each other by movement.

Dimension of the Space: The member of independent parameters needed to describe the change in position of an object in space.

N-Cross-Polytope: The convex polytope of dimension n in which opposite related of centrum edges do not have connection of edge.

N-Simplex: The convex polytope of dimension n in which each vertex is joined by edges with all remain vertices of polytope.

Parallelogon: A polygon capable of filling a two-dimensional plane without gaps during its translation, with adjacent polygons adjacent to each other on the whole sides of the polygons.

Parallelohedron in Three-Dimensional Space: A polyhedron capable of filling a three-dimensional space without gaps during its translation, with adjacent polyhedrons adjacent to each other along whole flat faces of the polyhedrons.

Polytope: Polyhedron in the space of higher dimension.

Chapter 9
Hierarchical Filling of N-Dimensional Spaces

ABSTRACT

The hierarchical filling of the n-dimensional space with geometric figures is studied, accompanied by a process of discrete similar changes in their sizes, that is, process of scaling. The scaling process in these fillings does not depend on time and is determined only by the geometric characteristics of the figures, which are preserved when their size is changed. Two possible ways of hierarchical filling of space are defined, under which the original figure incrementally increases its size fills the space. Investigations of the hierarchical filling of concrete geometric figures of a plane, three-dimensional space, four- and five-dimensional spaces are carried out. The denominator of geometric progressions characterizing sequences of figures in the process of scaling are determined depending on the shape of the figure and its dimension.

INTRODUCTION

For the first time the question of the hierarchical filling of the plane by geometric figures was posed by Kepler (Kepler, 1611). A hierarchical arrangement of any triangles on the plane was considered of Zhizhin (Zhizhin, 1993) and hierarchical arrangement special rhombs on the plane was considered of Penrose (Penrose, 1974). The hierarchical arrangement of regular pentagons on the plane was considered of Mackay (Mackay, 1981, 2006). In these works, the hierarchy of a structure was understood in the sense that at each stage of its construction, an exact replica of some initial shape (not just the pentagon) is used from the subnets. At the same time, the structures from sub - sources are not dense, i.e. between the subordinates there is an empty space of some form. However, more accurately the notion of hierarchy corresponds to structures in which there are several levels corresponding to the forms of a certain figure that differ in size (Zhizhin, 2010, 2012, 2014). In these hierarchical structures, the question of scaling naturally arises, i.e. scale change in the shape of some original figure. This concept of scaling differs from the process of scaling (or scale invariance) in physics, since in this concept there is no temporal coordinate. The scaling process in physics consists in preserving the equations of the process when the

DOI: 10.4018/978-1-7998-6768-5.ch009

distances and time intervals are changed by the same number of times (Prokhorov, 1992). The scaling process in this case is only a variant of the long - known self - similarity of the process (Barenblatt, 1982). One will consider hierarchical structures that have different levels of organization, differing in the sizes of such figures. Before considering the hierarchical filling of n - dimensional spaces, it is reasonable to turn to the hierarchical filling of a two - dimensional plane (Zhizhin, 2010).

The presentation of the question about the hierarchical filling of n - dimensional space in this chapter is based on previously published works of the author. This presentation is necessary to consider the connection between the hierarchical filling of space and the normal translational partition of space, which is discussed in the next final chapter of the monograph.

HIERARCHICAL FILLING OF TWO - DIMENSIONAL FLAT

Let the figure that one fill the plane be a regular pentagon $A_0B_0C_0D_0E_0$ (Figure 1).

Figure 1. The regular pentagon

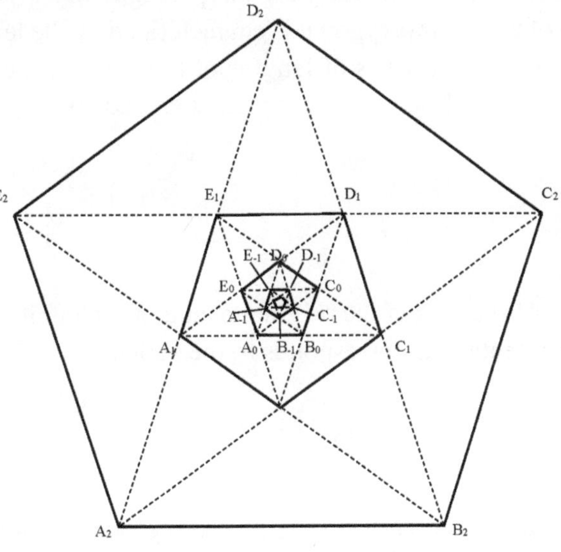

Continuing its sides before crossing into the outer region, can get a pentagon similar the original one $A_1B_1C_1D_1E_1$. Having done the same with a pentagon $A_1B_1C_1D_1E_1$, can get an even larger regular pentagon $A_2B_2C_2D_2E_2$, and so on. By joining the adjacent vertices of a polygon $A_0B_0C_0D_0E_0$, skipping one next vertex, can obtain a polygon $A_{-1}B_{-1}C_{-1}D_{-1}E_{-1}$ inside the original and then an even smaller polygon $A_{-2}B_{-2}C_{-2}D_{-2}E_{-2}$, and so on. Considering such triangles formed from the intersection of lines, can conclude that $\Delta\ D_0D_{-1}C_0 \sim \Delta E_1D_1C_1$, and hence

$$\frac{D_0C_0}{2E_1D_0 + D_0C_0} = \frac{D_{-1}C_0}{D_1C_1}, \tag{9.1}$$

but

$$\frac{D_{-1}C_0}{D_1C_1}=\frac{D_{-1}C_0}{D_1C_0+B_0C_0}=\frac{D_{-1}C_0}{D_0C_0+E_1D_0}=\frac{D_{-1}C_0/E_1D_0}{1+D_0C_0/E_1D_0}=\frac{D_0C_0/E_1D_0}{\left(1+D_0C_0/E_1D_0\right)^2}. \tag{9.2}$$

We denote the relation $E_1D_0/D_0C_0=\tau$, then, substituting (9.2) in (9.1), after the transformations there is

$$\tau 2-\tau-1=0 \tag{9.3}$$

The value has only the positive root of equation (9.3)

$$\tau=\frac{1+\sqrt{5}}{2}=1,618033989\ldots. \tag{9.4}$$

This root characterizes the golden section of the segment E_1C_0 into two parts E_1D_0, D_0C_0. It is obvious that such divisions of segments exist in every regular pentagon that arises on the plane as a result of construction. Since the size of a regular polygon it is characterized by the length of its side, the sequence of pentagons can be compared with a series of lengths of the sides of the pentagons. The ratio of the sides of regular pentagons that are close in size is constant and equal

$$\frac{E_1D_1}{D_0C_0}=\frac{E_1D_0+D_0C_0}{D_0C_0}=\tau+1. \tag{9.5}$$

If we consider the length of the side of the original pentagon as a unit, then the sequence of lengths of the sides of the increasing pentagons corresponds to the series

$$\ldots(\tau+1)^{-3},(\tau+1)^{-2},(\tau+1)^{-1},1,\tau+1,(\tau+1)^2,(\tau+1)^3,\ldots, \tag{9.6}$$

where the terms of the series after 1 correspond to pentagons spaced outside the original pentagon, and the terms of the series to one correspond to the pentagons located inside the original pentagon. The sequence (9.6) is an increasing geometric progression with the denominator $\tau+1$.

Such a construction models, for example, the growth of petals of flowers, which, as a rule, have a smaller size at the center and increasing in size away from the center, or the growth of cabbage leaves that are small size in center and large in size at the periphery. Thus, with the help of a regular pentagon it is possible to construct, so - called (Zhizhin, 2010), an increasing geometric variety modeling the growth of living systems. Between the pentagons in this manifold, a discrete scaling is performed and the similarity coefficient $\tau+1$ can be considered a scaling function. A regular pentagon is not the only polygon possessing this property. Figure 2 shows the growing geometric variety with an example of a regular heptagon (a flower - a seven - color).

Figure 2. The increasing geometric variety on bases of regular heptagon.

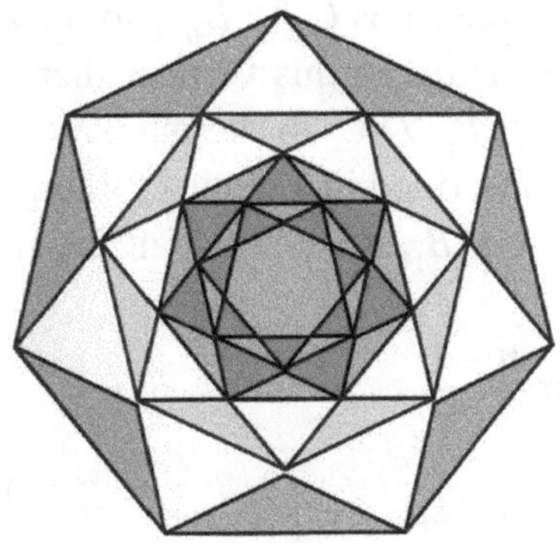

Theorem 9.1: Each regular n – angle for $n \geq 5$ can fill the plane with a growing geometric manifold with division of distances between adjacent vertices in the polygon with respect to

$$d_n = 0,5\left(2\cos^2\frac{\pi}{n} - 1\right)^{-1},$$

and the sides of the polygons thus form an increasing geometric progression with the denominator

$$q_n = \frac{\cos(\pi / n)}{\cos(2\pi / n)}.$$

***Proof*.** Assume that the original regular n - angle ($n \geq 5$) has vertices $B_0, C_0, D_0, E_0, \ldots$ (Figure 3).

Figure 3. Filling the plane with the correct polygon in the "growing" mode

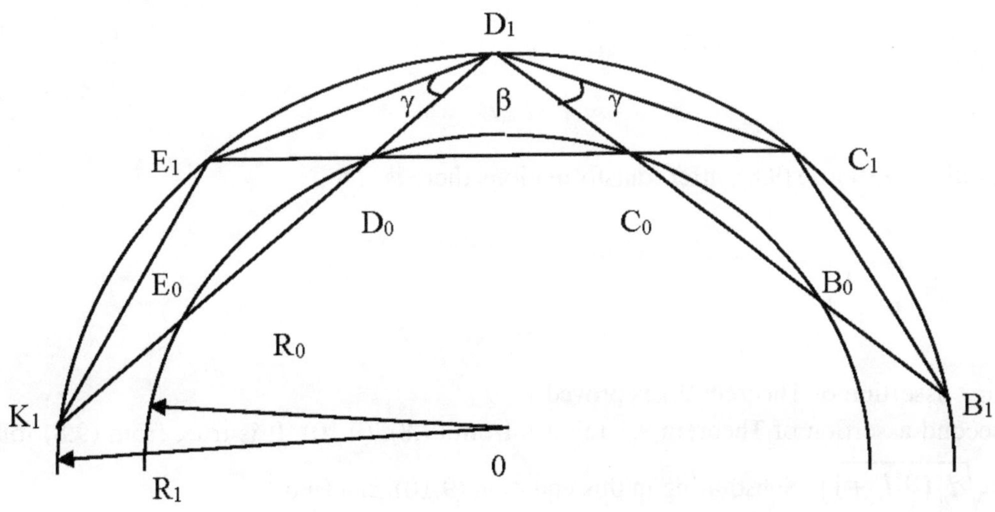

Through the vertices B_0, C_0, D_0, E_0 there passes a circle with a radius R and the center O into which this n - angle is inscribed. Continuing the sides C_0B_0, D_0E_0 to the intersection, can get the vertex D_1 of the polygon of the nearest larger size. Through this vertices there passes a circle with a radius R_1 and center O. Continuing the sides C_0B_0, D_0E_0, D_0C_0 to the intersection with circle with radius R_1 can get also vertices C_1, B_1, E_1, K_1 of n – angel inscribed in a circle of radius R with center O. The angle $K_1D_1B_1$, as an inscribed angle, resting on an arc K_1B_1, is equal to half the central angle, resting on the same arc,

$$\angle K_1D_1B_1 = \beta = \frac{1}{2}\left(2\pi - 4\frac{2\pi}{n}\right) = \pi\frac{n-4}{n}.$$

Thus,

$$\angle E_1D_1D_0 = \angle C_1D_1C_0 = \angle D_1E_1D_0 = \angle D_1C_1C_0 = \frac{\alpha-\beta}{2} = \frac{\pi}{n} = \gamma. \tag{9.7}$$

Triangles $E_1D_1D_0$, $D_1C_1C_0$ are similar of triangle $E_1D_1C_1$. Therefore,

$$\frac{E_1D_1}{D_0C_0 + 2E_1D_0} = \frac{E_1D_0}{E_1D_1},$$

that is, by introducing $d_n = \frac{E_1D_0}{D_0C_0}$, can obtain

$$\left(\frac{E_1D_1}{D_0C_0}\right)^2 = d_n\left(2d_n + 1\right). \tag{9.8}$$

But, taking into account (9.7) $E_1D_1 = E_1D_0 2\cos\frac{\pi}{n}$ or

$$\frac{E_1D_1}{D_0C_0} = 2d_n\cos\frac{\pi}{n}. \tag{9.9}$$

Substituting (9.9) into (9.8), after transformations there is

$$d_n = \frac{1}{2}\left(2\cos^2\frac{\pi}{n} - 1\right)^{-1}. \tag{9.10}$$

The first assertion of Theorem 9.1 is proved.

The second assertion of Theorem 9.1 follows from (9.8), (9.10). It is true, from (9.8) follows that $\frac{E_1D_1}{D_0C_0} = \sqrt{d_n\left(2d_n + 1\right)}$. Substituting in this equation (9.10), can find

$$\frac{E_1D_1}{D_0C_0} = \frac{R_1}{R_0} = q_n = \frac{\cos(\pi / n)}{\cos(2\pi / n)}. \tag{9.11}$$

Consequently, the sides of the n - angles form an increasing geometric progression with the denominator $q_n = \frac{\cos(\pi / n)}{\cos(2\pi / n)}$.

It is of interest to compare the arbitrary term of the geometric progression defined by the theorem 1 with the two preceding members of this progression. Denote by $S_{n,k}$ the length of the side of the regular n - angle at the step k of the increase from the initial n - angle. By the geometric progression, we have $S_{n,k} = q_n S_{n,k-1}$. Imagine

$$S_{n,k} = S_{n,k-1} + (q_n - 1)S_{n,k-1} = S_{n,k-1} + q_n(q_n - 1)S_{n,k-2}. \tag{9.12}$$

It follows from (9.12) that the terms of the geometric progression $S_{n,k}$ represent the Fibonacci series (i.e. each term of the series is the sum of the two preceding ones) only then, when $q_n(q_n - 1) = 1$, i.e. then when

$$q_n = \frac{1+\sqrt{5}}{2} = \tau = 1{,}681033989.$$

But can there be such a value of q_n? Table 9.1 shows the results of calculating d_n and q_n as a function of n from the obtained formulas (9.10), (9.11). It follows from the table that the sequence of lengths of sides of regular n - angles for any n is not a Fibonacci series. In addition, it follows from the table that as n increases, the denominator of the geometric progression decreases, tending to unity.

This means that the second term in (9.12) decreases rather rapidly, asymptotically tending to zero. From Figure 9.1, the growth of regular pentagons implies that the Fibonacci series consists of the sides of regular pentagons and, alternating with them, the sides of star pentagons (for example, $E_0E_{-1} + E_{-1}D_{-1} = E_0D_0$, etc.). However, it remains unclear the meaning of the summation of clearly inconvenient figures, which are convex and star - shaped pentagons.

In all other cases, for any n sequences of sides of convex polygons, sequences of sides of star polygons, sequences of alternating sides of convex and star polygons do not lead to the widely known Fibonacci series (Vorobiev, 1984). Instead of the Fibonacci series, for a sequence of sides of convex regular polygons of interest, a more complex series is realized, characterized by a decrease in the influence (memory) of the selected member of the series, lagging behind it by two steps of the other member of the series.

For a regular triangle and a square, can suggest a slightly different way of obtaining a growing geometric variety. Consider, for example, the right triangle. From the center of the triangle draw the bisectors of the angles of the triangle and continue them outside the triangle (see Figure 4). To each of the bisectors one draw perpendiculars passing through the vertices of the triangle.

Table 1. The functions d_n and q_n at number sides of polygon

n	d_n	q_n	$q_n(q_n - 1)$	q_n*
3	-	-	-	2
4	-	-	-	1,41421356
5	1,6180033989	2,618033989	4,236067	1,2360679
6	1	1,73205	1,2679472	1,15470538
7	0.801937735	1,445041867	0,6431041	1,10991626
8	0,707106781	1,306562965	0,4005438	1,0823922
9	0,652703647	1,2266816	0,27806614	1,06417772
10	0,618033989	1,1755705	0,2063955	1,05146222
11	0,594351144	1,140551777	0,16030658	1,04221711
12	0,577350269	1,115355071	0,12866186	1,03527618
13	0,56468078	1,096544366	0,10586518	1,02992783

Figure 4. The increasing geometric variety on bases of regular triangle.

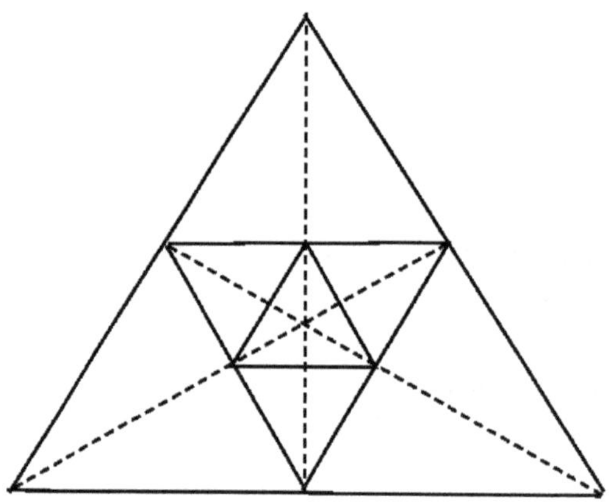

To each of the bisectors can draw perpendiculars passing through the vertices of the triangle. Three perpendiculars, intersection, form the right triangle of the new generation. Having carried out the same procedures with a new triangle, can get the right triangle of the next generation. Obviously, such constructions can be carried out with a regular polygon for any number of sides. Let us study this method of obtaining a growing geometric variety in the general case. Assume again that the original regular n - angle ($n \geq 3$) has the vertices $B_0, C_0, D_0, E_0, \ldots$ (see Figure 5).

Through the vertices B_0, C_0, D_0, E_0 there passes a circle with radius R_0 and center O, into which this n - angle is inscribed. We draw the bisectors of the angles of the n - angle, which naturally pass through the center of the circle. To these bisectors one draw perpendiculars passing through the vertices of the polygon. Crossing, the perpendiculars form a n - angle of a new generation with the vertices $B_1, C_1, D_1, \ldots$. They lie on a circle with radius R_1 and center O. We consider two triangles ΔOB_1B_0 and ΔOFB_0. These two triangles are rectangular with a common angle $\angle B_0OB_1 = \pi/n$. From the similarity of these triangles it follows that

$$\frac{B_1B_0}{FB_0}=\frac{C_1B_1}{C_0B_0}=\frac{1}{\cos(\pi/n)}=q_n^*=\frac{R_1}{R_0}. \quad (9.13)$$

Figure 5. Filling the plane with the correct polygon in the "growing" mode of second way

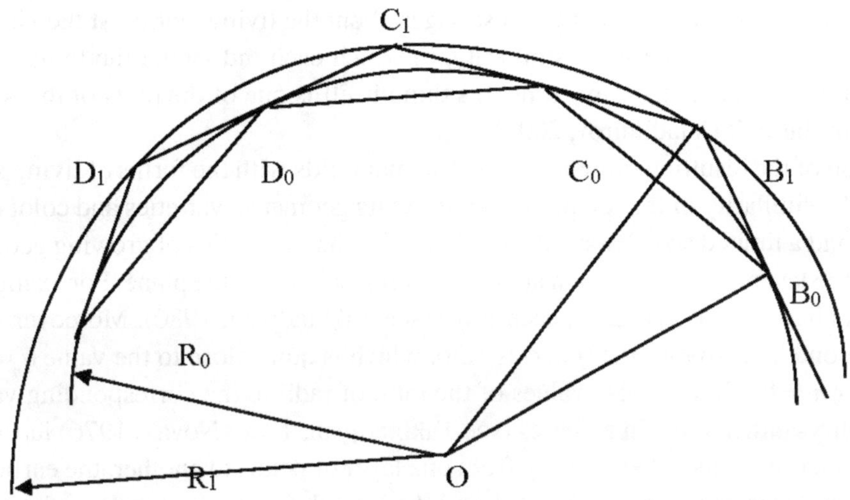

In the last column of Table 9.1, the values of the geometric progression denominator q_n^* are shown for this variant of constructing a growing geometric manifold as a function of the number of sides of a regular polygon *n*. From the consideration of Table 9.1 it follows that the denominator of the geometric progression q_n^* increases *n* very rapidly with increasing 1. This means that, with $n > 5$ the increasing geometric variety practically ceases to expand; remains in the finite domain. This property is also observed in the first method of constructing a growing geometric manifold, but the restriction of its growth in this case occurs for sufficiently large *n* ($n > 12$). Therefore, it is better to use the second method for constructing a growing geometric manifold for small values. It can be seen that when constructing a growing geometric variety in the second way, the lengths of the sides of polygons of different generations also do not form a Fibonacci sequence.

It should be emphasized that the introduced concept of a growing geometric variety, in spite of its relative simplicity, has a number of properties that allow us to consider a growing geometric variety as a model of growing living systems. Any regular *n* - angle generates a growing geometric variety. In this case, at each growth stage a new regular *n* - angle of a larger size is formed and triangles forming a new *n* - angle, similar to the triangles making *n* - angle in the previous stage of growth, also appear. A growing geometric variety generates an infinite discrete series of proportions, in which the golden section is only a particular case. It is interesting that, in spite of the fundamental possibility of filling the entire plane with the growing geometric variety with an unlimited increase in the number of stages of growth, the rate of increase in the size of figures decreases with increasing number, and the growing geometric variety remains in the finite domain. The center of a growing geometric variety is a singular point, since when moving towards the center, there is an unlimited reduction of figures that can not continuously turn into a point. The concept of singular points in tiling of a plane it was introduced by Grunbaum and Shephard (Grunbaum & Shephard, 1987).

Each proportion in an infinite number of proportions corresponds to some geometric progression, and, and this is essential, the denominator of the progression depends on the number n. The algorithm for filling the plane with an infinite number of similar figures (regular n - angles) increasing their size for moving away from some point of the plane (center of growth) is constructed. This algorithm is consistent with the views of the ancient Chinese sages about the living world. At the views an organism possessing the principle of scale self - similarity, in which each individual the thing consists of parts similar to it as a whole, and at the same time this thing itself is one of the parts of the whole world and is structurally similar to it (Vladimirov, 2002).

A comparison of the resulting growing geometric manifolds with the forms of living systems such as flowers shows the similarity of the regularities of growing geometric varieties and colors having several layers of petals and a limited size. Especially close are the characteristics of growing geometric varieties to the colors of the layers of petals, which are located almost in the same plane. For example, Eupomatia bennettii flowers have about 8 petals in each layer (see Takhtadjyan, 1980). Moreover, the ratio of the radius of the second layer to the first (later) is 1.36, which is quite close to the value q_n=1,306 given in the Table 9.1 for $n = 8$. Similar close values of the ratio of radii to the corresponding values are found in dahlias and chrysanthemums. In all cases (see Takhtadjyan, 1980; Novak, 1976) in flowers of plants with a large number of petals, when passing from one layer of petals to another, the earlier, the diameter of the layer of petals increases insignificantly, and the less, the more the number of petals in the layer. This completely agrees with the dependence of the basis of the geometric progression q_n on n, presented in Table 9.1. This is completely inconsistent with the assumption of the subordination of the laws of plant growth to the Fibonacci series and the associated golden section, since in this case the ratio of the diameters of the petal layers should be a significant value of 2.618033989 and should not depend on the number of petals in the layer.

HIERARCHICAL FILLING OF THREE - DIMENSIONAL SPACE

Hierarchical Filling of Space with a Tetrahedron

In space, the faces of the tetrahedron do not intersect when they continue to form a larger tetrahedron. Therefore, in order to construct a similar tetrahedron, it is necessary to extend the first method of constructing such a polygon to spaces. The bisectors of angles in the plane case are transformed in space into planes that cut the faces along the bisectors of the angles passing through the two vertices of the tetrahedron. Three such planes pass through each vertex of the tetrahedron. They form four perpendiculars, coming from each vertex perpendicular to the opposite face.

Draw to each of these four perpendiculars a perpendicular plane passing through each vertex, respectively. Crossing, these planes form a new tetrahedron of a larger size. This is a consequence of the duality property of the tetrahedron to itself. Continuing this process, we obtain a sequence of increasing tetrahedron with a single center.

Can to denote the vertices of the original tetrahedron O_1, O_2, O_3, O_4 and the vertices of the larger tetrahedron A, B, C, D in this construction (see Figure 6).

Figure 6. Hierarchical filling of space with a tetrahedron

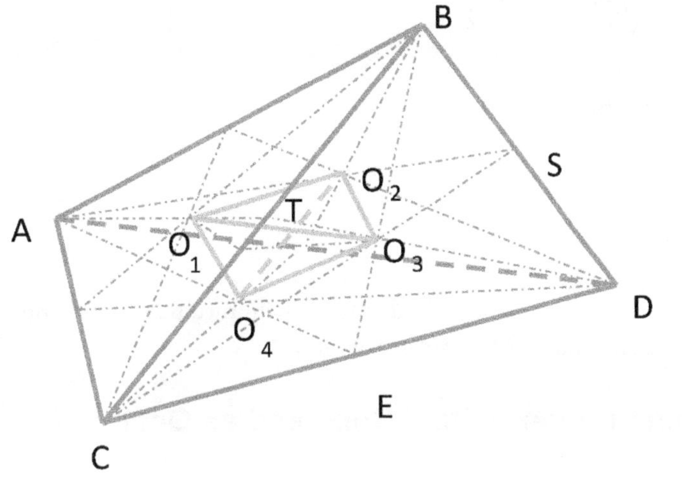

We denote the midpoints of the edges of the outer tetrahedron *BD*, *CB* and *CD*, respectively, *S*, *T* and *E*. Let the length of the edge of the outer tetrahedron be "*a*".

We consider the triangle *AEB*. The length of the side *AB* of this triangle is, by assumption, "*a*". The lengths of the sides *AE* and *EB* are equal to $a\frac{\sqrt{3}}{2}$ as heights in the right triangle. The O_4O_3 segment in the triangle *AEB* is parallel to the side *AB*. The length of the segments $O_3B=AO_4$ on the sides *AE, EB* in the triangle *AEB* is found from the triangle *CBD*.

Considering triangles TBO_3 and *CBE* in triangle *DCB*, can find

$$\frac{BE}{TB} = \frac{CB}{BO_3}.$$

Since $CB = a$, $TB = a/2$, $BE = a\frac{\sqrt{3}}{2}$, from the last equality can find

$$BO_3 = AO_4 = \frac{a}{\sqrt{3}}.$$

Now, considering similar triangles EO_4O_3, *EAB*, there is

$$\frac{O_4O_3}{AB} = \frac{EO_3}{EB}.$$

Since

$$AB = a, EO_3 = a\frac{\sqrt{3}}{2} - \frac{a}{\sqrt{3}}, EB = a\frac{\sqrt{3}}{2}$$

from the last equality can obtain

$$O_4O_3 = \frac{a}{3}.$$

Thus, the lengths of the edges of tetrahedrons that hierarchically fill the space form a geometric progression with the denominator of the progression $q_T = 3$.

Hierarchical Filling of Space with a Cube and an Octahedron

The edges of the cube, when extended, do not give additional edges, which could lead to the formation of a larger cube. Therefore, to construct a polyhedron external to the cube, on the basis of the characteristics of the cube alone, the first method of construction is applicable.

We draw through each vertex of the cube perpendicular to the diagonal of the cube passing through this vertex, the plane. Crossing, these planes form a polyhedron. Since the vertices of the cube are 8, the resulting polyhedron will be an octahedron that has exactly 6 vertices, i.e. as much as the faces of the cube. This is a manifestation of the duality property of a cube and an octahedron. Therefore, if a polyhedron is constructed outside the octahedron by the same principles, then a cube is obtained. Thus, in this case the sequence of polyhedron consists in alternating cubes and octahedron: a cube gives rise to an octahedron, an one gives rise to a cube. In order to determine the changes in the lengths of the edges of polyhedron in this sequence, can to turn to Figure 7, which depicts an octahedron whose focal points should be the vertices of an inscribed cube.

Suppose that in the upper half of an octahedron with an edge the faces of the faces are points O_1, O_2, O_3, O_4. Consider the triangle *EDC* passing through the vertex of the octahedron *D*, the points O_1, O_3 and the midpoints of the two parallel edges of the octahedron of the point *E* and *C*.

Can find the distance O_1O_3, which is the diagonal of the square that is the face of the inscribed cube. From consideration of two similar triangles O_3LC and *DCO* follows that

$$\frac{LC}{OC} = \frac{CO_3}{DC},$$

But

$$O_3C = \frac{a}{\sqrt{3}}, OC = \frac{a}{2}, DC = \frac{\sqrt{3}}{2}a.$$

Hence,

$$LC = \frac{a}{3}, O_1O_3 = a - 2\frac{a}{3} = \frac{a}{3}.$$

Therefore,

$$O_2O_3 = \frac{a}{3\sqrt{2}}.$$

Figure 7. An octahedron

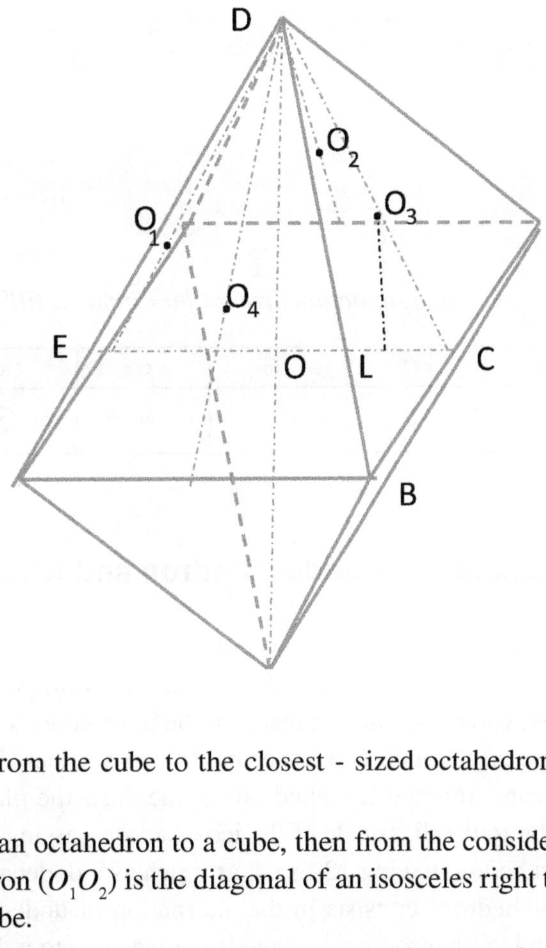

Thus, at the transition from the cube to the closest - sized octahedron, the length of the edge increases by a factor $3\sqrt{2}$.

If the transition is from an octahedron to a cube, then from the consideration of Figure 8 it follows that the edge of an octahedron (O_1O_2) is the diagonal of an isosceles right triangle with legs $a / 2$ if a is an edge of the described cube.

Thus, when a cube is formed from an octahedron, the edge increases by $\sqrt{2}$ times. If we take as the initial figure a cube with an edge 1, then the sequence of edges with alternating cubes and octahedrons looks like Table 9.2.

From Table 9.2 it follows that if from this sequence to separate the sequence of cubes and the sequence of octahedrons, then both sequences form a geometric progression with a denominator of $q_{c,oc}$=6, differing by an initial element of the sequences: for the cube -1, for the octahedron - $3\sqrt{2}$.

Figure 8. Hierarchical filling of space with a cube and an octahedron

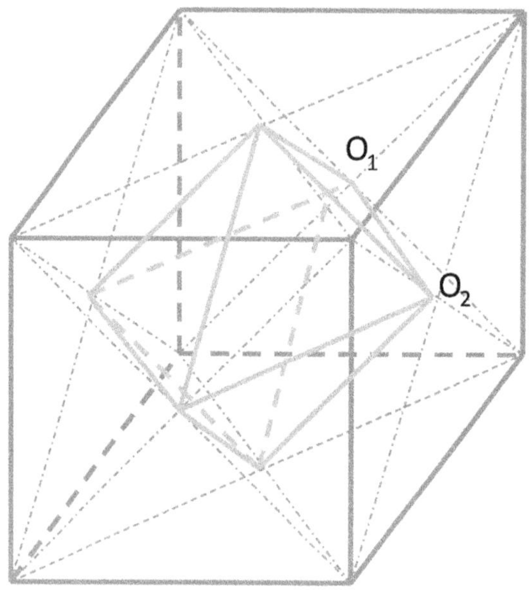

Table 2. The length of edges a cube and an octahedron at hierarchical filling of space

——	C	Oc	C	Oc	C	Oc	C	oc
length of edges	1	$3\sqrt{2}$	6	$18\sqrt{2}$	36	$108\sqrt{2}$	216	$648\sqrt{2}$

Hierarchical Filling of Space with Dodecahedron and Icosahedron

For the dodecahedron and icosahedron, only the second method of hierarchical filling of space is also applicable. Drawing a plane through the vertices of the dodecahedron perpendicular to the diagonals passing through these vertices, can to obtain as a result of the intersections of the planes a closed figure - an icosahedron having 20 faces, as much as a dodecahedron of vertices. This is a consequence of the duality property of the dodecahedron and icosahedron. If one draw the planes through the vertices of the icosahedron, perpendicular to the diagonals of the icosahedron passing through these vertices, then can to get as a result a dodecahedron having 12 faces, as much as a dodecahedron of vertices. Thus, in this case the sequence of polyhedrons consists in the alternation of dodecahedrons and icosahedra: a dodecahedron gives rise to an icosahedron, an icosahedron gives rise to a dodecahedron.

Let the icosahedron be inscribed in the dodecahedron (Figure 9).

Then the middle of the regular pentagons, which are the faces of the dodecahedron, are connected by the sides of one regular triangle, as one of the faces of the icosahedron. To determine the ratio of the sides of a pentagon and a triangle, calculate the dihedral angle in the triangular corner formed by the faces of the dodecahedron. Between the plane angles of the triangular angle *abc* and the dihedral angle *A* there is a relation of the spherical geometry (Zalgaller, 1967; Zhizhin, 2009)

Figure 9. Hierarchical filling of space with dodecahedron and icosahedron

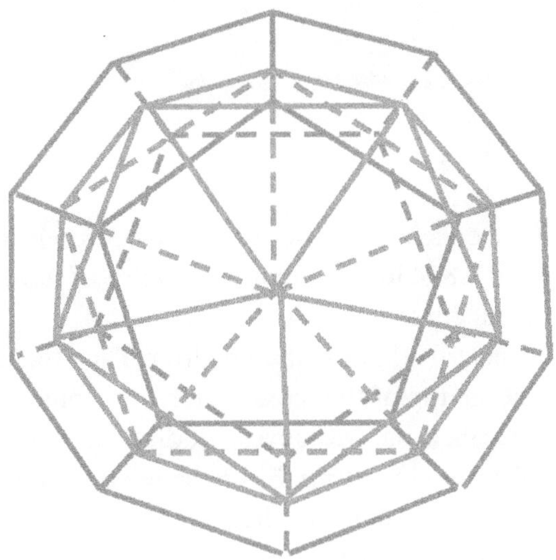

$$\cos a = \cos b \cos c + \sin b \sin c \cos A \tag{9.14}$$

where the dihedral angle A lies opposite the angle a. Since in the regular pentagon all angles at the vertices are 108°, hence $a = b = c = 108°$, then from (9.14) can find the cosine of the dihedral angle between the faces of the dodecahedron

$$\cos A = \frac{\left(\cos 108^{o}\right)\left(1-\cos 108^{o}\right)}{\left(\sin 108^{o}\right)^{2}}. \tag{9.15}$$

From (9.15) can find that $A = 116{,}5°$.

Consider the triangle *BOL* formed by the edge of the icosahedron and the perpendiculars *OL* and *OB* dropped from the centers of the faces of the dodecahedron to the edge of their connecting (see Figure 10).

Figure 10. The dihedral angle in the dodecahedron

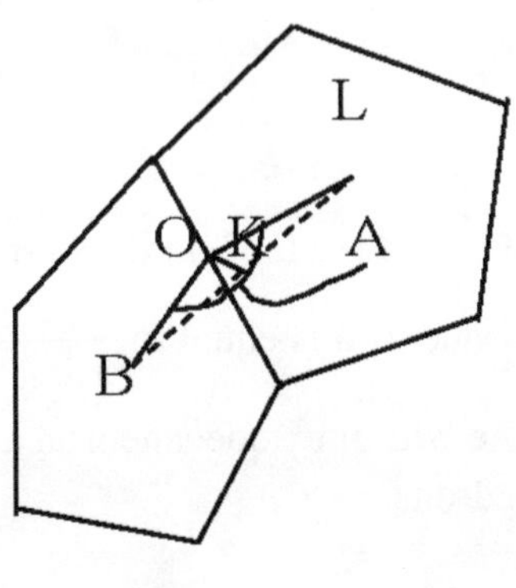

Since $BO = OL = a / 2\ tg\ 54^o$, where a is the length of the edge of the dodecahedron, edges of an inscribed icosahedron

$$2KL = a\sin\frac{A}{2}tg54^o = a1{,}170409428. \tag{9.16}$$

It follows from (9.16) that if the edge of the inscribed icosahedron is taken as 1, then the edge of the described dodecahedron is equal to 0.854401866, that is, the edge of the described dodecahedron is less than the edge of the inscribed icosahedron!

Now let the dodecahedron be inscribed in the icosahedron. Then the middle of the regular triangles forming the pentagonal corner, at the top of the icosahedron, are connected by the sides of the regular pentagon, like the faces of a dodecahedron. Moreover, all dihedral angles in the pentagonal angle are equal (pentagonal angle is completely symmetric). From the relations connecting dihedral angles to the pentagonal angle (Zhizhin, 2009) it follows that the dihedral angles in a completely symmetric pentagonal corner with faces from regular triangles are $138.2^o = \alpha$.

One consider the triangle DO_1C formed by the edge of the dodecahedron DC and the heights O_1C and O_1D from the centers of the faces of the icosahedron (Figure 11).

Figure 11. The dihedral angle in the icosahedron

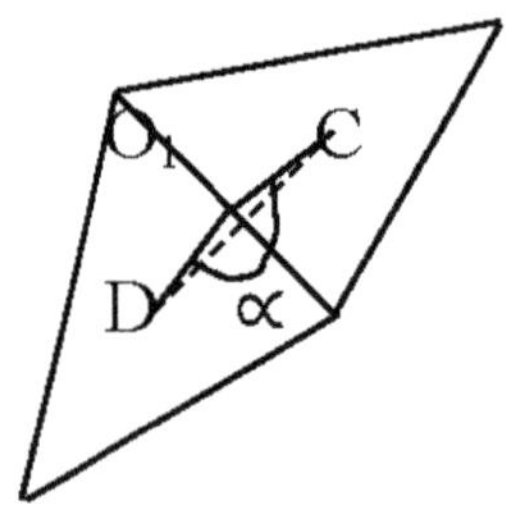

Since, $O_1C = O_1D = \dfrac{a}{2\sqrt{3}}$, where a is the edge of the icosahedron, the length of the edge of the inscribed dodecahedron

$$DC = 2\frac{a}{2\sqrt{3}}\frac{1+\sqrt{5}}{2\sqrt{3}} = a\frac{1+\sqrt{5}}{6}.$$

It follows from the last equality that if we take the length of an inscribed dodecahedron as 1, then the length of the edge of the described icosahedron is equal to $\dfrac{6}{1+\sqrt{5}} = 1{,}854038233$.

Thus, the hierarchical filling of space with the dodecahedron and icosahedron leads to a sequence of alternating dodecahedrons and icosahedrons.

The lengths of the edges of polyhedrons and their volumes with increasing number of steps "k" in the sequence are presented in Table 9.3, provided that the initial figure ($k = 0$) is a dodecahedron with a length of edge 1.

From Table 9.3 it follows that, if from the sequence of alternating dodecahedrons and icosahedron isolate the sequences of dodecahedron edges and separately the sequence of edges of icosahedron, then these two sequences form geometric progressions with the denominator $q_{d,i} = 1.584093726$. However, these sequences have different initial edge values: for the dodecahedron edges sequences, the initial length of the edge 1, for the sequence of edges of icosahedron the initial length of the edge is 1.854038233.

It should be noted that, despite the decrease in the length of the edge, the volume of the described polyhedron is larger than the volume of the inscribed polyhedron from the icosahedron to the described dodecahedron for any generation (see Table 3).

Table 3. The length of the edges of dodecahedron and icosahedron at hierarchical filling of space

K	0	1	2	3	4	5
a dodecahedron	1		1,584093726		2,509352933	
a icosahedron		1,854038233		2,93697033		4,652436278
V Dodecahedron	7,66		30,44888481		121,0358468	
V Icosahedron		13,89352832		55,22747305		219,5319799

Hierarchical Filling of a Three - Dimensional Space by Convex Semi - Regular Polyhedrons

Among semi - regular polyhedrons (Zhizhin, 2014) there is no dual to itself (the numbers of vertices and faces are different) and there are no pairs of dual polyhedrons. Therefore, semiregular polyhedrons cannot hierarchically fill the space in a way with common points that increase their size by polyhedron. Semi - regular polyhedrons can hierarchically fill space only in the second way, i.e. not having common points with each other.

Variants of the hierarchical filling of space by semi - regular convex three - dimensional polytopes have been studied successively in Zhizhin's book (Zhizhin, 2014). The similarity coefficients for all semi - regular bodies are determined. More precisely, these similarity coefficients should be called self - similar coefficients. The values of the self - similarity coefficients determine the regularities of the hierarchical filling of space by convex semi - regular polyhedrons. It is essential that semi -regular polyhedrons, in contrast to regular polyhedrons, have as a rule several (from 1 to 5) similarity coefficients, i.e. several variants of hierarchical filling of space by each polyhedron. This leads to the possibility of simultaneously increasing the volume in several ways at each step of the filling, which considerably complicates the filling structure and contributes to the densest filling of the space. In addition, this leads to the possibility of a hierarchical chaotic filling of the space, since at each step of filling different ways of filling can be arbitrarily realized.

HIERARCHICAL FILLING OF HIGHER - DIMENSIONAL SPACE

Now, applying the methods developed in previous studies, one to consider the hierarchical filling of a multidimensional space in the process of enlargement of multidimensional convex bodies (Zhizhin, 2012). As was shown earlier when studying the hierarchical filling of a plane with polygons (Zhizhin, 2010), there can be two ways of filling it. In a multidimensional space is first way of hierarchical filling with polytopes, i.e. continuation of the edges of the figure previous stage before intersecting them outside the figure, forming the vertices of the figure in a new stage. For the tree - dimensional polytope the method of hierarchical filling of space was based on the transition of a polytope to a dual polytope at each step of filling the space. In this case, investigating scaling processes, one need to consider a method for filling a space of higher dimension with polytopes, at which only the scale of the figure would change at each step, and the polytope's background would be preserved.

On Figure 4 of Chapter 3 presented projection of the 4 – simplex.

Continuing the outer edges of the projection to the intersection with each other, connecting the resulting vertices with edges, can obtain an enlarged projection of the 4 - simplex (Figure 12).

Figure 12. The hierarchical filling of the space by a 4 - simplex

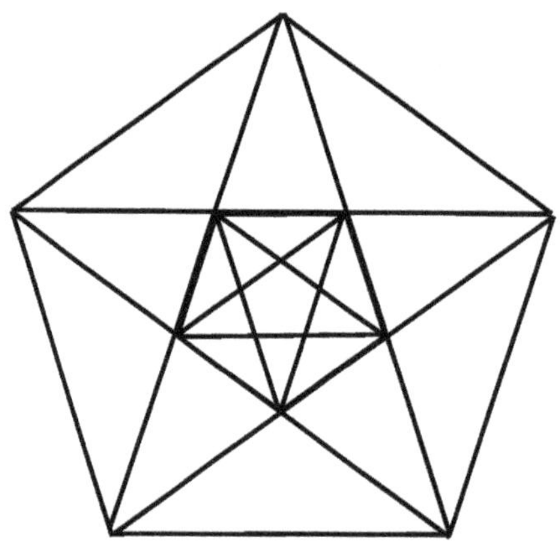

It should be emphasized that the lines that extend the edges of the figure are of an auxiliary character (however, as in the hierarchical filling of the plane). In this case, the edges of the figure of the next generation, due to the multidimensionality of space in the projection, can be partially superimposed on the auxiliary lines and edges of the figures of previous generations.

Obviously, the process of extending the 4 - simplex can be continued further. Each time the size (length) of the edges of the next 4 - simplex from the size of the edges of the previous 4 - simplex is increased by $1 + \tau$ times. Where τ is the golden section. Since the coefficient of the geometric progression of the discrete increase of each side of the regular pentagon in the hierarchical filling of the pentagon plane is $1 + \tau$ (Zhizhin, 2010), then each segment joining the vertices of the pentagon increases in the same progression. The difference in the lengths of the segments of the projection of a regular the 4 - simplex into a two - dimensional plane is related to the design process, but their images in four - dimensional space have the same length. Thus, the geometric progression denominator of the hierarchical filling of the space by a 4 - simplex is $1 + \tau$.

On Figure 3.10 of Chapter 3 presented known projection of the 4 – cube in symmetrical form.

Continuing the outer edges of the projection to the intersection with each other, connecting the resulting vertices with edges, can to obtain an enlarged projection of the 4 - cube (Figure 13).

Figure 13. The hierarchical filling of the space by a 4 - cube

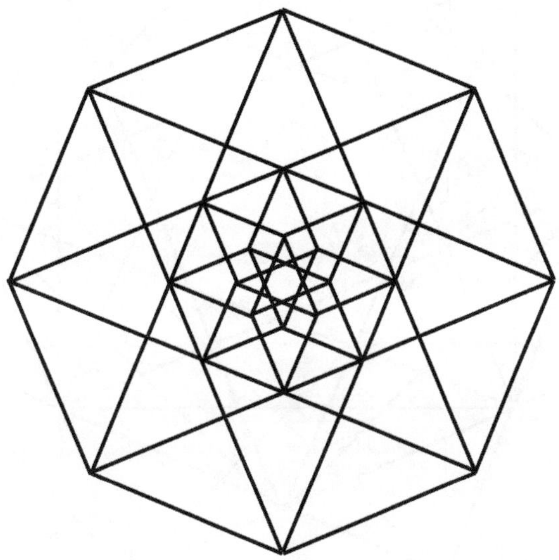

Obviously, the process of extending the 4 - cube can be continued further. Easy to see that each time the size (length) of the edges of the next 4 - cube from the size of the edges of the previous 4 - cube is increased by $1+\sqrt{2}$ times. Thus, the geometric progression coefficient of the hierarchical filling of the space by a 4 - cube is $1+\sqrt{2}$.

On Figure 9 of Chapter 2 presented projection of the 4 – cross - polytope in symmetrical form.

Continuing the outer edges of the projection to the intersection with each other, connecting the resulting vertices with edges, can to obtain an enlarged projection of the 4 - cross - polytope (Figure 14).

Obviously, the process of extending the 4 – cross - polytope can be continued further. Easy to see that each time the size (length) of the edges of the next 4 – cross - polytope from the size of the edges of the previous 4 – cross - polytope is increased by $\sqrt{1+\frac{1}{\sqrt{2}}}$ times. Thus, the geometric progression denominator of the hierarchical filling of the space by a 4 – cross - polytope is $\sqrt{1+\frac{1}{\sqrt{2}}}$.

In the same way, it is possible to construct a hierarchical filling of a space of dimension greater than four starting from some initial figure of this dimension. For example, let this figure be the 5 - simplex. Its image on the plane is shown in Figure 3.2 of Chapter 3.

This polytope has 6 of the vertices, 16 of the edges, 20 of the two - dimensional faces, 15 of the three – dimensional faces, 6 of the four - dimensional faces. Continuing the outer edges of the projection to the intersection with each other, connecting the resulting vertices with edges, can to obtain an enlarged projection of the 5 - simplex (Figure 15).

Figure 14. The hierarchical filling of the space by a 4 – cross - polytope

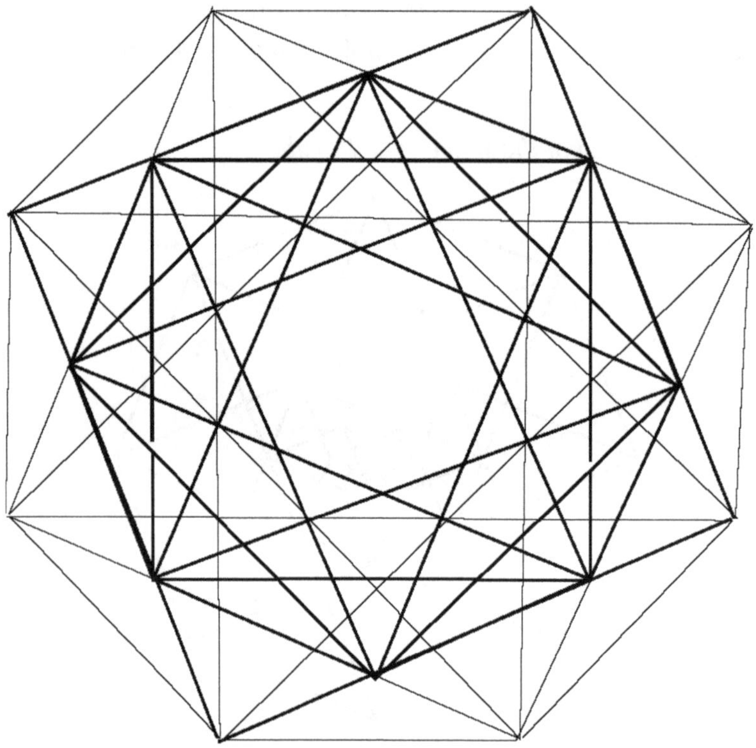

Figure 15. The hierarchical filling of the space by a 5 - simplex

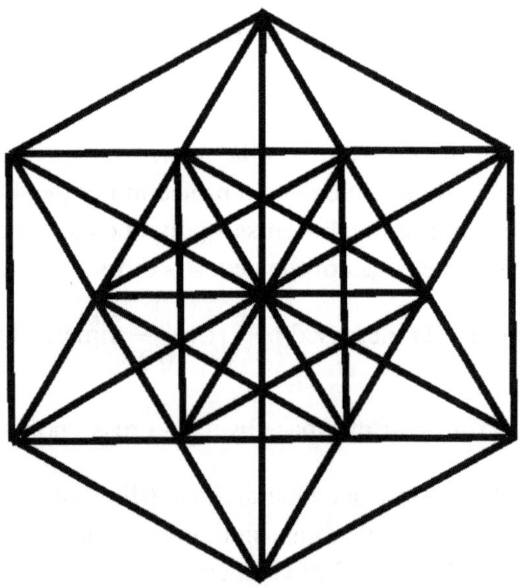

Obviously, the process of extending the 5 - simplex can be continued further. Easy to see that each time the size (length) of the edges of the next 5 - simplex from the size of the edges of the previous 5 - simplex is increased by $\sqrt{3}$ times. Thus, the geometric progression denominator of the hierarchical filling of the space by a 4 - simplex is $\sqrt{3}$.

CONCLUSION

The process of scaling (zooming in) has attracted the attention of researchers in recent years in analyzing various phenomena. In this chapter, we study the geometric characteristics of the scaling process in the hierarchical filling of spaces. The investigation was started from the hierarchical filling of the plane by regular polygons in the "growing geometric variety" regime. This mode simulates the growth of biological objects, for example, flowers. Two ways of the scaling process are discovered, which also manifest themselves in spaces of higher dimension. The denominator of geometric progressions characterizing a discrete series of polygon sizes in the process of hierarchical filling of the plane are determined. These denominators can be regarded as the values of the scaling function of the corresponding processes. It can be noted that changing the size of the figures during the hierarchical filling of the plane does not obey the Fibonachchi series for any number of sides of the polygon. The hierarchical filling of a three -dimensional space by regular and semi - regular polyhedrons is investigated. In the hierarchical filling of space by three - dimensional polyhedrons at each level of filling, the type of the polyhedron changes by a dual type. When the space is filled with semi - regular polyhedrons, each variant of a semi - regular polyhedron can have different variants of changing the size of the figure. This leads to possible chaotic hierarchical filling of space. The value of its scaling function is calculated for each type of three - dimensional polyhedron. The processes of hierarchical filling of a space of higher dimension by convex polytopes are also considered. These fillings are constructed for a number of polytopes of higher dimension, and the values of their scaling functions are determined.

REFERENCES

Barenblatt, G. I. (1978). *Similarity, self - similarity, intermediate asymptotes*. Hydrometioizdat.

Grünbaum, B & Shephard, G. (1987). *Tilings and Patterns*. New York: A Series of Books in the Mathematical Sciences.

Keppleri, J. (1611). *Strena, seu de nive sexangula*. Tampach: Francofurti ad Moenum.

Lord, E. A., Mackay, A. L., & Ranganathan, S. (2006). *New Geometry for New Materials*. Cambridge University Press.

Mackay, A. L. (1981). De Nive Quinquangula: On the pentagonal snowflake. *Kristallografiya., 26*, 910–919.

Novak, F. A. (1976). *Illustrated encyclopedia of plants*. Artia.

Penrose, R. (1974). The role of aesthetics in pure and applied mathematical research. *Bulletin - Institute of Mathematics and Its Applications*, *10*, 266.

Prochorov, A. M. (Ed.). (1983). *Physical encyclopedic dictionary*. Soviet Encyclopedia.

Takhtadzhyan, A. L. (Ed.). (1980). *Life of plants. T.5. Flowering plants*. Enlightenment.

Vladimirov, Yu. (2002). *Metaphysics*. Moscow: Binom. Knowledge lab.

Vorobiev, N. N. (1984). *Fibonacci numbers*. Science.

Zalgaller, V. A. (1966). *Convex polyhedrons with correct sides*. Science.

Zhizhin, G. V. (1993). *Parquetages of equal triangles, adjoining on whole sides*. Polytechnika.

Zhizhin, G. V. (2009). *Convex Right – Faceted Polyhedrons*. Saint – Petersburg: Polytechnika – service.

Zhizhin, G. V. (2010). *Geometric bases of the dissipative structures*. Polytechnika.

Zhizhin, G. V. (2012, October). *Hierarchical filling of spaces with polytopes*. Paper presented at "St. Petersburg Scientific Forum: Science and Human Progress". 7th St.- Petersburg meeting of Nobel Prize laureates, St. Petersburg, Russia.

Zhizhin, G. V. (2017). *Chemical Compound Structures and the Higher Dimension of Molecules: Emerging Research and Opportunities*. IGI Global.

KEY TERMS AND DEFINITIONS

Denominator of Geometric Progression: The relationship between the previous and subsequent elements of a geometric progression.

Fibonacci Series: A series of numbers in which a member of a series is the sum of the two previous members of the series.

Golden Ratio: The section of the segment into two parts, in which the ratio of the length of the larger part to the length of the smaller part is $\tau = \frac{1+\sqrt{5}}{2} = 1,618033989....$

Hierarchical Filling of Space With a Figure: Filling a space with a figure with a similar increase in the size of the figure around a certain center according to some law related to the geometric characteristics of the figure itself.

Scaling: Scale change of shape.

Chapter 10
Joint Normal Partitions and Hierarchical Filling of N-Dimensional Spaces

ABSTRACT

The structures arising in spaces of various dimensions with simultaneous normal partitioning of spaces and their hierarchical fillings are considered. The conditions for the appearance of translational symmetry in these structures are investigated. It is shown that simultaneous hierarchical filling and normal tiling in three-dimensional spaces do not lead to the formation of translational symmetry. Such consistent transformations lead to many elements of translational symmetry in spaces of higher dimension. The higher the dimension of space, the more complex the emerging structure and the more symmetry the elements.

INTRODUCTION

In the previous chapter, when studying the hierarchical filling of n - dimensional spaces, we made sure that the distance between the lattice points of such a filling increases unlimitedly with distance from the center of the hierarchical filling (expansion center). This of course should lead to the decay of matter into individual atoms. This does not happen in nature. To avoid this, nature employs a mechanism of expansion (or hierarchical filling) from each point in space (the principle of uniform expansion of the Universe). Moreover, as it is not surprising, translational symmetry appears, leading to a normal partition of n - dimensional space. This chapter explores the mechanism leading to this partition.

DIFFRACTIONS PATTERNS OF QUASICRYSTALS

Finding in 1982 of ordered structures deprived (as it seemed) of translational symmetry (Shechtman et al., 1984), next called "quasi - crystals", had marked the beginning of numerous cycles of papers and books devoted to the experimental and theoretical study of these unusual materials. Later it was found

DOI: 10.4018/978-1-7998-6768-5.ch010

that the diffraction patterns of quasicrystals have a latent periodicity (Zhizhin, 2014), if we consider the diffraction pattern as a projection of a structure from a space of higher dimension. Figure 1 shows a typical diffraction pattern of intermetallic compounds.

Similar structures have other intermetallic compounds involving transition metals, for example: Al_6Mn (Shechtman et al., 1984), $Al_{70}Fe_{20}W_{10}$ (Mukhopadhyay et al., 1993), $Ti_{54}Zr_{26}Ni_{20}$ (Zhang & Kelton, 1993).

Figure 1. Electron diffraction pattern of compound $Al_{72}Ni_{20}Co_8$ (Eiji Abe et al., 2004)

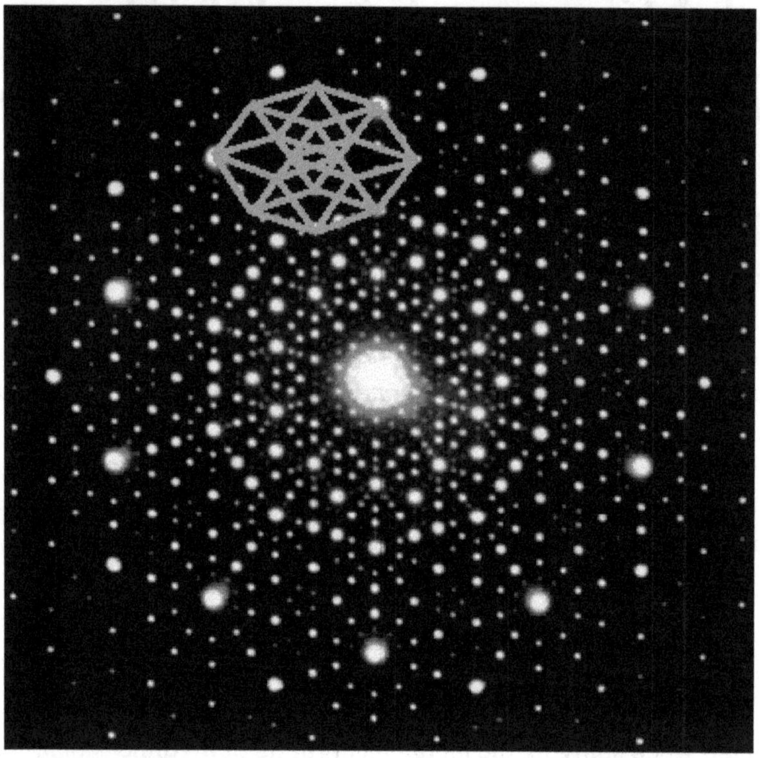

Figure 10.1 shows that the luminous points, which are a reflection of light from the impact of the electron beam, form five families of parallel lines oriented with respect to each other at angles of a multiple of 72 degrees. The distances between the parallel lines and the angles are determined by the golden section. A geometrical model of the structure of the diffraction patterns of quasi - crystals was constructed (Shevchenko, Zhizhin & Mackey, 2013a; Shevchenko, Zhizhin & Mackey, 2013b; Zhizhin, 2014; Zhizhin & Diudea, 2016). It was shown that the elementary cell of this geometric structure is a polytope of dimension 4, which was called a gold hyper - rhombohedron. This cell is plotted on the diffraction pattern in Figure 1 with solid segments of light lines. I can see that it passes through the luminous points of the diffraction pattern observing its geometry. This cell fills the entire space with the translation reflected by the diffraction pattern. To determine the regularities of this cell, it is depicted in Figure 2 on an enlarged scale.

Figure 2. Golden hyper - rhombohedron

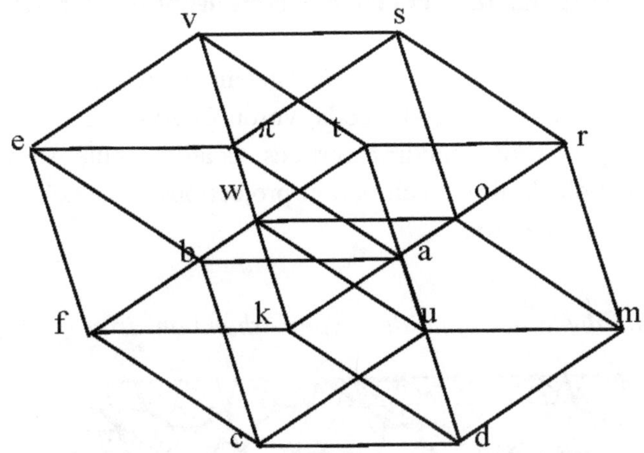

The polytope in Figure 10.2 consists of 8 three-dimensional polyhedrons whose faces are determined by construction with a golden section. Namely, all two - dimensional faces represent the same diamonds with angles

$$\varphi = \arccos\frac{1}{2\tau}, \gamma = \pi - \varphi,$$

where *t* is golden section. The three -dimensional faces (rhombohedra) are adjacent to each other along the flat faces, so that all planar faces are common to some two rhombohedrons, which is necessary for the polytope to exist. The polytope in Figure 10.2 has the same number of vertices (16), edges (32), planar (24) and three -dimensional (8) faces as 4 - cub. Therefore, substituting these quantities of elements of different dimensions into the Euler - Poincaré equation, can get that this equation is satisfied for a value of dimension equal to 4. The difference from the 4 - cube is that its flat faces are rhombs, and not squares. Therefore, the golden hyper - rhombohedron is a semi - regular polytope, since its two - dimensional faces are not regular polygons, although all vertices are compatible with motion. The periodic part of the structure of quasi - crystals is the translational filling of a four - dimensional space by a golden hyper - rhombohedron. The edges emanating from its vertices carry out the product of the rhombohedron into a segment. Just like a 4 - cube is obtained by multiplying the cube by an edge. As a result, the number of elements of different dimensions in the gold hyper - rhombohedron is the same as in the 4 - cube. The difference from the 4 - cube lies in the fact that all its metric relationships are based on the golden section. It was shown (Zhizhin, 2014, Zhizhin& Diudea, 2016) that the lattice of the vertices of the golden hyper - rhombohedron during its translation highlights the latent periodicity of the diffraction patterns.

POLYTOPES OF VARIOUS DIMENSIONS AND SHAPES ON THE LATTICE OF VERTICES OF THE GOLDEN HYPER - RHOMBOHEDRON

On this lattice can be built all the Platonic bodies, the Bravais (Bravais, 1848) and Delaunay (Delaunay, Padurov& Alecsandrov, 1934; Delaunay, 1937) cells, Voronoi (Voronoi, 1908) cells. Figure 3 representing the lattice of golden hyper - rhombohedron vertices, as an example, shows how from the vertices of the golden hyper - rhombohedron one can obtain projections of Plato's bodies: tetrahedron, cube, octahedron, icosahedron, dodecahedron.

Figure 3. Plato`s bodies on the lattice of golden hyper - rhombohedron

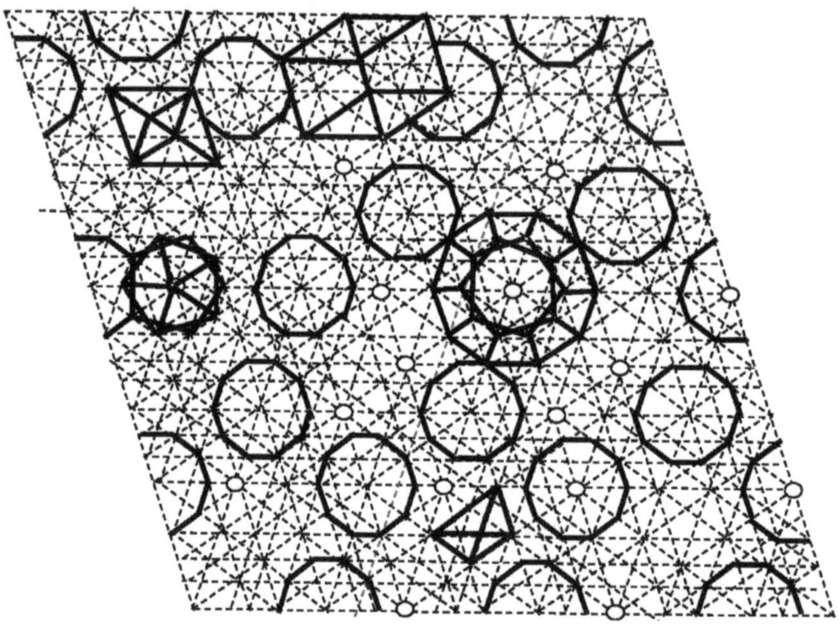

Moreover, these figures can vary in size. It is essential that in addition to the already built golden hyper - rhombohedron of certain size, these polytopes of other dimensions can be built on vertices lattice (see Figure 4).

On the Figure 3 and Figure 4 the vertices of golden hyper - rhombohedron big size is indicated of a white circle. On Figure 4 besides golden hyper - rhombohedron little size is indicated of a black segment.

The physical nature of the process of scaling in phase transitions of the second kind and in quasicrystals is a change in the structure during the phase transition of the second kind (Kadanoff, 1966), and in the processes of formation of quasi - crystals during annealing. However, if Kadanoff assumed for phase transitions the merging of smaller cells to larger ones, here the analysis of golden hyper - rhombohedron vertices lattice shows that the process of scale changing occurs hierarchically, i.e. a smaller hyper - rhombohedron cell is not combined with similar ones but simply increases in hierarchical mode, absorbing more and more of lattice sites. On the diffraction pattern (Figure 4), as an example, the projection of golden hyper - rhombohedron with the edge of a certain size is applied. It's clear that hyper - rhombohedrons of other sizes can be distinguished on the diffraction pattern. The analysis of diffraction pattern geometry

Figure 4. Golden hyper - rhombohedrons of different scale on the lattice of golden hyper -rhombohedron vertices

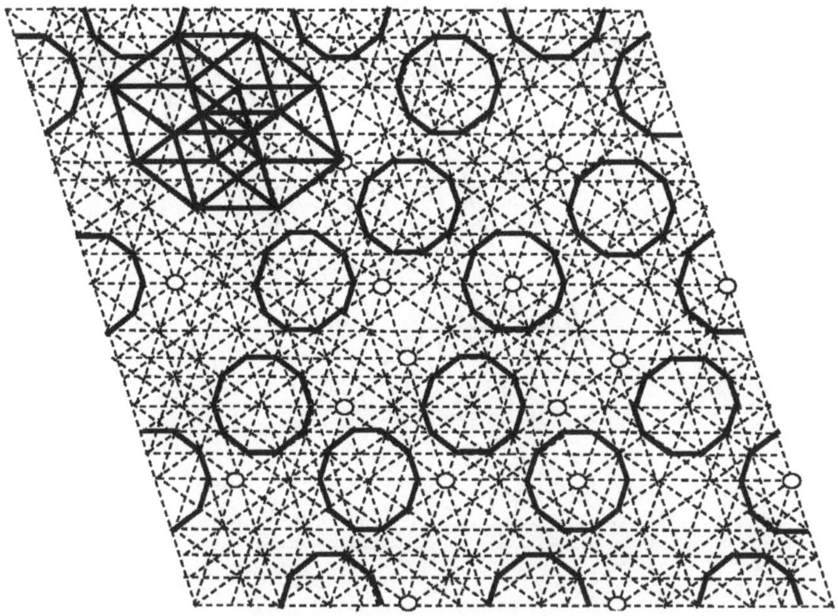

shows that the similarity coefficient of the hyper - rhombohedrons is the golden section t. Raising t in positive and negative integer powers one can receive the hyper - rhombohedrons of appropriate sizes, which can be identified on the diffraction pattern in the range of its solvability. It is essential that the coordinates of the vertices of the golden hyper - rhombohedron are expressed in forth - dimensional space in whole numbers. Let's choose the top *a* in Figure 10.2 as the beginning of coordinates system *(x, y, z, h)*. Let's direct the axis *x* along the edge *ar,* the axis *y* along the edge *ap*, the axis *z* along the edge *ad,* the axis *t* along the edge *ab*. Let's assume that the length of hyper - rhombohedron edge is a unit. Then the coordinates of the vertices of the hyper - rhombohedron *apebdkfc* are expressed in whole numbers *a(0, 0, 0, 0), π(0, 1, 0, 0), e(0, 1, 0, 1), b(0, 0, 0, 1), k(0, 1, 1, 0), f(0, 1, 1, 1), c(0, 0, 1, 1), d(0, 0, 1, 0)*. Now let's to change the coordinate x from 0 to 1. Then the coordinates of a*pebdkfc* hyper - rhombohedron vertices also to change v*(1, 1, 0, 1), t(0, 0, 0, 1),w(1, 1, 1, 1), u(1, 0, 1, 1), s(1, 1, 0, 0), r(1, 0, 0, 0), o(1, 1, 1, 0), m(1, 0, 1, 0)*. It's obvious that at hyper - rhombohedron translation the coordinates of all lattice vertices formed by hyper - rhombohedron vertices are integers. Filling the space of the hyper - rhombohedrons of other sizes also forms integral lattice, if the size of hyper - rhombohedron edges is mistaken for one. It's interesting that on the lattice of golden hyper - rhombohedron vertices in the space 4D it is possible to construct the projections of the golden rhombohedron from a space of even more dimension. Figure 5 shows a projection of the golden rhombohedron from the space 5D on the plane 2D.

It is clear that this projection can be laid on the lattice of golden hyper - rhombohedron vertices in the space 4*D* in Figure 10.3. The 5 – rhombohedron is the product of hyper - rhombohedron by one-dimension segment. If length of the segment equal of length edges of the hyper - rhombohedron, so the 5-rhombohedron is semi - regular polytope too. The 5 - rhombohedron can also fill the space of quasi - crystal displayed by the diffraction pattern in Figure 1, i. e. one can say that the quasi - crystal has a

Figure 5. The golden rhombohedron in space 5D

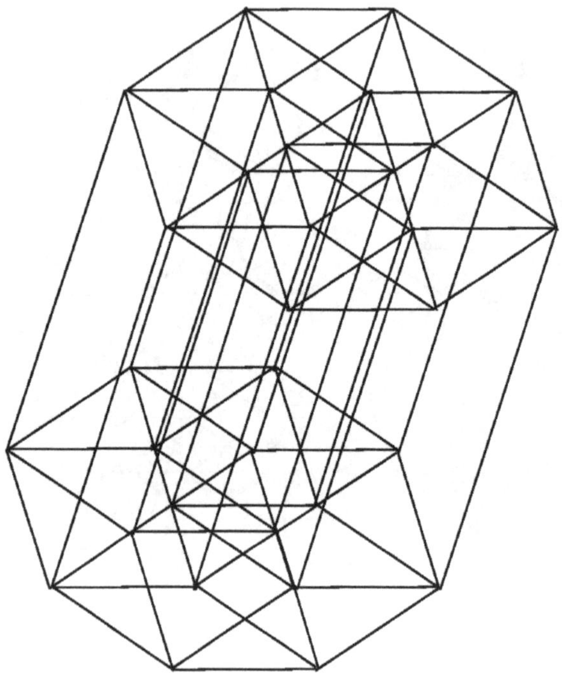

subspace of dimension 5. Similarly, a rhombohedron of even more dimension can be built on the diffraction pattern. This leads to the conclusion that quasi-crystals have subspaces of any integer dimension.

From the said above it follows that the lattice of the vertices of the golden hyper - rhombohedrons have the universal nature. On it you can find the projections of the polytope 4 - simplex and other polytopes of higher dimension.

It is also important to note that on the diffraction patterns of intermetallic (Figure 1) one can see not only the latent periodicity of the structure in spaces of higher dimension, but also the hierarchical filling of spaces in the vicinity of almost any point of the structure. It is expressed in an increase in the size of the figures around an arbitrary point of the structure while maintaining the shape of the figure.

NORMAL PARTITION OF A TWO - DIMENSIONAL PLANE AS A RESULT OF ITS HIERARCHICAL FILLING

In the previous chapter, the hierarchical filling of a two - dimensional plane with a regular triangle was considered (Figure 9.4, Chapter 9). Obviously, continuing the sides of the original triangle in Figure 9.4 in Chapter 9, as well as connecting the intersection points of these lines, we get the normal partition of the two - dimensional plane by a regular triangle (Figure 6).

Figure 7 shows the hierarchical filling of a two - dimensional plane with a square.

Obviously continuing the lines to the intersection and connecting the intersection points of the lines with line segments, we get the normal partition of the two - dimensional plane by a square (Figure 8).

Figure 6. The normal partition of the two - dimensional plane by a regular triangle

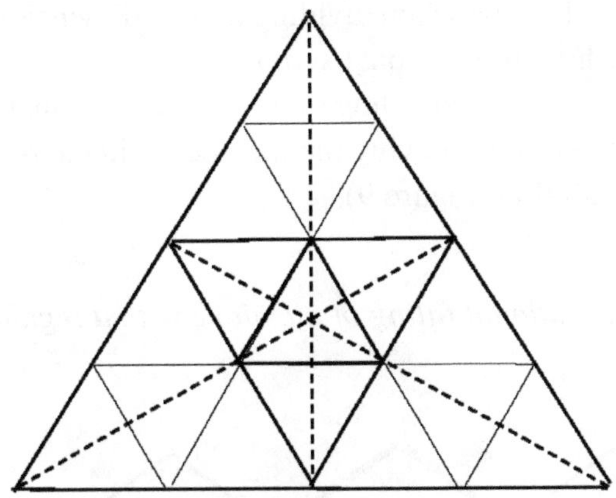

Figure 7. Hierarchical filling of a two - dimensional plane with a square

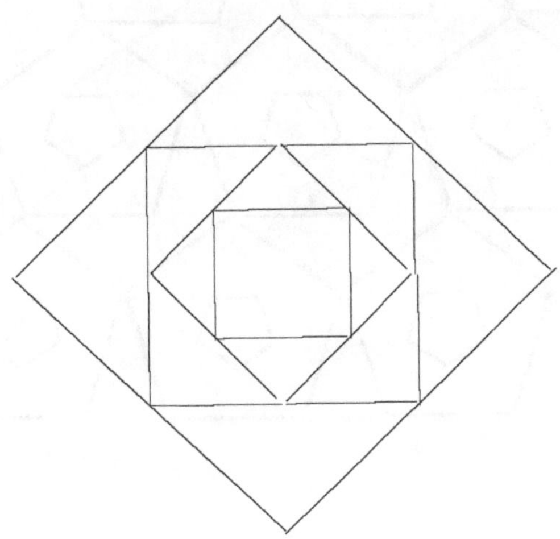

Figure 8. Normal partition of a two - dimensional plane by a square.

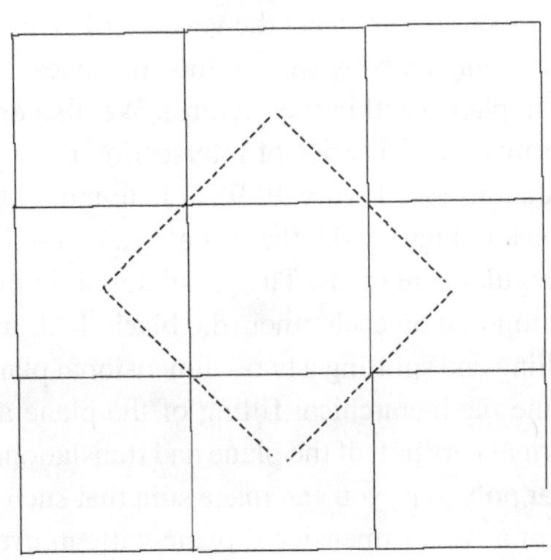

A parallel triangle when dividing a two - dimensional plane, a regular triangle is a rhombus composed of two triangles. The parallelogon when dividing a two - dimensional plane is the square itself. It is obvious that these partitions have translational symmetry.

In the previous chapter, the hierarchical filling of a two - dimensional plane with a regular pentagon was considered (Figure 1, Chapter 9). Following the method outlined in this chapter, continue the sides of the red pentagon to the intersection (Figure 9).

Figure 9. Hierarchical and translational filling of the plane with a regular pentagon

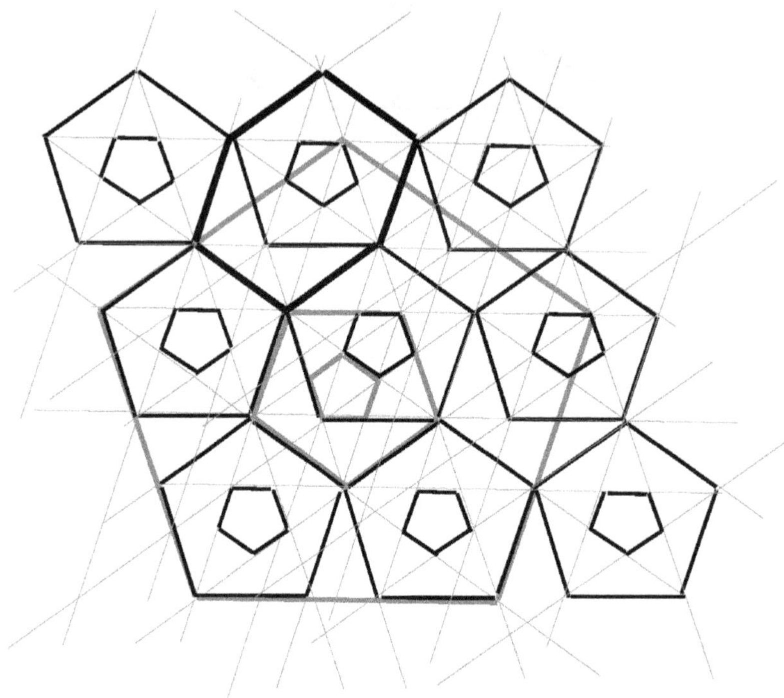

These intersections are the vertices of another regular red pentagon of a larger size. Continuing the sides of this pentagon to the intersection, can to get the vertices of an even larger pentagon. This is how the plane is filled with a regular pentagon. Now we continue the sides of the regular red pentagons into the whole foreseeable part of the plane (within the picture). We also draw part of the lines parallel to these sides. This leads to the formation of a series of intersection points that can be considered as vertices of other regular black pentagons (see Figure 10.9). It is found that these regular black pentagons translationally fill the plane. Thus, in Figure 10.9 there is also a hierarchical and translational filling of the plane by the same figures: regular pentagons. The parallelogon in this case is a hexagon composed of two regular pentagons superimposed on each other (the black thick lines).

In the considered cases of filling and splitting a two - dimensional plane, regular polygons were used. This made it possible to combine the hierarchical filling of the plane and the normal partition of the plane, so that in this case the normal partition of the plane had translational symmetry. If you try to make such constructions with irregular polygons, you can make sure that such a combination is impossible to obtain. The hierarchical filling of a two - dimensional plane with an irregular polygon is accompanied by a normal partition of the plane, but such a partition does not have the property of translational symmetry. This occurs as a result of the fact that figures of various shapes and sizes appear on the plane as a result of the irregularity of the original polygon.

NORMAL PARTITION AND HIERARCHICAL FILLING OF A THREE - DIMENSIONAL SPACE

In the previous chapter, hierarchical fillings of three - dimensional space with three -dimensional figures were considered. It follows from this consideration that even regular three -dimensional figures cannot lead to normal partitions of space with translational symmetry when they hierarchically fill the space. This is due to the fact that when hierarchically filling three -dimensional space with regular three - dimensional figures, many three - dimensional figures of various shapes and sizes are formed. For example, when hierarchically filling the space, a cube turns into an octahedron, an octahedron turns into a cube, an icosahedron turns into a dodecahedron, a dodecahedron turns into an icosahedron. When hierarchically filling the space, the tetrahedron becomes a tetrahedron, but the space between the tetrahedra is filled with various irregular shapes. We can say that the hierarchical filling of three - dimensional space with three -dimensional polyhedrons is accompanied by a normal partition of the space, but this partition does not have the property of translational symmetry.

NORMAL PARTITION AND HIERARCHICAL FILLING OF A HIGHER - DIMENSIONAL SPACE

The most successful combination of hierarchical filling of the space and its normal partition occurs in spaces of higher dimension. This was already noted at the beginning of this chapter by the example of intermetallic diffractograms. Let's consider simpler examples. Let there be a simplex of dimension 4. Choose the origin at its leftmost vertex located at the base of the simplex (Figure 10).

Figure 10. Hierarchical filling and normal partitioning of a four - dimensional space 4 - simplex.

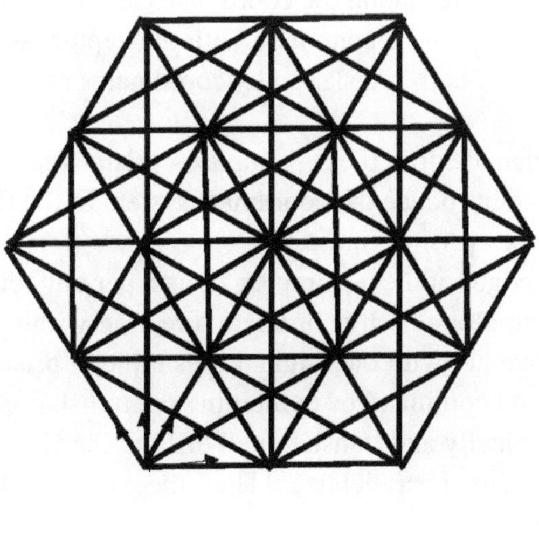

We direct the coordinates along the edges of the 4 - simplex and we will translate the 4 - simplex along each coordinate to the length of the edge. Then, in addition to the original 4 - simplex, we get four more 4 - simplices of dimension 4, which together with the original 4 - simplex form a large 4 - simplex. In the center of this 4 - simplex, as can be seen from Figure 10, there is a small polytope 4 - simplex. Thus, the translation of the 4 - simplex along the coordinates led to the hierarchical filling of the space with the 4 - simplex. This process can be continued with the capture of an ever-increasing volume of four - dimensional space, in which both translation by coordinates and hierarchical filling of the space will occur simultaneously.

Let there be a simplex of dimension 5. Choose the origin at its leftmost vertex located at the base of the simplex (Figure 11).

Figure 11. Hierarchical filling and normal partitioning of a five - dimensional space by 5 - simplex.

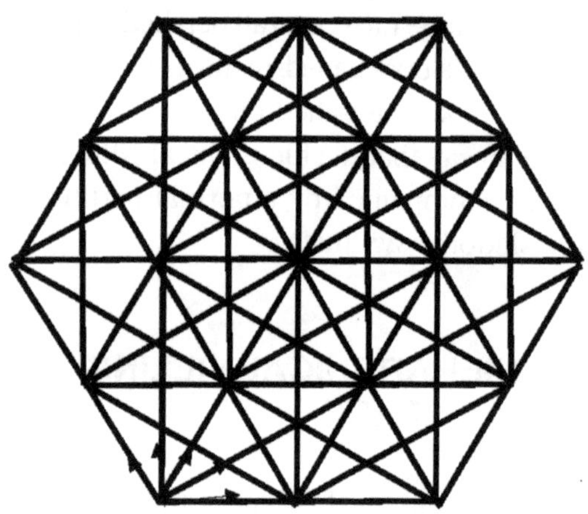

We direct the coordinates along the edges of the 5 - simplex and we will translate the 5 - simplex along each coordinate to the length of the edge. Then, in addition to the original 5 - simplex, we get five more 5 - simplices of dimension 5, which together with the original 5 - simplex form a large 5 - simplex. In the center of this 5 - simplex, as can be seen from Figure 10.11, there is a small polytope 5 - simplex. Thus, the translation of the 5 - simplex along the coordinates led to the hierarchical filling of the space with the 5 - simplex. This process can be continued with the capture of an ever-increasing volume of five - dimensional space, in which both translation by coordinates and hierarchical filling of the space will occur simultaneously.

Let there be a polytopic prismahedron the 5 - simplex - prism (Figure 5.3 in Chapter 5). This prismahedron has dimension 5. Can introduce the coordinate system of the five - dimensional space in one of the vertices of the 5 - simplex - prism (Figure 12).

Assuming the length of all edges of the 5 - simplex - prism per unit, one will translate the 5 -simplex prism by the length of the edge along each coordinate direction within the figure. There are four the 5 - simplex – prisms, which together with the original 5 - simplex - prism form a 5 -simplex - prism of a larger size. This process can be continued by combining the translation of a 5 – simplex - prism with growth its size. This is hierarchically and translationally filling the five - dimensional space with a 5 - simplex - prism on the same system of parallel lines. Thus, there is no contradiction between hierarchical and translational filling of spaces.

Figure 12. Hierarchical and translational filling of the five - dimensional space by the 5 – simplex - prism

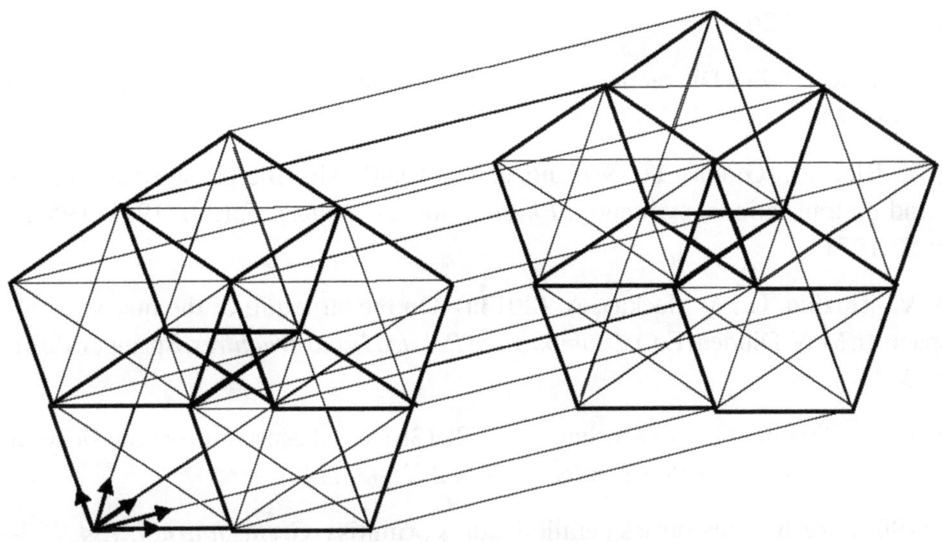

CONCLUSION

At the same time, a hierarchical filling of the space and its normal partitioning is possible for any space dimension greater than or equal to 2. It is significant that in three - dimensional space this does not lead to translational symmetry. The simplest structures with translational symmetry during hierarchical filling and normal partitioning arise in a two - dimensional plane with regular polygons. The most interesting structures with simultaneous hierarchical filling and normal partitioning are formed in spaces of higher dimension. Moreover, due to translation over a large number of coordinates, with hierarchical filling of the space, more and more complex structures are formed with a large number of symmetry elements. Diffraction patterns of intermetallic compounds are a striking example of the natural formations of spaces of higher dimensions with simultaneous hierarchical filling of the space and normal partitioning with various translation elements.

REFERENCES

Braveis, A. (1848). *Abhandlung über die Systeme von regelmässing auf einer Ebene oder im Raum vertheilten Puntsten*. Ostwald`s Klassiker.

Delaunay, B. (1937). Geometry of positive quadratic forms. *Uspekhi Matematicheskhi Nauk, 3*, 16–62.

Delaunay, B., Padurov, N., & Aleksandrov, A. (1934). *Mathematical foundations of the structural analysis of crystals*. ONTI.

Eiji, A., Yanfa, Y., & Pennycook, S. J. (2004). Quasicrystals as cluster aggregates. *Nature Materials, 3*(11), 759–767. doi:10.1038/nmat1244 PMID:15516956

Kadanoff, L. P. (1966). Scaling Laws For Ising Models Near τ_c^*. *Physics*, *2*(6), 263–272. doi:10.1103/PhysicsPhysiqueFizika.2.263

Mukhopadhyay, N.K. (1993). Diffraction studies of icosahedral phases in $Al_{70}Fe_{20}W_{10}$. *Journal of Non-Crystalline Solids, 153 – 154,* 1193 – 1197.

Shechtman, D., Blech, I., Gratias, D., & Cahn, J. W. (1984). Metallic phase with longerange orientational order and no translational symmetry. *Physical Review Letters*, *53*(20), 1951–1953. doi:10.1103/PhysRevLett.53.1951

Shevchenko, V., Zhizhin, G., & Mackay, A. (2013a). On the structure of the quasicrystals in the high dimension space. In M. V. Diudea (Ed.), *Diamond and Related Nanostructures*. Springer. doi:10.1007/978-94-007-6371-5_17

Shevchenko, V.Ya., Zhizhin, G.V. & Mackay, A.L. (2013b). On the structure of quasicrystals in a space of higher dimension, *Izvestiya RAS. Chemical series*, *2*, 269 - 274.

Voronoi, G. (1908). Recherches sur les parallelloedres primitiss. *Grelle Journal*, *134*, 287–1908.

Zhang, X. & Kelton, K.F. (1993). High-order crystal approximant structures alloys $Ti_{54}Zr_{26}Ni_{20}$. *Journal of Non-Crystalline Solids*, *153 – 154*, 114 – 118.

Zhizhin, G. V. (2014). *World – 4D*. Polytechnic Service.

Zhizhin, G. V., & Diudea, M. V. (2016). Space of nanoworld. In M. V. Putz & C. M. Marius (Eds.), *Sustainable nanosystems, development, properties, and applications* (pp. 221–245). IGI Dlobal.

KEY TERMS AND DEFINITIONS

5-Simplex-Prism: The product of a 4-simplex by a segment.

Congruent Polyhedrons: Polyhedrons that are compatible with motion.

Hierarchical Filling of Space: The filling of space by a figure with its discrete resizing, preserving the resemblance of figures at each step of its change.

Polytopic Prismahedron: The product of polytopes.

Scaling: Scale change of shape.

Conclusion

The text of the monograph can be divided into two parts. The first part, consisting of the first six chapters, is preparatory. It contains the material necessary for the perception of the main part, consisting of the last four chapters, in which the problem of normal partitioning and hierarchical filling of n - dimensional space is solved actually. Consequently, in an extended version, the 18th Hilbert problem is solved.

In the first part, the necessary conditions for the existence of regular and incorrect polytopes of the highest dimension are determined. The main of these conditions is the fulfillment of the Euler – Poincaré equation for the polytope. Analytical expressions are obtained for calculating the number of faces of different dimensions and the incidence coefficients of low - dimensional faces with respect to high - dimensional faces. Certain values of these incidence rates are also necessary for the existence of polytopes. It should be noted that from the analytical expressions for the numbers of faces of various dimensions, the Euler – Poincare equation automatically follows.

Obtained expressions for calculating the dimension of multi - shell clusters. It is shown that the dimension of multi - shell clusters grows linearly with an increase in the number of shells. The shape of the shells can be in the form of any body of Plato. Shells have a common center in which an atom can be located or it is not there. It has been shown that gamma - copper clusters are cross - polytopes of higher dimension. The structures of clusters are investigated and their dimensions are determined if the metal atoms in the cluster form linear or nonlinear, as well as closed chains or form a polyhedral core. In all cases, the dimension of these clusters is higher than three. In the case of octahedral coordination, clusters consist of two polytopes of the highest dimension, touching along a flat section.

It was established for the first time that for any convex polytope there is an integral equality in the transfer of information from low - dimensional elements to high - dimensional elements and vice versa. This integral equality is called the law of conservation of incidents. In particular, this law can serve as the mathematical basis for the recently discovered epigenetic principle of transmission of hereditary information.

Based on the experiments of Mendel, a mathematical model is constructed that describes such experiments in a wide range of parameters. It has been shown that this coin has a balance between dominant and recessive forms, and therefore there is no need to use the Hardy -Weinberg model constructed logistically without taking into account real processes in plant populations.

It has been proved that nucleic acid contact occurs in a polytope of hereditary information of dimension 13 (13 - cross - polytope), on the coordinate planes of which flat nitrogenous bases can be located. It was found that the polytopic of hereditary information has a hidden degree of freedom, which leads to a unique correspondence of the sum of the number of nucleotides and the number of their locations in space with the number of amino acids. Thus, there is no need to use the genetic code entered logistically. The latent order of interaction of nucleic acids is also important when using analog nucleic acids, in particular, in xenobiology, as well as in the treatment of viral diseases.

The structure of polytopic prismahedrons, i.e. polytopes of higher dimension, which are the products of polytopes of lower dimension. The structure of these polytopes is considered as a function of the structures of polytopes of the factors of the product of polytopes. The geometric properties of the boundary complexes of polytopic prismahedrons are investigated.

Analytically considered are the products of molecules and clusters on polytopes of various dimensions. The results of these products are polytopes with a dimension equal to the sum of the dimensions of the factors, and the number of vertices is equal to the product of the numbers of vertices of the factors. In this case, copies of chemical compounds (molecules and clusters) are distributed over space depending on the type and dimension of the polytopes of the factors. If the factors are biological objects, then the works of political scientists in this case, in principle, allow us to build a quaternary structure of living matter. The presence of families of parallel edges allows you to fill the space without gaps as a result of their translation.

The second part of the book begins with a study of the geometric work of Delaunay and his school. The Delaunay theory of discrete systems is investigated in detail. It was found that this theory does not take into account many phenomena in the discrete world that have been discovered in recent decades, such as scaling, the existence of so - called quasi - crystalline systems, and higher dimensions of molecules and clusters. Delaunay theory does not take into account the fundamental features of figures of higher dimension, the necessary conditions for the existence of polytopes of higher dimension. The concepts of an empty ball introduced by Delaunay and the condition of contact of polytopes of the highest dimension only in facets are not related to the real geometric features of systems of higher dimension and have not been proved. It turned out that the well - known Delaunay classification of parallelohedrons of dimension 4, including 51 figures, does not take into account the main condition for the existence of figures of higher dimension of the Euler - Poincaré equation. It is proved that not one of the parallelohedra of this classification satisfies the Euler - Poincaré equation, and therefore cannot be a parallelohedron of dimension 4 (and no other dimension). The parallelohedron built later by Shtogrin was in the same position. Using concrete examples, it is shown that the geometry of discrete Delaunay systems leads to incorrect results. This is all the more important because this geometry is still followed in many countries of the world, considering it to be true. The theory of Delaunay stereohedrons, as a consequence of the geometry constructed by Delaunay, also turned out to be incorrect. Delone never managed to build a single concrete stereohedron of dimension more than three. Thus, Delaunay geometry cannot be used to solve the 18th Hilbert problem of constructing spaces of higher dimension from congruent figures.

The study showed that polytopic prismahedrons, due to their geometric structure, provide the possibility of creating high dimensional parallelohedra. The basis of these parallelohedra of high dimension are already well - known figures of higher dimension: n - cube, n - simplex, n - cross-polytope. The adjacent parallelohedra in the normal partition of n - dimensional spaces can have common elements of different dimensions (from n - 1 to 0). In the case of the existence of enantiomorphic forms in polytopic prismahedrons, parallelohedrons are formed from several enantiomorphic forms adjacent to each other over elements of different dimensions. Thus, using the developed method for constructing parallelohedrons of any higher dimension, it is possible to obtain concrete samples of these structures. Therefore, it can be argued that the 18th Hilbert problem of constructing n - dimensional spaces with the help of congruent figures is solved. It should be noted, however, that with large dimensions this task may turn out to be technically quite difficult. It is significant that the solution obtained corresponds to Fedorov's idea of the need to take into account the properties of figures when deciding to construct spaces using congruent figures. All works by Delaunay and his schools do not satisfy this condition. This was the reason for the lack of concrete substantiated results in their work. The resulting solution allows us to obtain new nanomaterials with new useful properties.

The geometrical characteristics of the scaling process are investigated for hierarchical filling of spaces with geometric figures (regular polygons, regular and semi - regular polyhedrons, polytopes of higher dimension). The corresponding coefficients of geometric progressions of a discrete increase in their sizes are determined depending on the shape of the figure and its dimension.

Structures arising in spaces of various dimensions with simultaneous normal partitioning of spaces and their hierarchical filling are considered. The conditions for the appearance of translational symmetry in these spaces are investigated. It is shown that translational symmetry arises under these conditions in spaces of higher dimension. In three - dimensional space, simultaneous hierarchical filling and normal partitioning does not lead to translational symmetry.

The system of knowledge about the structure of matter presented in the monograph is based on modern achievements in mathematics, physics, chemistry, and technology. In general, this knowledge system represents a new paradigm of the discrete world. It was this paradigm that made it possible to solve the problem posed more than 100 years ago to construct n - dimensional space using congruent figures (Hilbert's 18th problem).

DIRECTIONS OF FUTURE WORK

The published four books of the author (Zhizhin, 2018; 2019 a, b; 2020) together with this book (Zhizhin, 2021) represent a new worldview of the main natural sciences: chemistry, physics, mathematics, biology. The fundamental feature of this worldview is the description of living and inanimate objects of nature from the standpoint of Riemann's geometry, not Euclid's geometry. This means that in all cases when describing objects of nature, the a priori dimension of objects is not assumed to be equal to three or less. The dimension of objects is determined on the basis of a study of the geometric structure of an object according to the Euler - Poincaré law for an object of unknown previously arbitrary dimension n. The resulting spaces, as a rule of higher dimension, have a natural boundary, a natural boundary, i.e. they are not considered in advance to be infinite in any coordinate. The boundary of the object space is, in the general case, the boundary of spaces of different dimensions. It is essential that the concept of the highest dimension of space refers to the elements of earthly reality, i.e. of our world and the world inside us, and not to distant outer spaces, as in string theory.

The main material objects of earthly reality are atoms and molecules. In the author's monograph (Zhizhin, 2018), the geometries of various molecules are systematically studied, which include almost all elements of the periodic table of elements. It is shown that most of them have the highest dimension, i.e. more than three. This direction, i.e. the stereochemistry of organic and inorganic compounds in higher-dimensional spaces_should be further investigated, since many important chemical compounds have not yet been considered. Moreover, the more complex a chemical compound, the higher its dimension.

In the author's monographs (Zhizhin, 2018, 2019 a), it was proved that the Schrödinger equation has a solution in a space of higher dimension, in contrast to the previously accepted opinion. It follows from this that in addition to c, p, b, and f electron orbitals in the three-dimensional vicinity of the atomic nucleus, other electron orbitals can exist in the space of higher dimensions. Consequently, there may be other high energy levels of electrons, significantly increasing the chemical activity of the elements. This is indicated by the established facts of the chemical activity of elements exceeding the values of the chemical activity determined by the arrangement of the elements in the periodic table. It is necessary to develop a direction for studying the chemical activity of elements, taking into account the solution of the Schrödinger equation in a space of higher dimensions.

If molecules have a higher dimension, then physical and chemical interactions of atoms and molecules occur in spaces of higher dimension. Therefore, when studying the kinetics of chemical reactions, one should take into account the higher dimension of the spaces in which these reactions occur. Since chemical reactions are accompanied by the movement of atoms and electrons, it is necessary to study the elements of chemical kinetics in spaces of higher dimension with the analysis of the movement of particles (atoms and electrons) in spaces of higher dimension with the possibility of transition of particles from one space to another.

Due to the fact that molecules and their compounds have a higher dimension, it is necessary to develop the mathematical foundations of geometric shapes in spaces of higher dimensions (especially with a dimension greater than 4). These studies were started in the author's monograph (Zhizhin, 2019 a), but, given the novelty of these studies, they should be continued. In this case, one should not be limited to the correct figures since there are many examples of irregular figures in nature. This is, among other things, important for obtaining materials with new properties required in various fields of human activity.

We are talking primarily about nanotechnology and nanomaterials, which often exhibit unusual useful properties. Here, the problems of normal partitioning and hierarchical filling of spaces of higher dimension, considered in the author's monographs (Zhizhin, 2020, 2021), become important. It is necessary to develop the direction of investigation of normal partitioning and hierarchical filling of spaces of arbitrary dimension n in order to obtain new useful materials.

An important place in further research should be occupied by studies of geometric forms of life, from the moment of the origin of life on Earth to the present. In the author's monograph (Zhizhin, 2019 b), it is proved that biomolecules have the highest dimension. This leads to the need to create geometric biochemistry, taking into account the highest dimension of the constituent biomolecules. As shown in the author's monographs (2019 a, b), taking into account the highest dimension in the analysis of the interaction of nucleic acids and in the transmission of hereditary information is of great importance. In accordance with this, genetics should be presented taking into account the highest dimension of the constituent biomolecules. This is currently especially important due to the need to counter viral infections.

The listed areas of work on the development of the results obtained in these monographs by the author are important for human life on Earth. However, for the successful solution of these problems, a certain change in the complex of knowledge received by schoolchildren and students during their studies is necessary. In this regard, in textbooks on chemistry, physics, mathematics and biology, it is necessary to introduce ideas about the highest dimension of elementary objects of Nature and the methods of their study. In this case, the geometry of Euclid should be given historical significance, as a complex of greatly simplifying assumptions, often inadequate to reality.

REFERENCES

Zhizhin, G. V. (2018). *Chemical Compound Structures and the Higher Dimension of Molecules: Emerging Research and Opportunities*. IGI Global. doi:10.4018/978-1-5225-4108-0

Zhizhin, G. V. (2019a). *The Geometry of Higher-Dimensional Polytopes*. IGI Global. doi:10.4018/978-1-5225-6968-8

Zhizhin, G. V. (2019b). *Attractors and Higher Dimensions in Population and Molecular Biology: Emerging Research and Opportunities*. IGI Global. doi:10.4018/978-1-5225-9651-6

Zhizhin, G. V. (2020). *Nanotechnologies and Clusters in the Spaces of Higher Dimension: Emerging Research and Opportunities*. IGI Global.

Zhizhin, G. V. (2021). *Normal Partitions and Hierarchical Fillings of N - dimensional Space*. IGI Global.

About the Author

Gennadiy Vladimirovich Zhizhin: Head of the Russian Federation Branch of European Society of Mathematical Chemistry (2017-2020); Chief Researcher of the OOO "Adamant", Skolkovo, Moscow, Russia (2017-2020); Senior Fellow of the Institute of Silicate Chemistry, Russian Academy of Sciences, S.-Petersburg, Russia (2012-2014); Professor at North-Western State Technical University, Departments of Chemistry, Computational Mathematics, Computer Science, Applied Mathematics, S.- Petersburg, Russia (1994 – 2012); Professor of the Department of Mathematics of the National University of the mineral resource (Mining University), S.-Petersburg, Russia (2012-2013); Academic degree of Doctor of Technical Sciences (1991); Academic rank of Professor (1999); Corresponding Member of the Russian Ecological Academy (2006); Academician of the Russian Academy of Natural Sciences (2017); Member of the Scientific Council on Combustion and Explosion of the Russian Academy of Sciences (2010-2020); Member of the editorial board of the Journal Biosphere (2008-2020); Deputy editor-in-chief of the IJARB (2020); Member of the editorial board IJCCE (2019); author of 120 articles and 15 monographs.

Index

P

R

S

T

U

www.ingramcontent.com/pod-product-compliance
Lightning Source LLC
Chambersburg PA
CBHW081415220926
51613CB00010B/241